Speaking Beyond Earth

T0130744

Speaking Beyond Earth

Perspectives on Messaging Across Deep Space and Cosmic Time

Edited by PAUL E. QUAST *and*
DAVID DUNÉR

Foreword by Alice Gorman;
Afterword by Cornelius Holtorf

McFarland & Company, Inc., Publishers
Jefferson, North Carolina

This book has undergone peer review.

The Beyond the Earth Foundation is a non-profit public benefits, education, and research organization that principally investigates emerging human legacies, while also ensuring long-term custodianship of these essential human records. As part of this program, the foundation is investigating messages in deep space and cosmic time for the purposes of studying the methods employed within deep-time communication strategies, while also chronicling these intelligible legacies already "out there." We seek to apply this research toward developing our own introductory "Companion Guide to Earth" message and other archival applications to safeguard essential human records for posterity. For more information about our activities, please visit; www.beyondtheearth.org.

LIBRARY OF CONGRESS CATALOGUING-IN-PUBLICATION DATA

Names: Quast, Paul E., 1989– editor. | Dunér, David, 1970– editor. | Gorman, Alice (Alice Claire), writer of foreword. | Holtorf, Cornelius, 1968– writer of afterword.
Title: Speaking beyond Earth : perspectives on messaging across deep space and cosmic time / edited by Paul E. Quast and David Dunér ; foreword by Alice Gorman ; afterword by Cornelius Holtorf.
Description: Jefferson, North Carolina : McFarland & Company, Inc., Publishers, 2024 | Includes bibliographical references and index.
Identifiers: LCCN 2023057494 | ISBN 9781476690001 (paperback : acid free paper) ∞
ISBN 9781476649863 (ebook)
Subjects: LCSH: Interstellar communication. | BISAC: SCIENCE / Space Science / General
Classification: LCC QB54 .S6925 2024 | DDC 621.3820999—dc23/eng/20240129
LC record available at https://lccn.loc.gov/2023057494

BRITISH LIBRARY CATALOGUING DATA ARE AVAILABLE

ISBN (print) 978-1-4766-9000-1
ISBN (ebook) 978-1-4766-4986-3

Front cover images: Sol System clock graphic depicts the next closest alignment of all 8 planets (within a 90° arc) on 6th May 2492 CE. Illustration courtesy of the Beyond the Earth Foundation. Space Station flight over Hurricane Lane (22nd August 2018). Inverted photograph courtesy of astronaut Ricky Arnold/NASA.

Printed in the United States of America

*McFarland & Company, Inc., Publishers
Box 611, Jefferson, North Carolina 28640
www.mcfarlandpub.com*

To our most protracted human legacy now cast out into the annals of cosmos time, the Earth itself—without which, we would obviously not be having these very academic conversations. May these messages and relics, one faithful day, represent our (presently) lonely planet, rather than painting the idealized rendition from our minds. Humanity still has much to reflect upon, and hopefully change, for the benefit of distant posterity who cannot readily "have a say" today.

Acknowledgments

Writing a book is a very humbling, thought-provoking, and yet *challenging* experience. This is especially true when the subject matter is seldom explored in material culture investigations, or if it only possesses impoverished storehouses of evidence to support what will likely remain as our informed *theories* about ancient envoys from Earth. To my co-writers in this volume, I appreciate the depth of your innovative thinking, knowledge, expertise, and countless hours of spilled ink and coffee in giving your voices to this admittedly speculative, but substantial step towards understanding—*for ourselves*—the material profile of humanity emanating from Earth.

Of course, this collection would be a meager morsal were it not for all those who supported the project and chose to actively contribute to the treasure trove of visual materials donning these pages—most graciously donated to stimulate vital ruminations about the indelible legacies that we leave for posterity. The breadth of this volume, of course, would also not have been possible without the decades of generational studies conducted by a vast number of professional and academic researchers, much of which has proceeded apace in the formal and natural sciences under the auspices of SETI, alongside the disciplines contributing to space archaeology fields.

At the risk of unfairly naming a handful of these essential voices, and in the interest of brevity, I would like to thank all who shared their diverse thoughts, sometimes dissenting opinions, and pensive reflections for this volume—all of which is extensively cited. I would also like to thank the animate team at Atelier Éditions for initially pushing us on to this journey, and our publisher McFarland (and two mysterious peer-reviewers) for bringing these essays to page.

Paul: On a personal note, thank you to my former fiancée, now wife, for her loving support. I promise that our tables will be *less* cluttered. Thank you for your enduring encouragement, and never-ending patience. To my parents Pauline and David, and sister Aileen, I'm grateful for the invariable nudges towards the next adventure, with love and enthusiasm intended of course.

Table of Contents

Between pages 130 and 131 are 8 color plates with 14 photographs

Part 3: Reading Outside the Message

Foreword

Beyond Ourselves: An Archaeological Reflection on Sending Messages to Other Worlds

ALICE GORMAN

Making the Familiar Unfamiliar

I find myself coming to the topic of messaging extraterrestrial intelligence through two lenses: that of the space archaeologist, looking at the remains of human activities related to space exploration throughout the solar system, and that of the terrestrial archaeologist trying to discern ancient minds through the artefacts and traces they left behind in the deep layers of Earth. In both cases, the symbolic nature of material artefacts is a critical component of trying to understand how human life on Earth comes to be as it is right now.

In my research on space material culture, I have become engrossed by two artefact types that seem, superficially, to have little to say: zip-lock bags and cable ties (Gorman, 2016; 2017a). These mass-produced plastic objects are discarded by the thousands every day. They are so familiar that they generally go unnoticed. And yet both have become part of a space culture, on the International Space Station and other spacecraft. Although ostensibly functional and multipurpose, these bits of plastic also convey the social requirements around fastening and confining, marking these as activities that need to be quick, undertaken almost without thought, and easily undone. Their ubiquity is also their obscurity, as illustrated by Sherlock Holmes' famous conversation with Dr. Watson in "A Scandal in Bohemia" (Doyle, 1891). To Watson's startled attempt to remember the number of steps he mounted every day from the hall to their apartment, Holmes replied: "Now, I know that there are seventeen steps because I have both seen and observed."

This statement echoes one of the mantras of the archaeology of the contemporary or recent past: the need to make the familiar unfamiliar, in order that we may observe, record, and analyze it (Buchli and Lucas, 2001). This theme runs through this book. To even begin to formulate a plan to message the alien other, it is necessary to bring to the foreground everything about our own communications that has escaped notice because we are steeped in it, from the global and species level, to the local and personal level.

Objects of Stone and the Lithic Gaze

In my other archaeology, I was concerned with how humans developed symbolic behavior—and hence language—in the Palaeolithic era. My evidence was stone tools used on human bodies to shape their appearance: symbols written on skin and in hair (Gorman, 2001). These sharp pieces of stone were the equivalent of mass-produced plastic in that they were easily and cheaply manufactured, used, and quickly discarded. Everywhere, across the surface of Earth, stone tools are the most ubiquitous and durable evidence of human activity from 3.3 million years ago to the present day—billions of them found everywhere from archaeological deposits meters deep to land surfaces on every continent except Antarctica. There are even stone tools in space: on the *Voyager Golden Records* (VGR) the distinctive sound of knapping was recorded to convey the noise of an industry that was once heard in every human community, every day, for millions of years.

To the trained eye, the angles of a conchoidal fracture initiated by deliberate human action stand out like a sore thumb against backgrounds of gravel, broken rocks, and deceptive leaves. But, early in the emergence of archaeology as a discipline, the difference between "natural" and "cultural" stones was far from obvious. From the late 1800s into the early decades of the 20th century, a fierce debate raged about the status of eoliths, or "dawn stones." Eoliths resembled deliberately manufactured stone tools, and were interpreted as crude precursors to the art of knapping by minds and bodies less "evolved" (O'Connor, 2003). Eventually, experimentation resolved the debate, proving that crushing by slow-moving glaciers had created these simulacra. If it is so difficult to discern the traces of sentient action within our own biological sphere, stripped of all but environmental context, what hope does the alien other have of understanding human messages?

Resemblance is key here and leads me to reflect on another class of stone artefact. In 1908, an 11 cm–tall figurine, which became known as the Venus of Willendorf, was excavated from a site in Austria. Her appellation as "Venus" was an instance of the male gaze, her corpulence something to be wondered at and accounted for, so thoroughly confounding was her divergence from the expectations of female beauty of the time. She was part of a symbolic tradition expressed in female representations from Siberia to Britain, from 35,000 to 10,000 years ago—a common symbolic language that lasted over 20,000 years.

In 1981, a rough pebble dated to 230,000 before present was found at the site of Berekhat Ram on the Golan Heights in Israel. The dimpled rock superficially resembled the shape of Willendorf. The archaeologists proclaimed this pebble a precursor to the Venus tradition (D'Errico and Nowell, 2000). But to me, the pebble evoked rather the eolith, buying into an assumption that what is earlier must be "cruder." If the Venus of Willendorf had not already been such a visual icon, I have my doubts that this pebble would have been recognized as cultural.

The other burning question is what happened in the intervening 200,000 years or so separating Berekhat Ram from the Venus of Willendorf. If Berekhat Ram was an early emergence of symbolic behavior, its presence was almost meaningless. Davidson (1992; 2020) argues that in order for a symbol to be identified, there must be more than one of it. Here we find relevance for messaging extraterrestrial entities. Quast notes the difficulties in establishing a taxonomy of messaging objects for space; and Traphagan and Smith (this volume) point out that of the 31 attempts to send signals to other worlds, most have been "one time" transmissions. A symbolic system requires more than one utterance or more than one example so that the arbitrary pattern of meaning can be perceived as something other than index or icon. If the relationship between object and signifier is not arbitrary, we cannot be sure it is not natural.

Some have argued that the past is unknowable, as we have no lens but the present through which to scry its depths. Yet the METI/SETI project depends on the existence of an external reality: there's no point if it's just us expecting to hear from ourselves. Opinions about the comprehensibility of symbols sent into space are divided along disciplinary lines, as Quast and Capova say: "hard" scientists believe mathematical concepts will be understood universally, while humanities scholars argue that this is impossible. However, views which acknowledge a reality independent of human observation—for example, those articulated by Australian ecofeminist philosopher Val Plumwood (e.g., 2001)—have been gaining currency as we learn more about the universe. It would be vastly disappointing to confront the alien other and find it only a reflection of ourselves, our deep cultural anxieties, in the ultimate postmodern deferral of meaning. Communicating with aliens is ontological as much as epistemological. If there is no possibility of ever knowing anything outside ourselves, then we can never find them. We need the external referent to relieve the burden of solipsism.

Context and Continuity

So, I will take the optimistic path of assuming an external, knowable reality. Even so, without the context of production of symbols, neither we nor ETI have any way of knowing what they mean, only that they (potentially) exist. Quast and Capova (this volume) talk of the "displaced": the objects and messages sent to ETI which are meant to convey meaning, but in the sending become detached from their "natural" context in the setting of Earth, much like the countless stone tools which end up in obscure drawers in museum storerooms with nothing to indicate their origin, but a number written on their surface in ink pen.

The problem, Quast and Capova argue, is "when we expect an essentially arbitrary, symbolic medium-format (for example, a modulated EM signal or inscribed metal plaque of text) to lend an interpretive context or allude to decipherment clues." Without having seen a record player, I'm not sure I would be able to interpret the instructions for playing the VGR which were inscribed on them. The medium and the message may not be distinguishable, and the stripped-back context of the space vehicle, or structure of the EM signal itself, becomes an exercise in interpretation, an archaeological process of divining meaning in the interstices and relationships between material, symbols, and location.

While Traphagan and Smith (this volume) point out the problems in assuming ETI have a visual culture that in any way equates to human, Gillespie (this volume) breaks this down even further by examining "visual ecology." He notes the antiquity of the eye and color perception in the evolution of terrestrial life; and the complexities of both perceiving and interpreting images. Anthropologists have demonstrated the cultural dependence of the senses, even for something as apparently basic as color perception (Saunders, 1995); and this is before we take account of physiological phenomena such as synesthesia. How would beings, raised in the light of other suns, the density of other atmospheres, the geometries of other gravities, perceive the narrowness of "vision" bequeathed to us by the constraints of terrestrial environment?

Perhaps it may be useful to focus on *place,* rather than a one-size-fits-all message "cast into the interstellar darkness" (Capova and Quast, this volume). Once upon a time there was no option but to broadcast; but, as both observations and analytic techniques for characterizing exoplanets increase, narrowcasting enters the realms of the possible. If we consider the qualities of an exoplanetary landscape (McTier and Kipping, 2018), we might arrive at some conclusions about how to connect with the sensorium of putative inhabitants by targeting the message to senses evolved in those conditions. The first step is to mark the message as non-natural (Quast and Capova, this volume). Blend in too much, and the message will not be noticed; too little, and it becomes indecipherable. Effectively, it's the same process as reaching into a pile of gravel to pull out the one artefact which is a deliberately manufactured stone tool. Even then, as heritage court cases in Australia have demonstrated, the interpretation is still open to dispute: geologists and archaeologists will disagree about the difference between culturally and naturally shaped stones.

Mathematics is often held to be a universal language of sentience, providing the bridge between natural and cultural; but sadly, it's not quite as easy as that. On Earth, mathematics is both gendered and racially charged; once women were not supposed to be capable of such abstract thought, and colonized Indigenous peoples were assumed to have "primitive" notions of numbers and be incapable of counting beyond 10. Traphagan (this volume) notes how decimal-based mathematics is rooted in the human body, with its ten fingers and toes. And, as Mary Douglas argued in *Natural Symbols: Explorations in Cosmology*, "even the human physiology which we all share in common does not afford symbols that we can all understand" (Douglas, 1970). Nor can the concepts of logicality and rationality encoded in mathematics be assumed to be value neutral (Capova and Quast, this volume).

On Earth, the interpretation of messages is constrained by the social processes and the sensory limits of the human body, but also by gravity. Gillespie (this volume) alludes to this in referencing balance, uprightness, motion, and relative angles as components of the perceiving body and the represented body. Those accustomed to other gravities might take very different meaning from such relational factors. Gravity at 1G is a taken-for-granted background condition which structures everything about life on Earth (Gorman, 2019; 2009a). Twenty years of continuous occupation of the International Space Station is probably too short a time to detect the emergence of a different symbolization of the body in microgravity, so we have to work harder to foreground gravity's role in how we communicate. Future lunar or Martian people might come up with novel ideas about communication with ETI based on their emotional and social dispositions with regard to divergent gravities.

"Making Silent Stones Speak"[1]

A major theme of this book is "speaking for Earth"—who, to whom, and how—as a project. The material and immaterial artefacts sent out into space (even the voices speaking greetings in many languages on the VGR, and the multitude of languages and works of literature inscribed on long-term storage media, such as the disk on the European Space Agency's Rosetta spacecraft), only allow some the opportunity to be heard.

Capova and Quast (this volume) propose an interesting thought experiment. They ask not what the alien other would be able to understand from the VGR, but what they would miss. Death is notably absent from this heavily curated record; and the portrait of humanity is digitally sanitized, more aspirational than realistic. It's a deceptive overview effect for the alien: a vision of united humanity under the veil of the blue sky, concealing all of the horrors of extinction, pollution, and destruction closer to the surface.

We assume, perhaps, that self-appointed speakers for Earth are always of good intention, and *so far, as we know,* that seems to be true. There have been no threats, at least, unless the concealment of terrestrial dystopia is a sort of threat in itself. (However, messages of ill-will have an equally poor chance of decipherment as those of good intent.) Having said that, in 2020 a group of young witches on the social media platform TikTok reputedly tried to hex the Moon (Panecasio, 2020). As Arthur C. Clarke said in 1962, "Any sufficiently advanced technology is indistinguishable from magic." While I can't endorse the colonialist assumptions underlying this much-quoted statement, there is something to consider in how magical practice uses symbolism to reconcile fundamental differences in worldview.

Many of the messages sent out as physical objects could be argued to fulfil the role of talismans or fetishes. An alien observer might incorporate them into their own sensory and memory world in unpredictable ways. I am reminded of how, from the 10th century CE in Europe and well into the 19th century in some places, flaked stone points from the Mesolithic and Neolithic eras were interpreted as "elf-shots," the weapons of a faery world with its own logic. They were both dangerous, as the source of disease and misfortune for humans and animals, and their own antidote. Such arrow heads were sometimes set in silver to be worn by a person to ward off the danger they promised. The talisman can catalyze action, just as the monolith in *2001: A Space Odyssey* did (Quast, "A Cultural History of 'Speaking' for Earth," this volume), without its purpose or meaning needing to be known. Such an object can reveal other worlds, but they may bear no relation to the moment in time, or the place where the object originated.

Archaeologists take a flake of stone, a sherd of pottery, a grain of quartz sand, and make of it a scrying glass to see deep into a past, hoping to resolve details through the haze of more recent times. It was another class of stone object that Elizabethan sages Dr. John Dee and Edward Kelley used to decipher the Enochian language of the angels—a polished obsidian mirror taken from Mexico sometime between 1527 and 1530 CE during the military expeditions of Hernán Cortés. Currently in the collections of the British Library, the mirror is a featureless black disk which betrays nothing of its potential. Every vision reflected on its surface is in the eye of the beholder. In his prologue, Quast invites us to see images of Earth from the outside as such scrying tools for a much more recent past. The Pale Blue Dot that Carl Sagan persuaded NASA to photograph from the Voyager 1 deep space probe in 1990 becomes a "mental artefact of the Anthropocene." Through the mediation of space technology, Earth itself is terraformed twice over, in action and in perception.

Where you are situated in space and time makes a difference. There is a before and an after, separated by distance. The rite of passage, as Quast ("A Cultural History of 'Speaking' for Earth," this volume) describes the practice of sending national emblems into space, transforms states of being. Van Gennep (1960 [1909]) likened the rite of passage to moving through the rooms and corridors of a house. Quast's metaphor can be extended further, to apply to crossing the boundaries between mundane reality and magical worlds, human and machine, Earth, and space, present and past, ourselves and other. Effectively, we are always looking for the thresholds between the cultural and natural. Thus far, we are still on the landing of a promised room, counting the steps over and over. On the other side, objects and symbols generate new meanings, performing the work that they were sent to do. These are messages that must traverse time, space, bodies, and senses. Echoing L.P. Hartley's aphorism "the past is a foreign country," we could equally say "the future is an exoplanet." Sometimes, that exoplanet is Earth.

Note

1. Quote from the title of Schick, K.D. and Toth, N. (1993) *Making Silent Stones Speak: Human Evolution and the Dawn of Technology.* New York: Simon & Schuster.

A Brief History
and an Imminent Future

A BRIEF HISTORY

1. c. 3.6m BCE—Laetoli *Australopithecus afarensis* footprints in Tanzania
2. c. 3.3m BCE—Paleolithic era and earliest known stone tool developments
3. c. 3.2m BCE—"Lucy" remains from *Australopithecus afarensis* species
4. c. 2.588m BCE—Start of Quaternary Period (to present). Start of Pleistocene Epoch (to c. 9700 BCE)
5. c. 2.1m BCE —Assumed emergence of *Homo habilis*
6. c. 2m BCE —Assumed emergence of *Homo erectus*
7. c. 430000 BCE—Potential evidence for *Homo Neanderthalensis* (Neanderthals)
8. c. 280000–250000 BCE—Berekhat Ram, if human material culture, roughly created
9. c. 217000–122700 BCE—Age estimates for *Denisovan* hominids in Siberia
10. c. 168000 BCE—Departure for light now reaching us from Supernova 1987A
11. c. 117000–108000 BCE—Last known population of *Homo erectus* [soloensis]
12. c. 100000 BCE—Blombos Cave pigment workshop utensils
13. c. 71000 BCE—Blombos Cave "oldest known drawing" (73000 BP)
14. c. 50000 BCE—Reliable limit for Radiocarbon (C-14) dating
15. c. 37300 BCE—Oldest handprints in El Castillo complex
16. c. 34000 BCE—Oldest paintings at Altamira Cave (36000 BP)
17. c. 23000 BCE—Venus of Willendorf created (25000 BP)
18. c. 15000 BCE—Oldest dated Lascaux cave paintings (17000 BP)
19. c. 9700 BCE—Holocene epoch begins (11650 BP) (to present)
20. c. 6300 BCE—Departure for light now reaching us from NGC 1952
21. 4241 BCE—Earliest calendar record for Egyptian bureaucratic-priestly lore
22. c. 3500 BCE—Earliest known Mesopotamian pictographs and Indus script use
23. c. 3200 BCE—Earliest Babylonian cuneiform and Egyptian hieroglyphs
24. c. 2700–2200 BCE—Campo del Cielo earliest impact (4200–2700 BP)
25. c. 2000 BCE—Last surviving population of Woolley mammoths on Wrangel Island

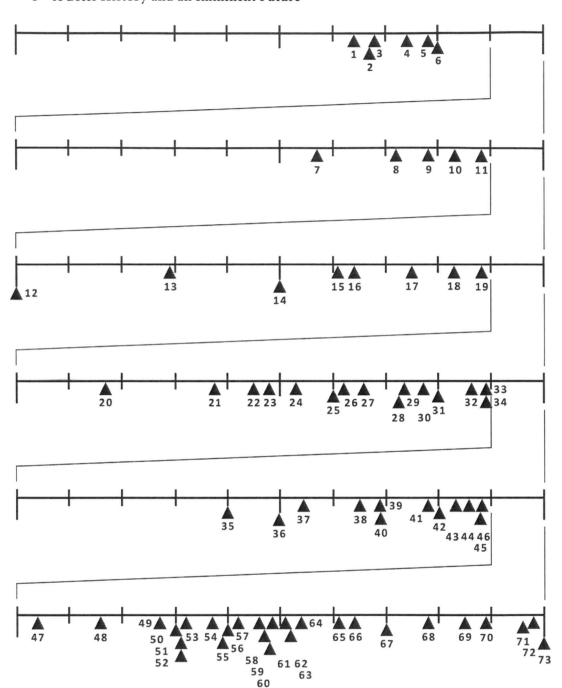

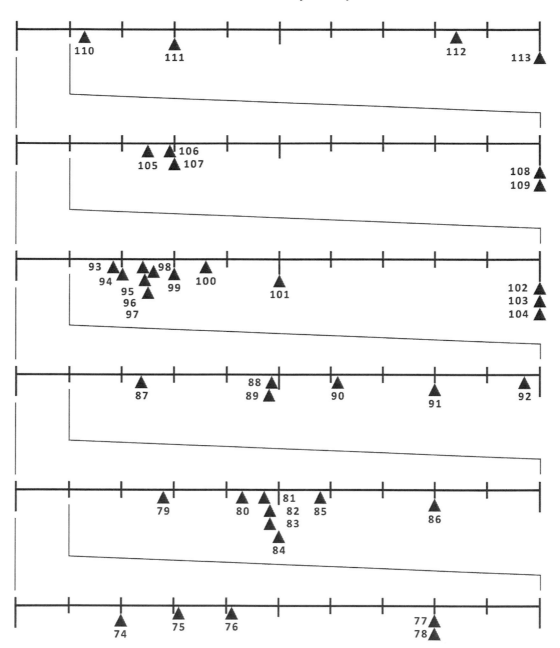

Opposite and above: A timeline illustrating historical events and future projections discussed throughout the book, in addition to some contextual markers along this arrow of time (illustration by Paul E. Quast).

26. c. 1800–1400 BCE—Earliest Linear A and Byblos script samples
27. c. 1425–1200 BCE—Earliest Linear B script samples
28. 762–631 BCE—Sargon II, Sennacherib, Esarhaddon, Ashurbanipal, Ashur-etil-ilani lifespans
29. c. 644–544 BCE—Buddhas of Bamyan constructed
30. c. 285–246 BCE—Library of Alexandria established
31. 0 BCE—Anno Domini (AD)/Gregorian calendar year divide
32. 622 CE—Hijri era/1st year of Islamic calendar
33. 949 CE—1 Feb, Last closest alignment of all 8 planets, and Pluto (within a 90° arc)
34. c. 975–1025 CE—Oldest surviving copy of Epic of Beowulf
35. c. 1400 CE—Oldest surviving Tsunami Stones erected
36. 1492 CE—12 October, Columbus lands in Americas, starting European colonialization
37. 1543 CE—Copernicus publishes *On the Revolutions of the Heavenly Sphere*
38. c. 1650 CE—Proposed start for Age of Enlightenment
39. 1686 CE—Bernard le Bovier de Fontenelle publishes *Plurality of Worlds*
40. 1687 and 1704 CE—Newton publishes *Principia* and *Opticks* respectably
41. c. 1780 CE—Proposed start of Anthropocene. Accepted start of Industrial Revolution
42. 1822 CE—Jean-François Champollion translates Egyptian hieroglyphs
43. 1859 CE—Darwin publishes *On the Origin of Species*
44. 1878 CE—Exposition Universelle world's fair in Paris
45. 1898 CE—H.G. Wells publishes *The War of the Worlds*
46. 1901 CE—12 December, Marconi's Transatlantic transmission test
47. 1924 CE—23 November, Hubble's observations expand scale of universe
48. 1936 CE—1 August, Adolf Hitler addresses a televised Berlin Olympics
49. 1947 CE—12 March, Cold War era general starting point
50. c. 1950 CE—The "Great Acceleration" of the Anthropocene
51. c. 1951–1953—Watson and Crick uncover DNAs double helix
52. 1951 CE—15 October, *I Love Lucy* first TV broadcast
53. 1952 CE—1 November, Ivy Mike thermonuclear test
54. 1957 CE—4 October, Space Age begins with launch of Sputnik. 3 November, Launch of dog Laika
55. 1959 CE—13 September, Luna 2 probe impacts Lunar surface
56. 1960 CE—Freudenthal developed *Lincos* language
57. 1962 CE—9 July, Starfish Prime thermonuclear test
58. 1966 CE—1 March, Venera 3 allegedly impacts Venus's surface
59. 1967 CE—20 April, Surveyor 3 lands on Moon. 10 October, Outer Space Treaty entered into force
60. 1968 CE—24 December, *Earthrise* taken during Apollo 8
61. 1969–1972 CE—Apollo Lunar Landing operations begin (Apollo 11, 12, 14, 15, 16 and 17)
62. 1971 CE—27 November, Mars 2 lander impacts Martian surface
63. 1972–1973 CE—2 March and 5 April, Pioneer 10 and Pioneer 11 launched
64. 1977 CE—20 August and 5 September, Voyager 2 and Voyager 1 launched
65. 1981–2011 CE—U.S. Space Shuttle operations begin

66. 1984 CE—SETI Institute established
67. 1990 CE—14 February, *Pale Blue Dot* photograph taken
68. 1998 CE—20 November, First ISS module launched
69. 2005 CE—14 January, Huygens module lands on Titan
70. 2009 CE—7 March, Kepler Space Telescope launched
71. c. 2016 CE—Predicted earliest response for Greetings to Altair (from 1983)
72. 2018 CE—6 February, Musk launches Tesla Roadster into space
73. 2021 CE—July, Bezos and Branson spawn "Billionaire Space [Tourist] Race"

* * *

AN IMMINENT FUTURE

74. c. 2040 CE—J002E3 may next return to Earth orbit vicinity
75. c. 2051–2069 CE—Cosmic Call question arrives to Sagitta and Cygnus targets
76. 2061 CE—28 July, Halley's Comet next approach
77. 2100 CE—Targeted timeline for United Nations climate actions
78. c. 2100 CE—Approximated arrival time for *Message to Qo'nos*
79. c. 2300 CE—Voyager 1 reaches boundary of Oort cloud
80. c. 2450 CE—*A Simple Response to an Elemental Message* anticipated arrival at Polaris
81. 2492 CE—6 May, Next closest alignment of all 8 planets, and Pluto (within a 90° arc)
82. c. 2500 CE—Svalbard Global Seed Vault specimen viability limit (from today)
83. c. 2500 CE—microfilm projected longevity (from today)
84. 2514 CE—500-year microbiology experiment ends
85. c. 2600 CE—All RORSAT nuclear powered satellite orbits likely to have decayed
86. 2800 CE—29 February, Gregorian calendar and Revised Julian calendar will no longer be synchronized
87. c. 4385 CE—Hale-Bopp next closest approach
88. 6939 CE—Westinghouse I grand opening
89. 6970 CE—Expo '70 grand opening
90. 8113 CE—Crypt of Civilization "omega" opening
91. 10000 CE—Projected lifespan of the Clock of the Long Now. WIPP nuclear waste depository goal
92. c. 11700–13727 CE—Vega becomes the North Star again due to Earth's axial precession
93. c. 18500 CE—Pioneer 11 likely to pass within 3.4 ly of Alpha Centauri
94. c. 20000 CE—Languages may retain only 1 out of 100 currently-spoken words

95. c. 24000 CE—Radiation in Chernobyl Exclusion Zone return to normal levels

96. 24100 CE—Half-life of Plutonium-239 isotope (from today)

97. c. 25000 CE—*Arecibo Message* arrival in region of M13

98. c. 26000 CE—*Nancay Message* arrival to galactic center, 26,000 ly away

99. c. 30000 CE—Voyager passes through Oort cloud boundary

100. c. 36000 CE—The star Ross 248 will come within 3.024 ly of Sun

101. 50000 CE—*KEO* (if launched) planned retrieval era

102. 100000 CE—Onkalo nuclear waste depository goal

103. c. 100000 CE—Stellar constellations in night sky become unrecognizable

104. c. 100000–150000 CE—10 percent of anthropogenic CO_2 remains in atmosphere (if now cut)

105. c. 250000 CE—Minimum time for WIPP plutonium to cease being lethal to life

106. c. 296000 CE—Voyager 2 may pass by Sirius at a distance of 4.3 ly

107. c. 300000 CE—Voyager 1 may pass within 1 ly of TYC 3135–52–1

108. 1m years CE—HUDOC and Memory of Mankind project goals

109. c. 1m years CE—Arnano sapphire disk projected longevity

110. c. 1.28m years CE—Gliese 710 will pass about 0.221 ly from the Sun, likely perturbing Oort cloud

111. c. 3m years CE—Earth's measure of day will be one minute longer than it is at present

112. c. 8.4m years CE—LAGEOS satellite re-enters atmosphere

113. c. 10m years CE—estimated biodiversity recovery after potential Holocene mass extinction

Part 1

WHY WE SPEAK

Prologue

Murmuring from *Earth: A Human Portrait of Representational Phenomena Across Deep Space and Cosmic Time*

PAUL E. QUAST

On February 14, 1990, having traveled a distance of over six billion kilometers, the Voyager 1 spacecraft pivoted its tired eyes back towards its homeworld one last time. The 1500 mm high-resolution, narrow angle vidicon camera would snap 60 frames at different millisecond exposure rates for each of the device's blue (0.72), green (0.48) and violet (0.72) color filters.

After storing these optical observations on magnetic tape for several months, the probe duly broadcast this data using its meager 22.4-Watt transmitter back towards its intended recipients over five and a half light-hours away[1]—not a trivial task, even with today's technological capabilities and global infrastructure. These frames, favorably captured within our species limited ocular sensitivity ranges, would become the final images to be relayed by this harbinger spacecraft as the cameras were thereafter deactivated to conserve power for the remainder of the probe's operational lifespan. At the time of writing, this journey of discovery continues through the interstellar darkness. Out of these frames, three clearly depicted one particular object of interest which, when recombined into a single-color mosaic, produced what is now known as the iconic *Pale Blue Dot* photograph (see color insert).

Parallel to many of my colleagues who have been actively involved within studying the evolving material relationships, mental impacts, and ensuing schisms between modern aerospace technologies and various disciplines in the humanities, I have been perpetually fascinated by the observations we gain from space exploration, and the many wondrous phenomena we are fortuitous to contemplate at this moment in cosmic evolution. The Hubble, Kepler and James Webb Space Telescopes, Planck space observatory and GAIA, amongst other great space observatories that serve as "telescopic time machines" (Benford, 1999), continue to passively expand the periphery of our celestial horizons outwards; furnishing astronomers with a humbling window into the early universe, while stoking our imaginations to critically address some of the most fundamental questions about our knowledge of cosmology, the evolution of matter, and the frequency of life's origins. Aside from these scientific, theological, and philosophical implications, our windows into cosmic evolution provide humanity with a context in

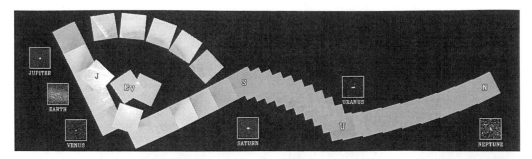

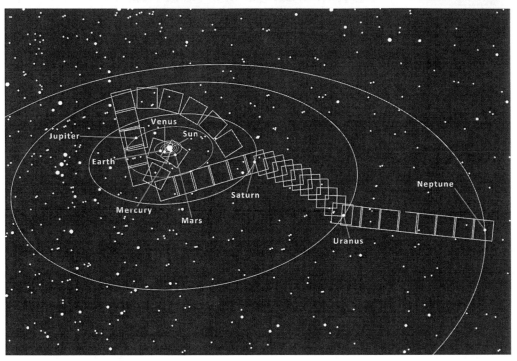

The 60 frames comprising the "Voyager family portrait," along with an annotated diagram for planet locations (photographic frames from NASA/JPL, processed by Carolyn Porco. Diagram based on a visual created by NASA/JPL. Re-illustrated by Paul E. Quast).

linear time, enabling us to position the big histories of various human societies, significant developmental stages for complex life, and Earth's geological chronology within the 13.799-billion-year timescale of the universe (Planck Collaboration, 2015). Time, in context with the human lifespan, seems to be a fleeting affair, measured primarily in the slow but predictable revolutions around our local star, itself only about 4.6 billion revolutions old.

By contrast, Earth portraits—taken during human spaceflight or distant robotic missions—are a unique class of remote observations that provide us with an inverted, ruminative *you are here* substrate outside their universal counterparts. The perceptual dissonance supplied by these non-traditional perspectives of our homeworld frequently serve to expose and disrupt many of the thin psychological pretexts we instinctively accept about our perceived position on this sphere. In the process, these images captivate the minds and energies of populations within cultural and cognitive shifts of awareness such as the notable "Overview Effect," and less-frequently experienced "Copernican Perspective" (White, 1987). These meditative photographs—as technical products, arising from these explorational activities—possess wonderfully counterintuitive relationships with our experiences of time, material connections with our planet, and our broader understandings of these subjects within the context of our local, cosmic neighborhood.

At the moment of their creation, these portraits physically peer back through *Earth-time* and enable us to reflect upon the immediate past of our homeworld, while also expanding our mind's eye by allowing us to observe our planet as it will appear to the futurescape of the cosmos. Using this mental time-travel process, the iconic *Earthrise* photograph, taken during the Apollo 8 mission from the far side of the Moon, enabled us to gaze back one second in time from lunar orbit. By stark comparison, the first image of our planet captured from space, taken by a converted V-2 rocket performing a sub-orbital flight (No. 13) in October 1946, enabled us to gaze back only about 200 microseconds. Similarly, the Spirit rover on Mars photographed Earth as it looked about 16 minutes into our past, while Cassini's recent *The Day Earth Smiled* portrait[2] served as a reflective window of our world as it appeared 80 minutes prior during a Saturnian solar eclipse. The perspectives of our planet do change, but these physical images seem to possess a timeless "snapshot" quality, one that becomes frozen for the ages, allowing any spectator with a viewing device to share in "the captured moment."

While these technologically mediated windows into our planetary past are rather limited to our immediate position within local space-time, as we can see, the further away a photograph is taken, the further back into this Earth-time *we can see ourselves*. By comparison, the Pale Blue Dot was created using photons that had departed the Earth system over 5.6 hours previously—a specter from our immediate past cast out into deep space which, much like our surrounding cosmic perspective, is faithfully preserved by the same quirk of physics; the finite speed of light.[3] The *Earth-frames*, taken by Voyager at 04:48 GMT on February 14, 1990, essentially captured light reflected by our planet that had departed around 23:12 GMT the previous day. This image remains as the most distant, poignant depiction of the Tellus system we've yet captured, but we do not need to look too far within mission catalogs such as the EPIC camera on DSCOVR, HDEV on the International Space Station, or elsewhere on the World Wide Web to find fascinating and beautiful photographs of our persistently changing planet. Our global telecommunication services are saturated with constellations of Earth imagery which, rather conversely, tends to dilute our relationship with the "only home we have ever known" (Sagan, 1994). Unfortunately, the now very trivial nature of these material images allows us to take for granted the many intricate ecosystems that have nurtured the development of not just our species, but also the diverse range of biota we share "Spaceship Earth"[4] with, and also the great, silent majority of life to yet come in this planet's future.

The Pale Blue Dot—as a technical concept and photograph—is one of my particularly favorite Earth portraits but, the more I inspect it, the more my opinion consistently vacillates on why this poor-resolution, 0.12 percentage of a single, pale pixel is significant to me. Is this mysterious preference due to the nostalgia I personally attach to a now time-faded, coffee-stained illustration of the image in Sagan's publication of the same name? It could be due to the traditionally unfamiliar vantage point of Earth as a meandering island corkscrewing through the vast cosmos, or perhaps the synthesis of representing "everyone that I have ever heard of," while concurrently featuring everyone

I have never known? Perhaps it is the bio-existentialist context lent by Sagan's resonant words, the rationalization of all nations separated by their ideals into one single point of pale light, the futurist and environmentalist connotations, popular science appeal, professional appreciation for the perceptual faculties required to *see* a conventional and minimalist picture, or even the absurdity that this image embodies the disparate and seemingly insignificant hubristic interactions of an entire species which once deemed itself as the center of the known universe? Perhaps it is simply due to the epistemic scaffolding supplied by this small terrestrial stage for the entire phylogenetic tree of life as we define it—and the single evolutionary branch tip where humanity precariously dangles, along this relatively short, transformative journey from the contested *Homo sapiens sapiens* subspecies to an envisioned "*Homo spacians*" (White, 1987).

The established biosphere and geochemical cycles on Earth may have continuously sculpted our planet for complex organisms to eventually emerge over eons, but the advent of organized human societies and technological dexterity, in concordance with "industrial time" as a conceptual barometer for our species' social and cultural proliferation under the inertia of our own feet, would substantially attenuate these ecologies for subsets of species (Lovelock, 1972). Informally known as the Anthropocene (Crutzen and Stoermer, 2000), this tentatively recognized planetary epoch of thermodynamic and environmental disequilibrium was coined to define an age in which the aggregate of all natural geomorphic and atmospheric processes on our homeworld began to denote measurable changes at an unprecedented rate, initiated by Earth's principle geological agent: humanity. As poignantly surmised by the environmental historian John Robert McNeill (2000), "In the twentieth century we became what most cultures long imagined us to be: lords of the biosphere." However, it is questionable how we have chosen to exercise this anthropocentric authority in adapting our homeworld to our built image, or indeed recognize the range of causal responses and positive feedbacks that are now arising from this amateur geoengineering experiment. Given this short-termism thinking, one could make the argument that we are falling short of reaching the potential mantle of "*Homo prospectus*"—defined by the psychologist Martin Seligman as a species whose behaviors are guided by its imaginative projections into the realm of its future, and sustainability therein (Seligman *et al.*, 2016).

Formal recognition of this epoch, along with a definitive "Golden Spike,"[5] is still subject to academic debate by the various international geoscience authorities. But the technical concepts of this human-manifest age have been philosophically embraced by many disciplines that continue to ignite ardent discussions in planetary stewardship praxis across all facets of organized human societies. Earth portraits certainly possess this studious, albeit relatively modern, custodial relationship with our past interactions, as seen through remote observation constellations and macroscopic research programs that monitor our dynamic planet; enabling us to position the fleeting steps that comprise the big, dovetailing histories of various contemporary human societies and recent global changes within vast overview composites of our world. However, in a counterintuitive sense, reflecting on these overview snapshots also allows us to conceptually peer *forward in time* and, perhaps, forecast the causality of humanities' growing, indelible reach through the coming ages.

If the Anthropocene has involved the artificial redistribution of material elements such as carbon, oxygen and nitrogen, alongside other compound molecules across the Earth system, this is also true of the spectral signatures of this accumulative matter, and the physical migration of cumbersome amounts of aluminum, silicon, titanium, carbon fiber, nickel, and cadmium from terrestrial environments into outer space (Gorman, 2014). In many ways, I suspect my underlying, deeper appreciation for the Pale Blue Dot is due to the symbolism of this photograph as a *mental artefact* for this Anthropocene era. It can be seen to draw into sharp focus the entwined cognitive and spatiotemporal dimensions of various active legacies that locally impact our world yet have now come to also represent modern human behavioral patterns distantly in material epiphenomena beyond the terrestrial biome—the unfolding futurescape, encoded in frequencies of light, that spacecraft currently take photographs of.

Our planetary system, and the aggregate of our interactions on it, does not end at the conceptualized Kármán line "boundary" of our atmosphere (Olson, 2010). To paraphrase Nigel Clark (2005), we reside on an integrated and open system in interchange with a dynamic cosmos in which the exchange of matter, information and energy is a frequent rather than rare occurrence—as plainly seen through simple optical phenomena such as *Earthshine*. By close association, the agency of humanities' physical planetary interactions also transpires much farther beyond the human and terrestrial realms—as unintended *messengers*. The "pale blue" atmosphere that has characterized our planet's oxygen-rich emission spectrum over the past several billion years, now radiates a gradually altering isotopic composition or, as the philosopher Frank White (1990) coined, "entropy pools." This is largely seen as due to our increasing techno-industrial respiration, but also arguably as an initial result of our species' gradual shift towards agriculture and practices of widespread land clearance over 10,000 years ago (Steffen *et al.*, 2011).[6] The Pale Blue Dot faithfully preserves a static, one-second snapshot of *our technosignature* legacy, or *Earth series, edition February 1990*; a collection of industrialized interactions within our biosphere, encoded within photons, that propagate outwards at the speed of light—epiphenomena that may take far longer than the entire phylogenetic history of the modern *Homo sapiens* genus to fully dissipate into the background noise of our galaxy.[7]

Photons that were sampled for this photograph are *now* over 30 light-years away. What stories about *us* do they carry? How might eavesdropping on the chronological sequence of atmospheric "strata," or ensuing editions of this *Earth compendium*, freely propagating outward from our planet, also supplement a potential external observers' interpretation of our world? Planetary custodianship—as rightfully advocated by those who wish to conserve our world for posterity—may also leave a distinctive *corrective* geoengineering imprint within this technosignature, thus adding to the scope of involuntary ephemera. Earth as *the artefact* of postmodern material culture is one of our species' most profound, enduring legacies to the cosmos, albeit an unintentional and relatively bewildering one, supporting often disordered, conflicting, incoherent, and illogical narratives about this ever-changing terrestrial stage.

The long-term paleo-archaeological record of life, what we define as intelligence, and signs of industrialized human terraforming activities, is already ingrained within the various naturally-disordered biosignatures and inadvertent technosignatures that scatter outward from the Tellus system; aging X-ray pulses emitted by atmospheric nuclear detonations such as the *Starfish Prime* shot (Dumas, 2015), the short-lived isotopic products and daughter by-products of radioactive decay from other nuclear testing activities, the carbon saturation of our atmosphere preserved in Earth's emissions spectrum alongside other attenuations in gaseous concentrations (Schneider *et al.*, 2010), gradually increasing thermal disequilibrium, and the artificial illumination emissions from global lighting infrastructure (Loeb and Turner, 2012). From an archaeological perspective, we may consider these synthetic, layered material legacies as "technofossils"; traces which serve as the *background context* for our more recent anthropological records in the cosmos, with varying technical degrees of observability.

In addition to these unintended, and debatably observable, background signals, the strata of material cultures arising from the Space Age exploration present us with another, distinctively new chapter in this exo-archaeological record. This accumulative process of active, defunct, and discarded space hardware presents evidence for how our species (or populations with access to these aerospace technologies) consciously interacted with their surrounding cosmic environment. But, perhaps more importantly, these aging heritage assets also indicate how the technical exploration of space, using significant artefacts of material culture, has demonstrably influenced the behavioral characteristics of modern human societies on Earth as mediated through orbital navigation constellations, surveillance systems, planetary exploration missions, and the interconnection of global communities through telecommunication services. We do not need to imagine distant observers to contemplate how significant some of this material heritage may be *for ourselves*—impacts range from "small-scale" behavioral changes (for example, the personal use of mobile devices, and how they impact familial relationships—see Miller *et al.*, 2021), to the macroscopic (e.g., how the proliferation of these devices changes how societies function and communicate and the resulting biological effects, such as hyper-timekeeping, dopamine-driven feedback loops etc.).

To date, these technologically-mediated legacies have cultivated an uneven cacophony of leakage radiation in electromagnetic (henceforth EM) frequencies over the past century (Scheffer, 2004)[8] which have already washed over 12,000–15,000 neighboring stellar systems (Gertz, 2016a). Along with these intangible legacies now representing Earth from afar, there are more physical elements abound in our planetary neighborhood, with examples including localized planetary debris fields, satellite constellations, interplanetary probes, hardware on other astronomical bodies, and an electric sports car haphazardly abandoned elsewhere in the Solar System. Voyager 1, the progenitor of the Pale Blue Dot, defines the outer corporeal boundary of this indelible material landscape, and will continue to (as Sagan stated) "plummet into the interstellar dark" indefinitely. Yet, this probe is still wonderfully connected with its terrestrial roots through its frequent and weak, electromagnetic murmurs *directed to Earth*. To listen out for these 13-watt whimpers, equivalent to the power for a refrigerator lightbulb, you need a large collection surface and sensitive electrical filters—technical constraints that may inhibit observation of our EM leakage from afar.

From Disorganized Signs to Material Signatures

The cultural phenomenon of *messages*,[9] otherwise colloquially known as *Postcards from Earth*, is an intriguing, purposeful subset of this archaeological landscape, arising as a direct result of social engagements with Space Age technologies, and conceptually stemming from age-old traditions and communicative aspirations still practiced on Earth today; time capsules, personal artefact deposits, and archaeological sites we partially leave undisturbed for future research opportunities (Jarvis, 2003). Much like their terrestrial counterparts, many of these ideological messages are aimed at transferring cultural properties to distant spatiotemporal recipients across differing material

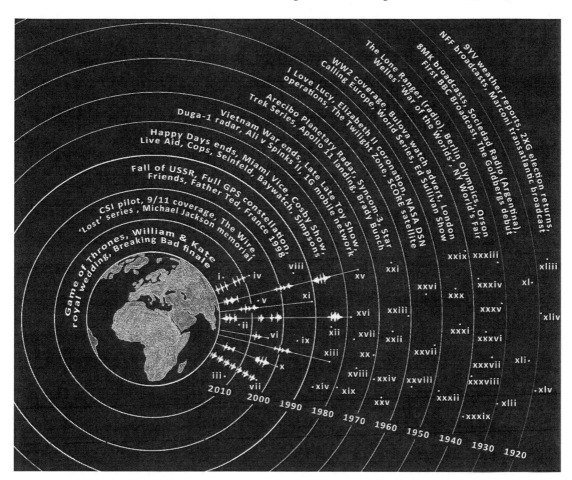

The background Radiosphere profile for Earth (marked in intervals of decades), along with unintentional types of interstellar message "layering" present across this leakage (i.e., patterns resulting from multiple transmissions sent towards the same target systems). The background consists of unassorted electromagnetic signals from radar search programs, deep-space exploration missions, satellite telecommunication services, and defensive missile tracking/early-warning systems, set against the historical backdrop of popular television and radio broadcasts for time context. Interstellar message "layer" types are described elsewhere (Quast, 2021). Message durations and star locations are exaggerated, but stellar distances are accepted measures. Featured Stars: Alpha Centauri (i), Barnard's Star (ii), Sirius (iii), GJ 273 (iv), Teegarden's Star (v), Altair (vi), HD 119850 (vii), Vega (viii), Fomalhaut (ix), Gliese 581 (x), HD 20630 (xi), Denebola (xii), Arcturus (xiii), Pollux (xiv), 47 Ursae Majoris (xv), 55 Cancri (xvi), Capella (xvii), Errai (xviii), TRAPPIST-1 (xix), Alderamin (xx), HD 7924 (xxi), Caph (xxii), Gliese 3293 (xxiii), Castor (xxiv), Gliese 676 (xxv), Aldebaran (xxvi), HD 197076 (xxvii), Larawag (xxviii), Regulus (xxix), Heze (xxx), Alphecca (xxxi), Aljanah (xxxii), Sabik (xxxiii), Ankaa (xxxiv), Gacrux (xxxv), Plecda (xxxvi), Algorab (xxxvii), Mizar (xxxviii), Alioth (xxxix), Alnasl (xl), Alpheratz (xli), Algol (xlii), Alkaid (xliii), Alnair (xliv), Edasich (xlv) (illustration by Paul E. Quast).

domains, including future human societies, post-biological entities, second-generation (sentient) terrestrial species, or an independently evolved Extra-Terrestrial Intelligence (ETI), for a plethora of communicative and votive applications.[10]

Understanding how such esoteric constructs of material culture may be interpreted by these presently unobservable recipients would require us to paradoxically know, in advance, whether an interlocutor may share in characteristics like our sense modalities, phylogenetic history, morphology, embodied mental faculties, and other properties of convergent evolution, or native cultural histories and epistemic conventions that we presently recognize as "human." Any semblance of "familiarity" we can project into this profoundly unknowable arena should ideally grant us pause for reassessing our capabilities to *intuitively know about something in a knowledge vacuum*. I would also argue that these properties are only *some* of the more definable obstructions, with more still likely to be secreted beneath a diverse range of other anthropocentric—or indeed earthly—partialities. Considering the likelihood of such convergences, a healthy skepticism is advised for such projects. I would encourage the readers to challenge any such anthropocentric assertions about imagined recipients being readily capable of interpreting our crafted messages, given the fact that we cannot possess any empirical evidence, or verifiable propositions, to support such *a priori* inferences.

Much like our Space Age window into examining recent changes across our biosphere, it is however possible to retrospectively investigate ourselves, as varyingly depicted within the material biographies of these modern messages—artefacts that serve as a unique storehouse of evidentiary information, documenting *how we understand and represent ourselves*. This focus may allow us to theorize how the authoring human populations consciously interacted with and, in turn, were mentally re-shaped by their surrounding celestial ecologies. However, it may also contribute crucial insights into evaluating how this residual material, alongside aerospace technologies and global satellite services in general, have profoundly influenced the transformation of modern human behavior. This is one of the prevailing topics now being explored today under the purview of space archaeological studies, a discipline which is no longer tethered to solely learning from the sediments and artefacts of the distant past.

One of the early pioneers of this modern research avenue, the behavioral archaeologist Michael Schiffer (1999), once opined that many social scientists have "ignored what might be most distinctive and significant about our species: human life consists of ceaseless and varied interactions among people and myriad kinds of things." Similarly, the archaeologist Lewis Binford (1962) commented that these material "things" function in the major subsystems of every society, such as technology, social organization, and abstract ideology; objects that are, as further argued by the archaeologist Carl Knappett (2005), "bodily and cognitive prosthetics … bound up in humans in their guises as biological, psychological, and social beings, as bio-psycho-social totalities." By treating this accumulated archaeological record as a unique source of enactivist, material evidence for studying processes of long-term change (Plog, 1974; Schiffer, 1975), *we could* feasibly construct inferences from the subtexts of purposeful artefacts that may inversely tell us more about the undercurrents of contemporary human space activities than we previously thought possible—in contrast to the frequent peripheral readings of these curated, representational biographies.

There is much to unpack from this inference. Our social lifestyles with myriads of things, and embodied mental interactions with others, are implicitly dependent upon such underlying, idiosyncratic behaviors, some of which may always remain obscure to us as *external voyeurs*—even if we live in the same eras, and possess all the fragments necessary for contemplating the symbolism, and origin points, for these period-pieces of material culture. Despite this, these messaging relics and signals from the early

Anthropocene enable us to raise provisional questions about the conscious relationships between people and material resources, cultural reinvention, personality, place, semiotic expression, memory, ritual exercises, and whether social attitudes correlate with selective patterns of cultural identity, alongside how the small groupings of project authors believe these artifices represent us (conceptually, ideologically, or literally) during a critical junction in the trajectories of human civilizations.[11] Regardless of whether we possess *enough* material evidence to fully support such inferences, at the very least, these approaches can reliably contribute new concepts, principles and heuristics to the behavioral toolbox to inform future material culture studies.

Thought is a nonmaterial, entangled bio-mental operation, projected outwards into the material world beyond the embodied mind and skin of any singular individual. We cannot physically touch it, or accurately measure it yet, at its inception, this rationalized substrate of information emerges as an intangible, organized concept, given shape from degradable organic matter (our brains and sensorimotor system) in response to its unique socio-cultural environment. These nonmaterial ideas and concepts are generated in accordance with the mental paradigms that selectively filter our living perceptions of an experienced world, and principally serve as a gesture towards inspiring interaction over different spatial settings (internal, interpersonal etc.) and experiential timescales (e.g., neural, biological, cultural, political etc.). But it can also—to varying degrees of questionable success—be symbolically packaged within constructs of material culture (and our extended relationships with these prostheses) for intervals of time longer than the individual lifespan of the intracranial structures that created them: the proverbial "ghost in the machine" or *memory-imbued matter* in this case. Such externalized objects of material culture also actively *reshape* the neural plasticity of our living minds in return, challenging our *ways of seeing* and thinking through matter usage, or cross-modal "thinging"—defined by the archaeologist Lambros Malafouris (2019) as the kind of thinking we do primarily within and through material things. (Thinking, by comparison, is defined as usually something we do in the absence of material things.)

These mnemonic objects and metaphysical EM emissaries, as representative extensions of their creator's minds, may therefore enable us (as fellow residents of these same socio-cultural subsystems) to provisionally infer how recent populations with access to aerospace technologies consciously reflected on, and reprocessed, their social interactions with outer space, before externalizing these thoughts within this material record of messages for a variety of explorative, communicative and psychological applications. It is frequently emphasized that these systematic processes function in a similar capacity to what Gerald Hawkins (1983) referred to as conceptual "mindsteps" for our species to the cosmos— characteristics that initially served as the mental foundations for our ancestor's numerous abstract beliefs, arts, languages and mathematical rationalizations of reality (and, thereafter, for notions of identity, heritage, value, trade and polity), which developed across different timescales, environments and social settings.

Given this unique lineage, it is perhaps not surprising to note that making sense of parietal art is often evoked as a metaphor for discussing the "incommensurability problem" (Vakoch, 1998a) which we face when transferring cultural[12] information to unknown recipients across millennia, while also correctly re-interpreting these dislocated artefacts of generational "distributed cognition"[13] (Benford, 1999; Paglen, 2012; Gorman, 2019). Framing the challenge in this way, however, broadly illustrates some disciplinary tendencies to misrepresent our capabilities for deducing meaning from isolated artefacts as

A negative handprint at the Cueva del Castillo archaeological site. Uranium/Thorium dating methods place the handprint at c. 39,000 BCE, implying that the vestige may in fact be material culture from Homo neanderthalensis. Featured in *A Simple Response to an Elemental Message* as image #4 (photograph courtesy The Cave Art Project. ©Pablo Herreros Ubalde. Courtesy Nuria Herreros Ubalde/Yo Mono Foundation).

a reduced *encoding-storage-reception-retrieval scheme*. "Making things mean," we may find, is a much more complicated affair that likely requires *living cultural and social settings* (and innate knowledge therein), rather than reading from isolated material signs, created by lives we verifiably know little about. Moreover, we should approach such antiquated signs as a form of *documentation for living thinking and thinging processes*, rather than aged records we may simply be able to read one day, and therefore understand, through the lens of *our own lived experiences*. We may never know exactly why a child's handprints were carefully impressed upon the walls of a cave. Yet, when we encounter such vestiges, we can understand to a degree the very human desire to create a mark upon our world, facilitated outwards through the extension of our extremities, and the articulative mental faculties that enabled this manifestation of purposeful creative thought as evidenced within our ancestor's ancient pigment workshop, imported ochres, and other material artefacts found in the Blombos Cave excavations. We can surely (or in some instances, at least *hopefully*) do better in interpreting our recent generation's intent and motives for leaving their marks across the Solar System (and beyond).

    One of the purposes of this volume is to expand upon the problematic analogy of "learning at a distance" (Finney and Bentley, 2014) when translating defined *neurocentric* information across intervals of deep time. To this end, we intend to query the seldom discussed problems inherent within our communicative media, while providing some fruitful insights into understanding the complexity of these interrelated challenges for various messaging practices. However, in context with the impracticality of near-future recovery for some (if not most) of these resources by human posterity or ETI, it can be argued that the distinctive legacy of these variegated messages may be tantamount to understanding the differences between the *a posteriori* debitage of hominid settlements (i.e., their "sustenance economics" actions), and their symbolic ochre effigies adorning global cavern walls. After all, it may become impractical to detangle the purposes and motives imbedded in the back end of a discovered unreadable message, not to mention whether it was indeed a purposeful material sign, and not simply a once-off "blip" in a much larger background technosignature. It is also worth highlighting that neither early hominid debitage, nor their ochre effigies were intended as messages *for us*.

Regardless of these challenges, the remnants may still provide rare glimpses into their creators' social, mental, aesthetic, and cultural attitudes, in addition to *that of the recipient*. As the archaeologist Bruno David (2017) surmised, "What *we* see in those caves … reflect[s] how we ourselves, as observers, have been conditioned to see": etiquettes we tend to subliminally and uncritically map onto the products of other minds to make sense of such vestiges. David's observation wonderfully highlights what we now recognize as *our social construction of knowledge about others*, which usually distorts any rational—or indeed useful—scholarly inferences about unknown ancestral cultures. In this, our old "art for art's sake" presumptions about ancient humans being too "primitive" for higher forms of cognitive reasoning have been discarded, in favor of now recognizing their cultural documents as rich, intellectual social histories utterly different from the lineage of our own customs. I would argue this to be perhaps one of the most significant discoveries we could logically glean from ancestral remnants, lessons which may prove fruitful for discussions about potential *messages from ETI*. Similarly, perhaps our exoatmospheric archaeological record, and the cultural documents we scrawl across the halls of gravity, when seen through the mind's eye of another, may someday capture how *we* naively saw our world—and our visions of a future living off it.

Space message deposits can be considered as excellent thought experiments, insofar as they serve as one of the highest forms of self-critical appraisal for terrestrial life and self-authored stories about human societies—conscious, introspective documentation that may be empirically valuable for future archaeologists examining the social lifestyles apparent within first-hand documentation from the Anthropocene. Many details about these messages, despite being fabricated recently within human history, remain obscure, yet they collectively describe our world in a plurality of ways, communicating narratives, philosophies, and poetic concepts, which expressively represent the ideals and core values of their author's cultural worldviews. As the idiom goes, putting "our best foot forward" is a natural process when considering contact with ETI, or visions of human posterity. We shall see how this factor has influenced many successive messages later though, in crafting these artifices, we inherently export our own personal biases, romanticized or abstract notions of reality, disciplinary partialities and other presumptions that may communicate more between the lines of our stories than we think. Such is the power of metadata, and its overarching context.

Not all messages are created equal. Readers of this particular volume may already be familiar with the purposeful analog message stowed on the outside of both Voyager space-craft, intentional rhapsodizing mosaics of our civilizations, which embody the philosophical and ideological threads for some of humanities' material heritage from the 1970s. Much literature is already devoted to the iconic *Voyager Golden Phonographic Records* (Sagan *et al.*, 1978; Nelson and Polansky 1993; Chen 2003; Macauley, 2006; 2010; Capova, 2008; Lemarchand and Lomberg, 2011; Schmitt, 2017; Ferris, 2017; Scott, 2019). This volume will also discuss the project's contribution to the early stages of what the Philosopher (and Jesuit priest) Teilhard de Chardin referred to as *the noosphere*.[14] However, these dual propaedeutic records are not alone in representing our world from afar. Since the dawn of the Space Age, small cohorts of humanity have independently transmitted *de novo* extraterrestrial messages to other stellar systems, deposited collections of questionable artefacts for ceremonial applications on other astronomical bodies (see color insert), and permanently incised our mnemonic marks upon the deep-time archaeological legacy of the Sol System.[15]

This *anthropo* mark is clearly evident across growing stretches of time and space, with an ever-increasing reach as we continue to add to this accumulative *a priori* legacy on a yearly basis. Indeed, at the time of writing, an inscribed metal plaque aboard NASA's Lucy spacecraft is one of the most recent additions to these expansive legacies. Given that we are now over 60 years into the Space Age, I believe it is now time to begin to look back and reflect upon the ideological portrait haphazardly crafted by aspects of humanity on behalf of Earth. Perhaps, more significantly, we will look at what this archaeological record—comprised of the aggregate of these aging narratives, concepts, analogies, conflicting ideologies, and justifications—actively reveals about the many social hegemonies, minority populations and the other biota inhabiting our shared planet at the time of their departure. Do these signatures compliment or contrast with the desultory of our pale blue legacy, and other background *messenger* markers we continually export as an archaeology of the future for ourselves—or distant ETI?

In attempting to rationalize this cultural mosaic of humanity in outer space for modern contemplation, it became necessary to consolidate a reliable chronology of communicative resources to serve as a foundation for documenting this exoatmospheric archaeological record. The *A Profile of Humanity* catalog (henceforth APOH) was compiled as the first active index for this application, before being published within the *International Journal of Astrobiology* (Quast, 2018a). This catalog has since grown remarkably, continually elucidating the historic and contemporary message artefacts that define what populations generally characterized as "humans and Earth" through the decades—or at least homogenous representations that embody our visual and conventional preferences, as well as widely-held paradigms that underpin our conceptions of experienced reality. An updated edition of this catalog has been faithfully reproduced in Appendix A of this volume for inspection. There have been several attempts by planetary scientists, heritage consultants and radio astronomers to characterize how detectable this sparse scattering of objects and tight EM beams may be for the futurity of observers: the majority of opinions suggesting that the likelihood of chance encounters to be extremely low, if not nil. Irrespective of the *still* frequently overblown odds of whether this asymmetrical smear of material culture is being—or even can be—contemplated by any external observers at a meaningful distance from Earth, the plurality of this material arising from the early Space-Heritage Age (Hays and Lutes, 2007) possesses crucial significance for the long-term cultural history of the human species in, and beyond, our local stellar system.

Technicians installing the Golden Record on Voyager 1 (photograph from NASA/JPL—Caltech [PIA21740]).

In light of this, the volume's authors consider all of these messages (composed of interstellar radio transmissions, artefact deposits or other, eclectic *space oddities*) as distinctive, social documents, heritage assets which varyingly aim to transfer cultural property or signified "meaning" across deep space and cosmic time to unknown denizens. The goal of this initial volume, therefore, is not to establish an overarching synthesis or grand narrative for how these messages "speak for Earth," by projecting idiosyncratic properties onto imagined minds. It suffices to state that in unaffiliated message contents, there is very little crossover or convergences in opinion, expression, tradition, or approach—*a consensus about the lack of a consensus*, if you will. For instance, the star HD 95128 will receive three interstellar messages in the forthcoming decades—two described as intelligible, the latter being a confectionary food advert. Rather, this volume will provide the first, consolidated look back at this rich storehouse of messages—*in our eyes*—and shine an interdisciplinary light upon the various human characteristics that are deeply embedded within our communication channels (i.e., what messages "say" outside of their delicately-curated contents). Subsequently, our hope is that this may optimistically reveal insights into how the many divergent narratives, concepts, and ideologies we employ in defining our world at a distance may be used to explore aspects of the sender's social philosophies and value judgements, committed on behalf of Earth's heterogeneous populations. Such articulated slices of life

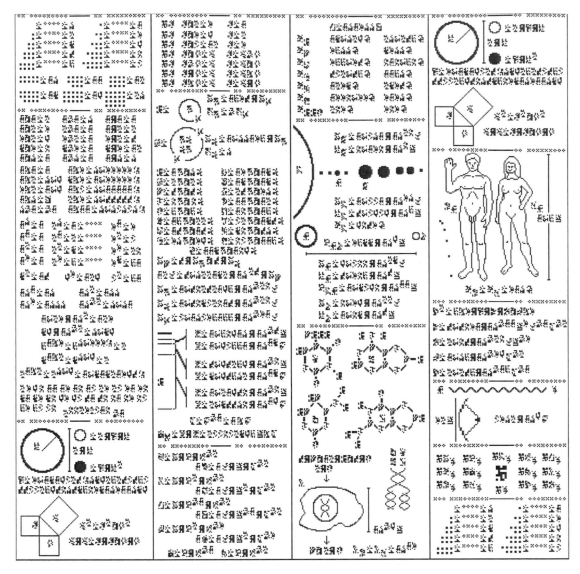

Sónar Calling bitmap tutorial for 14–16 May 2018 transmission series (bitmap is based on the Cosmic Call bitmap series by Yvan Dutil and Stéphane Dumas. Bitmap ©Advanced Music SL. Commissioned graphic developed by METI International. Re-illustrated from data by Paul E. Quast).

have previously revealed glimpses into some of the internal needs, hopes and virtues of humanities' psyche, alongside societal concerns of an era—truths packaged as stories about our relationships with things, places and life that are ironically embedded within out-bound media for contemplation from afar (Durrans, 1992; Vakoch *et al.* 2013; Quast, 2017; Schmitt, 2017).

The production of meaning from these relics and underlying behavioral patterns of the minds who crafted these elements of material culture is, of course, a maturing process under the aegis of several academic lenses that is subject to persistent revision, debates between intellectual traditions, and cross-disciplinary examination—often with competing outcomes. Much of this discourse, in turn, also centralizes around deconstructing the humanity in these complexities (given human attributes are often used as an archetype for envisioning the uncanny "other"). As such, the reader should expect some questions to arise that challenge conventional intuition, and presentation of contrarian viewpoints to the natural science and historical profiles of message which, frequently, are the dominant scholarly viewpoints available when reviewing "messages from Earth." As there are many facets to, and no singular approach for, analyzing the abundant aspects of this broad and intertwined mental landscape, the ensuing texts are comprised of several alternative, and sometimes contrasting, disciplinary essays which aim to condense the theories of our arguments into principal constituent approaches, while presenting diverse (and, at times, disparate) perspectives within epistemic, cognitive, linguistic and semiotic systems on how we might begin to interpret such an eclectic storehouse of material culture, as part of *our heritage futurity* (Harrison *et al.*, 2020).

Given this particular focus, the volume is conceptually divided into three distinctive parts which primarily address three key queries that form the bedrock of archaeological fieldwork studies when assessing uncovered traces of iconography, or residual inscriptions: why was the material created (*Why We Speak*), what does it mean (*What We Say*), and how does the surrounding material context affect this interpretation (*Reading Outside the Message*). In addressing "Why we speak," the volume begins by reviewing the social histories, underlying theories and motives for these material practices using aerospace technologies. Thereafter, part two transgresses into exploring the frequently arising aspects of anthropocentricity, cognitive or sensory discrepancies, and rational conflicts in representing "meaning" within our communication channels across spatiotemporal distances as enduring, ideological portraits for our planet. The third and final part of this volume is thereafter reserved for examining the more explicit contexts found outside of these autobiographies, including how we choose to write social documents about Earth, casually omitted subject matters, and found in-situ environmental conditions, before delving into the ethics and speculative future for our evolving relationships with these material ritual practices.[16] We suggest to the reader that they use these divergent viewpoints as an inductive bridge for informing their own exploration of this heritage legacy, and incorporate their own prerogatives into these arguments, in order to provisionally assess how humanity is manifesting discordant murmurs—*or mumbles*—*from Earth.*

NOTES

1. According to NASA's Jet Propulsion Laboratory Horizons tool, the distance between the Earth and Voyager 1 on February 14, 1990, was 6,054,587,000 kilometers, or 3,762,146,000 miles (40.472229 Astronomical Units).

2. The planetary scientist Carolyn Porco compiled this mosaic from the Cassini mission, in addition to previously combining Sagan's requested Voyager 1 frames into the Pale Blue Dot and Voyager family portrait series.

3. Otherwise known as "c," it is the maximum finite speed at which all conventional matter—and hence all known forms of information transfer—can travel through a vacuum. This velocity is defined as 299,792.458 km/sec.

4. The concept of Earth as a spaceship with limited resources was independently popularized by Kenneth Boulder (1965), Barbara Ward (1966), and later in *Operating Manual for Spaceship Earth* (Fuller, 1967).

5. The "Golden Spike" is defined as an observable marker within the geological record which clearly indicates a starting point for the Anthropocene epoch. Over 65 such spikes have been proposed across a broad range of ecological services, geological and atmospheric strata, as well as within celestial indicators (Grinspoon, 2016).

6. According to paleo-climatological and sedimentary records gained from Greenland and Antarctic ice core data, there have been several interglacial periods with increased CO_2 concentrations. However, recent recordings possess a distinctive departure from these natural cycles (since the Pliocene). The pale blue of the original photograph, alongside all other Earth portraits from the Space Age, varyingly reflect this characteristic shift in atmospheric CO_2 along with increases within CH_4, N_2O, HFCs and other isotopic concentrations. The influence of past human societies on these records is difficult to ascertain but should not be outright dismissed as ineffectual.

7. The visibility of such technosignatures at cosmic distances is contestable and often debated within the context of the brightness of background stellar emissions across a potpourri of EM frequencies, the historical strength of fluctuating leakage radiation, and future projections of sensitivity for observational technology and programs (Haqq-Misra *et al.*, 2013). The nature of this leakage really depends on its source at a particular point in time, but this inadvertent EM noise is likely only detectable out to approximately 100 light years if we assert ETI employs a receiver larger than the proposed (and overly ambitious) *Project Cyclops* array (Oliver and Billingham, 1972).

8. For the purpose of discussion, we shall treat this persistent EM profile from Earth or "radiosphere" as an active yet weak techno-geological strata propagating out into space which exponentially dissipates in accordance with the inverse square law of signal propagation (i.e., increasing spatial dilution over increasing distance). By stark contrast, narrow bandwidth, high-powered interstellar radio messages within this medium are recognized as distinctive phenomena, metaphorically comparable to notable artefacts "buried within" these strata.

9. By using the term "messages" here, I refer to the broad range of artefacts, inscribed content-bearing data carriers and encoded interstellar transmissions, alongside the array of media included within these artifices. The nondescript term "messages" is ubiquitous when discussing cultural objects used in space activities and is equitable with "depositories" as outlined within the book *Time Capsules: A Cultural History* (Jarvis, 2003).

10. By and large, many messages do possess a hypothetical addressee. However, it is worth acknowledging that other projects have been formulated as "gestures to eternity" or more accurately, *for us* as a human audience.

11. These prerogatives are also corroborated by interviews and source materials from living message authors.

12. The term "culture" is a rather ambiguous, nondescript phrase with numerous informal definitions. Herein, we define culture as a group of individuals or collective that share in an on-going process of reinvention within their knowledge, beliefs, art, customs, values, moral codes, and habitual systems as they manipulate their social and physical environments. It should be noted that boundaries between cultures, if there are any, are highly porous. "Material culture" denotes artefacts arising from these entwined social and mental interactions, with "cultural technology" denoting a collectives' aesthetic principles and other information-framing conventions.

13. The term "distributed cognition" alludes to socially stored knowledge outside of any single living brain (usually within inscribed or digitally encoded media), not conceptions of telepathy—though Stoneley and Lawton (1976) provide strange insights on the topic of ESP (extra-sensory perception) in context with space communications.

14. The Noösphere is broadly postulated as a "sphere of reason" or the highest stage of evolutionary development from the biosphere which is dominated by humanities' consciousness, the mind, interpersonal relationships, and rational activities. The Russian biochemist Vladimir Vernadsky co-defined a similar definition for this term.

15. In the interest of disclosing a potential conflict, this author was also responsible for one of these messaging projects; *A Simple Response to an Elemental Message* which was transmitted (to encourage outreach and engagement within this debate and exigent environmental ethics) to Polaris α UMi Aa on October 10, 2016. It is likely that the methods employed within encoding this content will ensure that the message remains illegible.

16. For considerations on the "found" environmental conditions associated with receiving human interstellar messages, I would recommend an external chapter on the semiotical implications for this subject (Quast, 2021).

A Cultural History of "Speaking" for Earth

A Taxonomic Approach for Exoatmospheric "Messages"

Paul E. Quast

Before we can delve too deeply into the enigmatic material practices that collectively contribute to the human ethos of "speaking for Earth," it's necessary to briefly understand some of the constituent theoretics, social arguments and broad technical discussions that serve as an ideological bedrock for this *purposeful* technosignature. The formal processes of establishing taxonomic categories—or separating populations of "things" into bifurcated classes—is a practice familiar to a diverse range of scientific, social, and archival professions, but we are all natural, and subjective, taxonomers at heart. We instinctively compartmentalize, and neatly order our surrounding material world and associated qualities into clean, value-instilled boundaries of living and non-living, edible and inedible, safe and dangerous, encompassing timescales and historical eras, movements, disciplines, traditions, objects, and cultural customs, among other categories, and subsets thereof. Yet, some social histories of things, in relation to other things, are difficult to demarcate; cultural objects whose informational benefit is deferred for unknowable posterity, or foreign audiences, being one blurry illustration of the obvious limitations and disparity in opinion for such orderly approaches. Taxonomic strategies do have drawbacks, as briefly summarized later, but let us provisionally explore this practice of speaking through time and space to imagined recipients under such a methodological framework for further context and discussion.

Throughout the history of our built-environment, humanity has persistently shaped our embodied minds through externalized material engagement practices, frequently for the gradual inventive production of technological tools to bootstrap lifestyle requirements and, quite often, for ostentatious desires—a cultural relationship with things that inherently reshapes us in return. In this respect, the crafting of "messages" or communicative artefacts that intend to speak beyond the senders' resident dimensions in space-time is not a new behavioral phenomenon to emerge from organized human societies. Since pre-antiquity, multitudes of empires, dynasties, religious orders, nomadic tribes, and other precursor civilizations have chosen to imprint pictorial, mathematical or proto-linguistic elements of their representative age on cavern walls, standing stones, clay tablets, monument facades, and other portable artefacts capable of carrying these *a posteriori* marks beyond their temporal horizon. The applications and addressees vary considerably throughout this antique material culture record.

While the intended recipients or degree of conscious awareness for undertaking representational material practices is still disputed (Malafouris, 2007), some of the oldest Upper Palaeolithic parietal paintings within a Borneo cavern in Lubang Jeriji Saléh date to over 40,000 years ago. These figurative, geometric and zoomorphic illustrations may have likely served as an expressive scaffolding for the painter to actively rationalize and abstractly think about their position with the surrounding environment, capturing a snapshot of the artisan's visual observations, tactile-thinking, social lifestyles, beliefs and desires within decorative effigies that could likely serve as intergenerational memory cues, or as pedagogic resources for their next seasonal visits (Lewis-Williams, 2002; Whitley 2006). This parietal art may have also served ceremonial[1] or "shamanist" purposes by representing sacred traditions that transcend the artisans from the human realm to a spiritual one. But, without knowing the intention behind these marks or relationship to larger symbolic systems, such inferences will remain as such. The negative spaces and lithographic features, which naturally form contours that are then used as aspects of these artworks, certainly lend some credence towards these imprints as spiritualized gestures that uniquely interconnect the mastery of the artisans with their surrounding ecologies. Equivalent European sites such as the Maltravieso, El Castillo, Coliboaia, Cosquer and Lascaux cave systems present us with a rich tapestry of late-Middle to Upper Palaeolithic impressions, created by successive hunter-gatherer generations who seasonally migrated to these regions—some of the artisans for the two former sites conceivably belonging to the *Homo neanderthalensis* branch of the hominid genealogical tree (Pike *et al.*, 2016; Hoffman *et al.*, 2018). Many researchers and visitors who are fortunate enough to encounter these archaeological wonders of prehistory regularly evoke a notional "time capsule-like experience" when observing these abstract legacies from the distant past.

Aurochs, horses, and deer in the "Hall of the Bulls" at Grotte de Lascaux near Montignac village, France, c. 2006. Featured in *The Last Pictures* as image #24 (photograph from Saša Šantić/Wikimedia Commons, CC BY-SA 3.0).

The establishment of integrated, regional societies with workforce specialization and surplus energy reserves (Smil, 2006) consequently led to the building of architectural marvels that are still discernible today, eventually including the purposeful development of votive messages for commemorative applications. The Neo-Assyrian king Sennacherib, son of Sargon II, regularly erected notable buildings and highly ornate monuments in conquered lands, but perhaps his empires' most resonant legacy can be found in the cuneiform inscriptions secreted beneath lineage structures by his heir Esarhaddon. These inscribed clay tablets provide a purposeful record (for descendants or posterity) of this dynasties' triumphs but also vital documentation from this period of antiquity (Ellis, 1968). We *still* discover such significant vestiges, including recently excavated cuneiform inscriptions, buried beneath the destroyed Tomb of Jonah in Nineveh. Esarhaddon's son, Ashurbanipal, perhaps inspired by these paternal building and cornerstone traditions, subsequently sought to forge his own legacy by consolidating the Library of Ashurbanipal—a treasure trove of over 30,000 cuneiform tablets from approximately 700 BCE.[2] Over the passage of millennia, this intentional practice of building within and through time, or compiling intentional *a priori* messages to unknown recipients, has served to project the values of these authors beyond their snapshot in space-time, providing a posthumous sense of immortality, while ensuring that their legacies are remembered by familial descendants, subjects, worshippers or, perhaps, as offerings to defined deities.

In his seminal publication on the cultural history of time capsules, William E. Jarvis (2003) established a multi-tiered framework for categorizing our contemporary definitions of time capsules, as he attempted to group together common archetypes, and distinguish these eclectic deposits for further discussion. These synoptic definitions range from non-intentional *a posteriori* deposits (experienced archaeological sites, such as the parietal art examples that *we impose this time capsule label upon*, in addition to discovered shipwrecks, and other "found" ancient monuments), to purposeful *a priori* deposits with unscheduled retrieval dates, such as our generational Assyrian-era cornerstone caches and archival projects. Jarvis also further delineates these earlier depositing practices from modern *time capsules* as initiatives comprised of intentional deposits, stored inside durable vessels, with targeted retrieval dates, marker strategies, prospective audiences, and elaborate send-off or recovery ceremonies[3] (examples include the enigmatic *Crypt of Civilization*, the *Expo '70*, and *Westinghouse Time Capsules*).

Outer space is not empty of the ideologies behind these material culture deposits. At the time of writing, human spaceflight—once romanticized as the next step in our cosmic evolution—seems to be logistically confined to geocentric orbit in favor of exploring our surrounding spatial environment by proxy of robotic avatars and EM signals—a fraction of which carry prosthetic extensions of our minds to local and extrasolar worlds. Many of these intentional messages are the consequence of age-old behavioral traditions, tool production customs, and other material-ritual practices that are uniquely intertwined with processes of mental externalization—projecting older narratives of human experience, proliferation stories, and prospective social exchange devices into new environments for imagined, exotic audiences. Several authors (Jarvis, 1992; 2003; Mehoke, 2009; Capova, 2013a; Harrison, 2014) have already recognized the parallels between terrestrial time capsules and these "space-time capsules," including their common lineage stemming from cornerstone depositories. However, Jarvis's publication does not extensively document these celestial projects (considering its focus is on the cultural history of *terrestrial* depositories). While the focus of this volume is uniquely on the examination of these exoatmospheric messages, it is worth provisionally discussing the confluences and divergences of these purposeful legacies in relation to Jarvis's considerable framework. Throughout the compilation of the APOH catalog (Quast, 2018a), three identifiable material-based categories emerge in which all *a priori* messages with a purposeful intent can be broadly segregated into:

1. Arbitrary objects placed aboard spacecraft, or on other astronomical bodies, as a variant of traditional cornerstone deposits. Examples of these material depositories can be seen on the Apollo lunar sites (memorials, photographs etc.), and several spacecraft depositories such as the New Horizons mementos. However, due to recent work within the legal recognition of space heritage sites, objects residing on other local astronomical bodies may also become subsumed under an alternative, *a posteriori* category (for archaeological sites that are deliberately left undisturbed). Preserving these archaic objects presents a unique paradox in that, while the physical artefact is preserved, the original cultural, utilitarian, and environmental context signified by the items may not be readily accessible for discoverers that do not share socio-cultural conventions, or other qualities, with the populations that launched these relics. These issues are also likely encountered with cultural items left as "Millennial" time capsules.

2. *A priori* "space-time capsules" designed to hypothetically provide an accessible, historical mosaic of information about present-day Earth for study by a future and likely foreign cultures. Jarvis notes that projects such as the *Voyager Golden Records* (henceforth VGR) and *Pioneer Plaques* tend to resemble traditional cornerstone repositories, as they do not possess a targeted retrieval date (and likely cannot be due to their remote locations). However, some of these deposits tend to also possess information capacities and content choices that correlate with trans-millennial time capsules,[4] alongside elaborate use of propaedeutic "primer" guides to promote comprehension. For comparison, the *Crypt of Civilization* included a unique "Language Integrator" mutoscope, invented by Thomas Kimmwood Peters to "teach English" to eighty-second-century posterity. As any interpretive context is always contingent upon the prior predispositions, value codes, and mutual understandings of shared referents by a recipient's culture, amongst other factors (see Traphagan and Smith, this volume), a large portion of these physical artefacts endeavor to provide accessibility for hypothesized recipients as a crude expression of altruism (even if some of these strategies will remain inaccessible).

3. Despite many overlapping characteristics with the prior deposits, I would argue that *de novo* interstellar radio messages constitute a separate category, as an EM signal's propagation speed and diametric trajectory (away from their author's location) will ensure that these monologues are permanently removed from future human consumption. These outbound *messages in a bottle* primarily service several psychological, artistic, novelty, and communicative applications, and are generally cheaper to fabricate than tangible *launched* archival projects. Despite being permanently removed from their creator's future influences, they may still impact posterity in several ways; either through a present-day reaction (contributing to the extraterrestrial contact debate, space ethics deliberations, popular culture references, purchasing affiliated products etc.), a conscious effort to archive the memory of these projects (Quast, 2018a; 2022; Welcher, 2018) within other depositories for posterity

(loosely related to Jarvis's retrospective, notional-contemplation category), or by stimulating a response by an external "participant-observer"—in this case, *purposeful reply messages* from an extraterrestrial intelligence. In expanding on the latter outcome, the partial re-use of these transmitted contents as a recognizable, artificial pidgin within a hypothetical extraterrestrial response signal[5] may also evoke a metaphorical time capsule-like experience, as defined under Jarvis's *a posteriori* category.

Table 1 below broadly integrates these three identified material-based space message categories into Jarvis's pre-established framework, alongside chronicling counterpart *a posteriori* variations of these practices for comparison. For practical context, I have included several examples that are discussed (or depicted) throughout this volume. The table is far from complete, with additional research likely to elucidate further avenues to incorporate, but it initially summarizes some of the prescribed trends, and documented aims for these projects, which may—intentionally or otherwise—be frequently assigned a future *messenger* status.

Culturing an "Insane," Intangible, and Indelible Legacy Through the Ages

Dispatching messages into outer space possesses a certain surreal intangibility, insofar as the vastness of our surrounding cosmic environment—even near-Earth orbit—is beyond the foreseeable grasp of future "terranauts," and the likelihood of a chance encounter with a *de novo* transmission by an unobserved interlocutor is minuscule, or a fantasy at best (Billingham and Benford, 2011). So, why create "insane" messages (Paglen, 2012) for other minds that paradoxically may never be recoverable? Why intentionally sacrifice matter and energy from our times, for enigmatic efforts to protect our cultural memory across an unpredictable futurescape? These questions are partly rhetorical, as they are broadly explained by our incessant, Cartesian desire to infuse human memory into matter and vice versa with matters transition into immaterial memory as elements of "distributed cognition"—purposeful intents shared by *at least* the last hundred generations of humanity, when fabricating *a priori* archival resources. While these mnemonic artifices leave Earth as concentrated packets of photons, and digitally encoded or etched matter, it can be argued that the perceptible threads of these prospective communiques remain in modern cerebral storage, as literary devices for fostering social cohesion and inter-cultural engagement with some of the most profound challenges facing contemporary global civilizations. As the scientist Barney Oliver acutely observed:

> There is only an infinitesimal chance that the [Voyager] plaque will ever be seen by a single extraterrestrial, but it will certainly be seen by billions of terrestrials. Its real function, therefore, is to appeal to and expand the human spirit, and to make contact with extraterrestrial intelligence a welcome expectation of [hu]mankind [Sagan *et al.*, 1978].

Ascertaining an overview of psychological reasons for why we choose to create seemingly absurd messages that may never be recoverable is a massive undertaking across schools of thought, worthy of extensive scrutiny in a separate, focused behavioral volume. Message contents, in turn, can be as diverse as their envisioned recipients, or anticipated recovery efforts. However, over the course of the last several years, I've somewhat patiently sifted through manuals, articles, project press statements, aging diskettes and newspaper snippets, alongside occasional message-author interviews and other ancillary literature in an attempt to at least catalog the exoatmospheric messaging heritage of our species (Quast, 2018a).

It is incumbent of me to state that there is no definitive, overarching logic or grand theory for why we (i.e., fragments of launch-capable populations, or subgroups with access to aerospace technologies) choose to create legacy messages. These artifices, much like their enigmatic creators, also rarely correspond to a precise either-or choice between the below categories.[6] As we still reside within the same global milieu that initially cultivated this Space Age material culture (and therefore share in, and may learn about, a number of desires, attitudes, conventions and behavioral characteristics), it is possible to deduce several dominant trends within these projects which broadly reflect the "extreme collecting" aspirations of terrestrial time capsules (Durrans, 1992), while encompassing additional urges for why we choose to figuratively speak beyond our terrestrial shores. Many of these observed motivations, trends in narratives, and concepts we employ within this emerging celestial noösphere are recognizable, and still warrant a deeper interdisciplinary analysis. However, it is hoped that this concise hierarchical listing will serve as a template for promoting further investigation into why we believe our world should be contemplated from afar.

Table 1. Types of Exoatmospheric "Space-Time Messages" and Retrievals

Recipients:		"A Priori" (Intentional Deposit)	"A Posteriori" (Non-Intentional Deposit)
Inferred for immediate or future humans, or as "gestures to eternity"	Scheduled Retrieval on a Target Date:	SPACE-TIME CAPSULES WITH PLANNED RETRIEVAL (KEO satellite proposal)	Hardware deposited into space and later retrieved by missions for materials tests (Surveyor 3 camera)
		Planned missions to recover test materials or artefacts in outer space (The Long Now Foundation's test sample on MISSE, LDEF)	Notational archival projects that document exoatmospheric messages and contents to maintain knowledge or reproduced space-time capsules
	Unscheduled Retrieval (happenstance; indefinite span):	"SPACE-TIME CAPSULES FOR FUTURITY" (Visions of Mars, The Long Now Foundation's Rosetta Disk, Arch Mission projects)	Defunct spacecraft, probes and rocket bodies in geocentric orbit and archaeological sites in outer space (Statio Cognitum—Apollo 12 site)
		SPACE-TIME "FOUNDATION" DEPOSITS (arbitrary artefacts left on spacecraft or archaeological sites in space, such as New Horizons mementos relics, Apollo deposits)	Defunct space hardware that occasionally returns to the vicinity of geocentric orbit (J002E3 S-IVB) or re-enters Earth's atmosphere and re-found (Tyazhely Sputnik pennant)
		"NAMES IN SPACE" or Kilroy-inspired projects (CHEOPS drawings, OSIRIS-Rex and MER Curiosity microchips). Possible subset of above category	Unintentional artefacts left behind on archaeological sites (Apollo 12 film rolls, microbial contamination of hardware), nano-strata of Earth

Recipients:	"A Priori" (Intentional Deposit)	"A Posteriori" (Non-Intentional Deposit)
Scheduled Retrieval on a Target Date:	DE NOVO INTERSTELLAR MESSAGES TO NEARBY EXOPLANETS that are confirmed to be inhabited. No candidates are yet confirmed	Notional contemplation as to how ETI may communicate with humans and our message search programs (SETA project, SETI, *Invitation to ETI*)
	CONTACT as; *de novo* messages from ETI, responses to our messages	ETI's partial reuse of recognizable "cribs" from prior human messages
Unscheduled Retrieval (happenstance; indefinite span):	"SPACE-TIMES CAPSULES FOR ETI" as intelligible messages (*Voyager Records, Pioneer Plaques, One Earth Message, StarChips*)	Interstellar spacecraft hardware that will eventually depart the Solar System for interstellar space
	DE NOVO INTERSTELLAR MESSAGES TO EXOPLANETS theorized to be inhabited (*Cosmic Call, Lone Signal, Sónar Calling GJ273B*)	Radiosphere (military/ astronomical radar searches, telecommunications broadcast in radio frequencies emitted since 1930s etc.)
	DE NOVO INTERSTELLAR MESSAGES TO DISTANT TARGETS (*Arecibo Message, Nançay Message*)	Other Bio and technosignatures (atmospheric spectrum, thermal disequilibrium, light emissions, etc.)
	NOVELTY INTERSTELLAR MESSAGES with low transmission power (artistic, outreach, etc., transmissions)	

(Left spanning cell: *Inferred for Extra-Terrestrial Intelligence (ETI)*)

Table 1 features types of messages, in context with examples discussed throughout the book (compiled by Paul E. Quast).

1. Space: The Final, Colonial Frontier

Prevailing human worldviews have always been tied to our perceived *weltanschau-ung*,[7] while our long history of looking outwards and contemplating the physical universe has served to persistently disrupt these deep-seated ideological, philosophical, and theological perspectives. For millennia, dominant worldviews and associated mental baggage were asymmetrically framed by the spatial resolution of known kingdoms, continents, and the Aristotelian-Ptolemaic geocentric system; humans, nations, and the Earth system occupied various privileged, divine niches within these traditional worldviews. Copernicanism would later overthrow this special relationship with the newer heliocentric model, positioning the Earth as a single planet within an interconnected system orbiting the Sun. Technological advancements which enabled observations by Harlow Shapley would, thereafter, decentralize our planetary schema further; casting the Earth as one insignificant world, orbiting an average star in the Orion Arm of a galaxy populated by billions of other stars (Palmeri, 2009). Due to Hubble's astute observations, we now know that the Milky Way is not even unique, as it is but one of over two trillion galaxies in the observable universe. We're still counting (see color insert).

Ideological vacuums, such as our shifting worldview, tend to be filled by familiar convictions capable of bridging the epistemic void, created when re-conceptualizing our uncharted surroundings.[8] This was certainly the case in the early Space Age and the exploration of the "final frontier," a concept arising at a propitious era which was influenced by the collapse of old European colonialism, the geopolitical musing of science fiction, and theoretical space exploration from the 19th and early 20th centuries,[9] not to mention the multi-centralized leaps in aerospace technologies that literally brought humanity to new heights (Osiander, 2009; Gorman, 2019). In the United States, space was seen as a "High Frontier" wilderness, with a moral and ideological vacuum, waiting to be occupied as part of a cultural custom of expansion and exploration—a traditional perspective inherited from the European colonialists of the "New World," which accorded well with the founding principles of the American state identity, manifest destiny and mission (Gorman and O'Leary, 2007). On the other side of the Atlantic, Russian Cosmism—a broad theory of natural philosophy which invoked elements of ethics, religion, and colonial expansion—served as a similar, yet characteristically distinctive ideological bedrock for the Soviet Union[10] (Groys, 2018), as this federal socialist state attempted to re-assert its national prestige and moral fortitude following the events of the Bolshevik Revolution and both world wars. However, the futurescape of humanity, and the defined ideological *vacuum* of space, could seemingly only be filled by the synecdochical tropes, mindset, hardware, and heroes of a single superpower (Billings, 2007).

The classical, condensed narrative of the Space Race is often remembered as a competition between these two opposing superpowers of the Cold War era,[11] instigated for the public perception of national superiority and a variety of opposing agendas and demonstrating technological supremacy, ambitious political propaganda, military acclamation, the conquest of communist or capitalist ideologies, and also the manifestation and enhancement of national prestige by imprinting this new frontier with very specific cultural connotations (Bryld and Lykke, 2000; Gorman, 2009b). Similar ideological friction could be readily observed in other arenas of the Cold War, but space represented a crucial, technological theater, with global security and defensive implications. After the injection of the first "artificial Moon" Sputnik 1 into orbit in 1957, many of the other early Space Race ideological triumphs would be claimed by the Soviets, who seemed to best Western nations at every milestone despite using a standardized probe hardware for these varying goals during this celestial competition. A curious incident in the 1960s involving the CIA "borrowing" a touring Luna probe certainly raises eyebrows in regard to this perceived *technological gap*. However, as this rivalry was essentially a surrogate battle for military superiority with propaganda weapons in this high frontier, both adversaries grew apprehensive of each other's expansionist capabilities to ideologically claim new celestial territories for strategic advantages (and how this translated into influence, and prestige, amongst undecided populations on a politically polarized world).

Since the dawn of interplanetary probes, the Soviet space program had traditionally stowed "pennants" aboard its spacecraft—objects depicting political and cultural iconography which would be subsequently deposited onto other astronomical bodies as part of international propaganda campaigns—in the process, laying ideological and pseudo-colonialist claims in this vacuous frontier. The Luna 1 (Mechta probe), launched in 1959 to impact the Lunar surface, ultimately missing the Moon by less than 6,000 kilometers; however, it carried a small sphere of pentagrams (embossed with Soviet insignia) designed to explosively scatter these medallions upon impact. This probe and its ideological cargo ended up in a heliocentric orbit (still carrying what some describe as a live fragmentation grenade), while its successor mission Luna 2 would hard-impact six months later with a similar sphere. Similarly, Venus would also receive its first pennants in either 1965 or 1967,[12] and Mars in 1971, alongside several regions of geocentric and heliocentric orbit (from scarcely documented, failed missions). As a result of this and other dubious activities planned in the high frontier, the *Treaty on Principles Governing the Activities of States in the Exploration and Use of Outer Space, Including the Moon and Other Celestial Bodies* (United Nations, 1967)[13] was expediently ratified to prohibit formal claims of sovereignty beyond Earth's surface. While the OST expressly prohibits formal sovereignty claims, this practice of transporting pennants would remain as a staple component of Soviet planetary missions[14] (Reeves, 1994), with the U.S. subsequently depositing its first political emblems[15] on another astronomical body in 1969, when the Apollo 11 astronauts raised the American Flag during the first human walk on the Moon—an act by state agents which is symbolic of claiming territory, as set by historic precedent (Platoff, 1993).[16]

Since the end of this Space Race era, other countries have continued this *rite of passage* tradition—transporting cultural property and national insignia on board techno-colonial objects—as a standard benchmark in advancing their statehoods under the set rubric of Western scientific paradigms.[17] Many of the interplanetary spacecraft today continue to overlay a meta-colonial stake in this frontier by carrying depictions of national flags on their exterior hulls,[18] while other missions carry cultural archives and intimate artefacts that symbolize their creators and country of origin. It is perhaps no coincidence that these old ideas of depositing nationalist motifs seem to correlate with the traditional practice of enclosing cultural media within terrestrial time capsules to preserve a snapshot of an author's

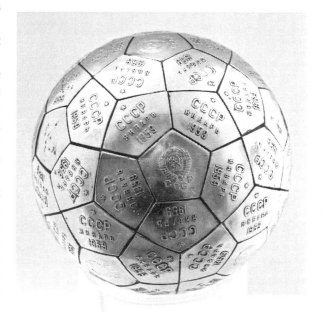

Genuine copy of the Luna 2 pennant (without explosive charge) on display in the Cosmosphere, Hutchinson, Kansas. The counterpart pennant, which was presented to Dwight D. Eisenhower by Nikita Khrushchev during his visit to the U.S. in 1959, is kept separately in the nearby Eisenhower Presidential Library (Abilene, Kansas). Pentagrams depict the Soviet Union coat of arms, and Cyrillic script "СССР Январь 1959" (USSR January 1959). Sphere designed by an unknown Soviet illustrator (photograph courtesy the Kansas Cosmosphere collection, cosmo.org).

encompassing national heritage for posterity. Examples of this practice abound: the Beresheet spacecraft which recently hard-impacted the Lunar surface, carried with it an "Israeli archive"; the New Horizons spacecraft is currently carrying an eclectic array of Americana into interstellar space; while a package containing a paper-printed copy of Saparmurat Niyazov's touted moral guidance book *Rukhnama*, is presently orbiting our planet to "conquer space" for about 150 years (BBC News, 2005). Examples of other techno-colonial expansions into outer space are also plentiful in the varying messaging enterprise, in addition to related explorational activities ranging from planetary science missions (as openly seen within the rhetoric used by popular media, when outlining a list of nations who chronologically land on other planetary bodies), to the observed peaks in nationalism associated with sending astronauts to space stations. Given this history, we may assume space is no longer a value-neutral "frontier"—if it ever was considered one.

2. Are We Alone? Making Contact in Context

The concept of the *Other*, and our desire to communicate with exotic entities that reside outside our immediate anthroposphere or empirical reality, predates academic disciplines by several thousand years. Examples of these other denizens range from deities, angels, and ancestral spirits to modern, conceptualized interlocutors such as our future descendants and ETI. However, these latter phantoms still need to manifest within our telescopes. The earliest recorded discourse about extraterrestrial civilizations can be traced back to the ancient Greeks in the musings of Epicurus, Democritus and Metrodus, with further abundance or rarity of sentient life arguments later advanced by Giordano Bruno, René Descartes and Bernard le Bovier de Fontenelle (among other contemporary philosophers, scientists, fiction writers and historical figures). All these voices have fundamentally contributed towards the modern characterization of extraterrestrials, and the nearby "plurality of worlds" they may inhabit (Dick, 1984; 1998; Crowe, 2003).

As Sagan (1994) once eloquently stated, "we continue to search for inhabitants. We can't help it. Life looks for life." This may be a truism of human curiosity, but there are no concrete reasons to suspect such life may be actively searching for us or providing us with indicators for finding them (Webb, 2002). Our initial planetary exploration of neighboring astronomical bodies during the Space Age revealed a lack of discernible evidence for these hypothesized "advanced" societies, subsequently forcing our theories about the abundance of multicellular life (intelligent or otherwise) to abandon our immediate Solar System. The ensuing modern scientific Search for Extra-Terrestrial Intelligence (henceforth; SETI[19]), which looks outside of our immediate stellar system, was initially founded by a convergence of events in the 1960s.[20] Today, SETI continues to passively sift through the "cosmic haystack" using sensitive receivers for tantalizing microwave or optical signals directed near our planet,[21] along with physically observing candidate star systems for a series of proposed technosignatures that are indicative of active, or extinct, civilizations. SETI—definable as a pre-emptive search for signs of prospective ETI material culture—is an excellent example of international cooperation amongst many research disciplines; however, it also serves as a nexus for investigating what it means to be human and our known limits of knowledge about life. As such, SETI forces us to examine some of the long-held assumptions about our own evolution, cultures, civilizations, epistemologies, and communicative technologies, while also addressing some of the most profound questions facing our emergent future—all of which are useful for exploring the range of developmental pathways a sentient ETI society may take "out there"—*or not.*

Contrary to this passive enterprise of searching for transient or beacon signals, there have also been numerous proposals put forth over the last several centuries to actively initiate communications with such "advanced" ETI (Raulin-Cerceau, 2010; Raulin-Cerceau and Cyrille-Olou, 2019; see also Oberhaus, 2019). The Serbian-American inventor Nikola Tesla may have been the first to initially consider the wireless transfer of energy as an adequate communicative medium between worlds in the onset of the 20th century.[22] However, the origins of the nascent *Messaging Extra-Terrestrial Intelligence* (henceforth; METI) enterprise using these same, eminently practical EM channels, can be traced to the immediate context of the Cold War and the technological demonstration activities of the Soviet Union in 1962 to stimulate a response from the (at the time, still plausible) inhabitants of Venus. This broadcast sent three Morse-encoded words; "MIR" (Peace), "Lenin" (i.e., Vladimir Lenin), and "CCCP" (known to Western audiences as "USSR"[23]). While documentation for the reasoning behind the message is scarce, this prototypical *de novo* transmission event is generally considered as a symbolic effort, or classified as "Pseudo-METI" as, although it was purposefully aimed at a theorized extraterrestrial audience, the signal possesses insufficient accessibility, content redundancy, or communication of intent, to be defined as a true METI attempt (Zaitsev, 2012). The technical feat simply lacks context, content, and access to transfer meaning to other minds.

Since this early Cold War experiment to stimulate a response from ETI with aerospace technologies, a number of successive METI practitioners have pioneered encoding stratagems that focused on relaying pictorial bitmaps and other visual-symbolism preferences within the *Arecibo Message* (1974),[24] *Greetings to Altair* (1983), both *Cosmic Call* series (2001 and 2003) and several other METI projects, alongside using syllogism in artificial languages such as *Lingua Cosmica* (henceforth; "Lincos"), executable code like Paul Fitzpatrick's *CosmicOS*, musical scores, and other semiotic or proto-linguistic media to test hypotheses, or provoke a response from the universe. All these cultural formats, to a degree, reflect human forms of expression (for instance, most are also visually orientated)

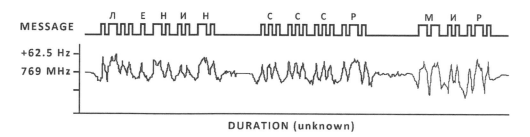

Signal sequence for second "Morse Message" from 24 November 1962, along with signal decipherment "Message" above. The signal reflected off Venus, and was received back on Earth 4 minutes, 44.7 seconds later. "Dots" were 10 second pulse durations, "dashes" 30 second pulses (re-illustrated by Paul E. Quast, based upon a graphic from the Soviet *Red Star* newspaper, edition 30 December 1962 [pg. 5]).

and material thinking, communication theories that, for the moment, cannot be experimentally verified as being capable of relaying the complex sign-relation histories embedded at the heart of *our* information-exchange systems. As argued by the astronomer Seth Shostak (2005) and other commentators on METI activities, transmitting a continuous, artificial sinusoidal wave at narrow band frequencies could suffice as an obvious communicative signal to ETI, but active search strategies have been far more elaborate. This distinction comes into sharper relief once we consider that the defined praxis of METI (Zaitsev, 2006) may be better understood as a prospective cultural exchange program using light-speed media and contents, that supposedly serve as mutually accessible pidgins with a common empirical basis. These transmissions may be best described as simple "postcards" written in hypothetically shared physical properties.

Our historical METI emphasis has so far hinged upon the *exchange* element of making contact, rather than an initial "we are here." Considering this general approach,

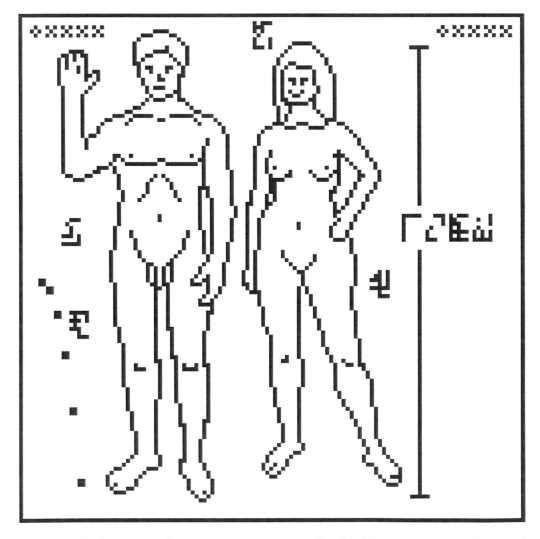

Cosmic Call 1, bitmap #15 (bitmap courtesy Yvan Dutil and Stéphane Dumas. Re-illustrated by Paul E. Quast).

the communicated media is theorized to elicit an equivalent response from a potential interlocutor to discuss their histories, cultures, religions, values, and ways of organizing stable, long-term societies—a reply generally assumed to contain beneficial information which our civilizations may also find practically useful, rather than simply answering that we are indeed *not alone*. As such, this prospective exchange program possesses psychological connotations of a "cargo cult" (White, 1990; Traphagan, 2018), or a similar *ask and you shall receive* philosophy. These outcomes are often sensationalized as "technologically mediated prayers" (Brin, 2019) in popular culture depictions of METI, in addition to depicting an innate desire to "elevate humanity" by metaphysically entering the accustomed "higher realm" historically occupied by gods (though this is usually associated with pseudoscientific contributions). METI, as we shall see later, remains a contentious subject in both academic and public spheres across a plethora of disciplines, with both sides of the isle contributing valid points to the enduring debate. However, it has been noted that calls for a moratorium on transmission "reconnaissance" efforts, as they have been described, may in itself be framed as an answer to the "Great Silence" (Brin, 1983), and numerous proposals put forth for the Fermi Paradox (Ćirković, 2018).

In addition to employing EM emissaries, there is also a historical precedent for crafting physical "Passive-METI" (Quast, 2018a) objects which are subsequently bolted to space probes that achieve escape velocity. The rocket pioneer Robert H. Goddard, in his report concerning rocket technologies for space exploration (Goddard, 1920), envisaged metal plates containing geometric shapes that could be used as a communication strategy for extraterrestrials. This concept was shared by early contact proponents such as Karl Friederich Gauss who (at least is attributed to have) proposed to create large geo-glyph diagrams of mathematical theorems, and other concepts, by repurposing vast expanses of forestry. Similarly, Joseph von Littrow proposed to ignite large canals of kerosene in the Sahara Desert for interplanetary communication (Vakoch, 1998a). Thankfully, these ecosystems were spared the disruptive groundworks, though I'm curious what nearby extraterrestrials would have thought about such massive, willful representations of ecocide. However, the concept of using spacecraft as physical messaging devices would eventually be adopted within passive-METI projects aboard four 1970s harbinger space probes, while the fictional *Friendship One* probe from the *Star Trek Voyager* series also technically conforms to this Passive-METI definition—though it presents an intriguing ethical conundrum that explores the notion of conflicting legacies arising from observed ambiguities within these intentional emissaries.[25]

In recent years, a growing number of other messaging projects have stated to support this Passive-METI application and line of intellectual enquiry[26] alongside recent developments within interstellar message construction. This momentum will likely expand in light of the Breakthrough Message prize to foster competition, alongside the *New Arecibo Message challenge*, and the SETI Institute's *Earthling Project* to promote engagement within this interdisciplinary enterprise. Outside of these initiatives, an increasing number of private and public ventures have been established to bequeath the cultural properties of humanity to profoundly unknowable minds, residing in postulated candidate stars, several of which are discussed throughout this volume. It has been asserted that such messages (and spacecraft) may also serve as "conquerors" of interstellar space by facilitating, as the late cultural geographer Denis Cosgrove (1994) surmised, the "expansion of a specific socio-economic order across space." Perhaps, Cosgrove's assertion is correct, in that these objects tend to draw into sharp focus the historical "frontier" origins of our messaging praxis and aerospace technologies as an extension of terrestrial socio-political power dynamics while, as we have seen, enabling us to contemplate how these undercurrents of messaging provide "new ways of understanding how colonisation is narrated … on an extraterrestrial scale" (Schmitt, 2017). Food for thought.

Ultimately, SETI and/or METI activities may either confirm or contradict our shifting cosmological perspective given *enough* time (Dick, 1989). However, in the meantime, both enterprises will likely continue to enjoy a high-publicity profile within outer space activities, regardless of whether we detect an extraterrestrial signal, or receive one in response to our crude messages to the stars. Whether we remain alone—or not—this intellectual exercise will likely continue to motivate the creation of these unidirectional communique as we fulfil a seemingly innate desire to connect with something bigger than the Earth system, raising an abundant amount of profound, existential questions about the evolution of matter into sentient minds (Mallove, 1987). Regardless of the immediate null results, SETI/SETILO should continue this enigmatic search for the *other*.

3. I Once Existed and Looked Like This

Messaging artefacts, by their very nature, are spatiotemporal snapshots of their creator's "distributed cognition," frozen in time, that immediately become representational antiques of these eras once removed from their host communities' gradual influences. Legacy-crafting of effigies that will far outlast the life-expectancy of their respective author(s) is an age-old memorializing vocation that has been undertaken by ancestral cultures since antiquity, but not all messages have imagined audiences or targeted recipients. In contemporary iterations of this "memory value" regime (Riegl, 1903), the protracted surrogate memory of an author is paradoxically interlinked with maintaining concealment of the deposit. Once found, the immortalized legacy of the author, conferred by these prosthetic artefacts, loses its designated significance before slowly fading into the indistinguishable background noise of our species' history (Durrans, 1992).

The late art critic John Berger (1973) once perceptively observed a defining trait of postmodern material culture practices: in the ever-perpetuating *age of the now*, "the more monotonous the present, the more the imagination must seize upon the future." Therefore, it is perhaps hardly surprising, that this sensationalist practice of self-expressing one's personal identity outside of collective experiences, in extreme "conquering" environments, has increasingly manifested within outer space messages—organized, inert matter containing aspects of cultural and social life, which is dispersed into an expanse that will likely permanently inhibit any recovery, or transcendence by peers. The recent addition of a Tesla Roadster to this archaeological landscape would seem to correspond with this ideological intention to poise an idiosyncratic legacy, or technologically mediated memory, on the edge of eternity (Gorman, 2019), in addition to meeting several other suspected virile and exorbitant intentions (David, 2018; Gorman, 2018). Additionally, posthumous iterations of this practice may also be seen within the recent Stephen Hawking memorial transmission,[27] and the remembrance transmission remit of the commercial entity *Sent Forever*; however, there are also a number of physical effigies dedicated to deceased astronauts, casualties of terrestrial war, and other individuals scattered throughout the inner Solar System (including stylized urns containing human cremains).[28]

Many of these autobiographical lifetime-remembrance surrogates are also comparable to the building cornerstone deposit behaviors, or the infamous *"Kilroy was here"* marks left behind by American World War II infantry troops on several ancient monuments. A similar practice may be observed in the ostentatious activities of the renowned Italian archaeologist Giovanni Belzoni who, in a sign of the times, kindly cut his appellation into structures from antiquity for modern visitors to enjoy. It is worth pointing out that Belzoni was considered a pioneering archaeologist of his era and, I would also argue, also one of the premier amateur graffiti artists.

Perhaps, the most visible manifestation of this *participation mystique* legacy can be found within the various *names in space* initiatives that several space agencies have adopted as simple outreach strategies, or promotional stunts for scientific missions.[29] Since the 1975 Viking lander's "microdot," lists of people's names have gradually become a staple inclusion within these missions. However, these lists have also appeared within other, more serious messaging efforts, such as the infamous list of U.S. Congress members inserted by NASA on the VGR (much to the record author's dismay). As the science fiction author Ray Bradbury once surmised (about human agents prevailing over practical challenges such as climbing mountains or reaching space), "if we reach the stars, one day we will be immortal," a desire which these simple proxy gestures metaphorically fulfil for the minds that contribute to them. Indeed, there is already a subculture of online messaging boards that documents individuals' various contributions—and opportunities to partake within—this distinguished record, projects which are now further promoted by NASA under a "passport to space" service as mission outreach. While these immortalizing activities can be amusing, and personally gratifying to partake within, it is debatable whether they add anything significant to the exoatmospheric record, or may support unintended insights into our modern fetishes within the contemporary "disposable culture" (Cooper, 2010). In archaeological terms, evidence is evidence, and it will always practically contribute *some form* of insights into an artefact and its authors. However, other parties consider such cultural extensions as nothing more than a variant of kitsch or graffiti, with many dissimilar opinions also dispersed along this long scale of value judgements.

4. *Those in the Future Would Want to Know More About Our Time*

In contrast to the variegated traditions of parietal art deposits, the practice of bequeathing an inheritance of information, or "passing the torch" of our experiences, is a relatively modern affair. The aforementioned Assyrian deposits were likely not the first intentional archives of information from antiquity destined for an undefined human posterity, but this familial practice of serial crafting informal time capsule deposits exemplifies many of the familiar principles and customs used today for casting messages and cultural information beyond our temporal shoreline. Whether the progenitors of such archival resources believed their descendants would want to know more about them may be seen as a temporal gamble; however, modern studies in history and archaeology have demonstrated how such seemingly inconspicuous *gestalt,* arising from human antiquity, may one day prove insightful and valuable for understanding our social histories and ranges of cultural lifestyle experiences.

The creation of secure, off-world information vaults or a "backup of knowledge" outside of our terrestrial environment[30] is not a recent proposal, and likely stems from the pioneering science fiction literature of the 20th century (Jarvis, 2003).[31] The reader may be familiar with Arthur C. Clarke's *The Sentinel* (1951), later adapted to become Kubrick's film *2001: A Spacetime Odyssey*. While not an information archive *per se*, the storyline explores the discovery of mysterious extraterrestrial monoliths, whose function is to seemingly guide humanities' evolutionary trajectory towards a penultimate transformation (not too dissimilar from how we believe our bequeathed information may enlighten successive generations). In a similar example, the premise of Asimov's *Foundation Trilogy* presents the reader with an ancient, pre-recorded simulation of the protagonist Hari Seldon who, over the evolution of civilizations, is capable of reappearing to dispense political, social, and historical advice to posterity.[32] Other, more recent speculations within the arena of celestial archives have postulated the cultivation of a "Matrioshka brain" as a computational super-object (Bradbury, 1997–2000; Sandberg, 1999), alongside a recent International Space University study (2007) on the feasibility of creating a communicating vista of knowledge on the lunar surface. Some private organizations have even questionably suggested that the Moon's natural environment could play host to a biological ark of genetic information to preserve the genus of endangered terrestrial biota, similar in scope to the earlier (terrestrial-based) *Library of Life* proposal (Benford, 1992), but this raises additional bio-ethical and moral concerns as discussed later.

The development of such tacit celestial archives of knowledge as part of an international, globally-inclusive endeavor, or as a private commercial enterprise, may be viewed as a generational or personal "central project"—a phrase coined by the philosopher Frank White (1987) to describe a "means of focusing the energies of a population during an evolutionary transition to a higher level of culture." Such projects *may* serve as vehicles for more abstract social and psychological aims of immediate or successive generations.[33] It should be observed from this definition, however, that the value of this transition remains firmly with the immediate generations who benefit from such steps and, in the case of archival development, perhaps secondly for distant posterity who may find abandoned deposits. One could infer that we are concurrently pursuing several central civilization projects in establishing "the archaeology of the future," and that development of these off-world "mindstep" (Hawkins, 1983) repositories is yet another manifestation of this perpetual drive to simply archive, conserve, and bequeath intergenerational knowledge as a postulated stage in the "grand strategy" (Michaud, 1982).

These projects tend to embody two, broadly identifiable mnemonic operations of conserving extensive amounts of records for futurity as a prosthetic memory device, while conversely presenting an account of how the author's era actively interpreted its own culture *for itself* (Stiegler, 2009). It may also be inferred that projects which follow this trajectory also possess strong, sensationalist associations with the recovery of human civilization after a planetary catastrophe (Spivack, 2003), creating "a tomb for history" (Harrison, 2014), or keeping cultural memory perpetually alive in the minds of future inhabitants of Earth—hallmarks of the various anxieties of "cultural amnesia" (Haskins, 2007), and also reminiscent of the evocations of death and rebirth cycles prevalent within terrestrial time capsule entombing and recovery ceremonies. Certainly, such conceptions of *rediscovery* are prevalent across popular media depictions which, broadly, explore the narratives of recovering advanced artefacts from ancient human antiquity or, in more fanciful scenarios, technologies from forerunner extraterrestrial species. (How many popular media stories now revolve around the notion of finding ancient artefacts or advanced weaponry?) However, it is intriguing to observe how such thinking has now become embedded in the varying subcultures now contributing to information preservation.

There are several proposed, historical, and realized projects that conform to this information-preservation desire. *KEO*, perhaps the most ambitious heritage-satellite proposal to date, was founded by the French artist Jean-Marc Philippe to cultivate a millennial time capsule of crowd-sourced heritage resources and other curated contents for the eventual discoverers in 50,000 years' time.[34] Similarly, Arthur Wood's *Orbital Unification Ring Satellite* was due to service this function from Low Earth Orbit, with the (missed) goal of launching the structure by the year 2000. Another recent initiative to contemplate the development of a celestial vault was *Lunar Mission One*, which propounded to store private and public information in "millions of digital memory boxes," alongside an encyclopedia of Earth with human hair samples, all of which would be deposited in a borehole through the lunar regolith. There are several similar projects that relate to the establishment of these celestial libraries such as the Long Now Foundation's excellent *Rosetta Disk*, which was forcibly deposited by the European Space Agency (ESA) Rosetta spacecraft onto the comet 67P/Churyumov—Gerasimenko. I can't confirm if this disk is now resting in pieces, or forms part of the comet's halo. The Planetary Society's *Visions of Mars* DVD—a radiation-hardened, silica glass disk containing a rich mosaic of popular science fiction media about Mars for future Martian explorers—was also deposited onto the Martian surface (see color insert). In fulfilling their stated mission "to preserve and disseminate humanity's most important information across space and time, for the benefit of future generations,"[35] the Arch Mission Foundation has also begun actively launching dense archives comprised of contents such as Asimov's Foundation Trilogy, the English edition of Wikipedia, Project Gutenberg, and the *Internet Archive* repositories, alongside copies of the *Rosetta Disk* project and the Central Intelligence Agency's World Factbook (Spivack, 2019).

While seldom explored within space messages to date, the practice of conserving "essential information"[36] for effective planetary stewardship across deep time possesses a historical precedent across various terrestrial settings. For example, in earthquake prone regions of Japan, dozens of *Tsunami Stones* can still be found dotted along the coastline to forewarn future inhabitants of risks posed by building dwellings below a designated elevation near the ocean. Similar mnemonic and memorial devices should, ideally, be installed at limnic eruption sites which may suffocate nearby residents. More modern developments, such as the Waste Isolation Pilot Plant project (Trauth *et al.*, 1993), have conceptualized how to mitigate future human intrusion within a deep geological fissile waste depository using symbolic, lingual, pictorial, and architectural features, or by establishing an atomic priesthood and local populations of radioactive cats to serve as intergenerational custodians for 10,000 years. The *Landmarkers* project, developed as part of an open competition by the architectural initiative *Arch out Loud*, demonstrate perhaps the most imaginative and creative musing in recent years to address this challenge (despite some of these proposals' debatable resilience to the geomorphic cycles of Earth over time). While efforts to host this information within

The tsunami stone near Aneyoshi Bay in Honshū, c. 30 October 2011. This stele was erected near Omoe village, alongside another 316 standing stones across the region, bearing warning messages following the 1896 and 1933 tsunamis. The inscription roughly translates to: "Homes on higher ground will guarantee the comforts of descendants. This is a reminder for the horror of tsunamis: do not build homes below this point. We suffered tsunamis in 1896 and also in 1933, only 2 villagers in the former disaster and 4 in the latter survived. Keep on guard even as years pass by." The 11 March 2011 tsunami would sweep inland through Aneyoshi bay to an altitude of 40 meters, washing away several other stones, but stopping before this stele (located at 55 meters altitude). Everyone who followed the tablet's advice survived the 2011 Tōhoku tsunami (photograph from T. Kishimoto/Wikimedia Commons, CC BY-SA 4.0).

space-time messages will likely serve as an intangible redundancy for terrestrial strategies, the long-term communication of this "essential" information is nevertheless of paramount ethical interest to several other global stewardship entities. The *Companion Guide to Earth* initiative, under development by the *Beyond the Earth Foundation*, is seemingly the first experimental enterprise dedicated to uniquely exploring this stewardship potential with a research program and planned micro-etched artefact for Geosynchronous orbit. The guide is principally designed as a semiotics experiment and to foster interdisciplinary engagement in this emergent field (Quast, 2018b).

5. Freedom of Expression, and Autonomy in Action

5.1 "Art is about taking something that is small but can represent the whole"

The late philosopher Ayn Rand once described art as "a selective recreation of reality according to an artist's metaphysical value-judgements" (Rand, 1971). I do not think there is a need here to demonstrate the veracity of this assertion and its relevance for many, if not all, aesthetic-inclined cultures. Artisans do not merely replicate our experienced reality, but rather annotate and emphasize particular features while challenging our standard ways of mainstream thinking, personal dispositions, and cultural customs as part of a process of social experimentation and reinvention. The capacity for creative thinking, as a situated, mental process of reasoning through, and interacting with, a diverse range of transformative matter, is a common feature of every culture and social history to live on and explore Earth's horizons. This concise summary raises the tantalizing specter of whether such biosocial forms of inventive reasoning, facilitated through our interactions with things, is an innate, cognitive skill set resulting from how humans continuously absorb, filter, rationalize, and categories our surrounding ecologies—though, a process which obviously creates many disparate outputs.

The historical musing of (pre–Space Age) international artists about space exploration and a human expansion into this celestial realm is extensively documented (Petersen, 2009; Triscott, 2016). These chronicle peculiar artist space race projects, impacts from the imaginative exploration of celestial regions on avant-garde movements, space as a metaphorical utopian dream for equality (in stark contrast with established Western notions of progress), and also artisan's broad involvement within commemorating aspects of space missions through commissions or occasional residency programs. In conjunction with the impetus of previous motivations for crafting messages, the Space Age itself presented a destabilization of this traditional Earth-space perceptual boundary, a new and now permeable membrane for re-conceptualizing our customary worldviews and positions beyond the technological realms previously available to generations of human experiences. While science fiction literature has certainly prospered from the imaginative and technological exploration of this new frontier, it is perhaps hardly surprising to also see other creative fields reacting to, and producing activities directly within, the exoatmospheric archaeological record. Contributions have, thus far, been highly sporadic and asymmetrical, largely due to the common perception that space is the traditional province for science, engineering, and technological ingenuity. However, in more recent years, cultural and social fields have come to also breach our planet's thin blue veil in the form of a burgeoning enterprise of "astronautical art" in—or about—this frontier.

Jon Lomberg's "taking something small" sentiment (Reynolds *et al.*, 2017) refer to the common perception of the VGR as a crescendo in human creativity and maxim in representing our diverse aesthetic and acoustic accomplishments, unique to a particular time, for a substantially foreign culture to contemplate. The message is often described in poetic and deeply meditative terms (Sagan *et al.*, 1978), and has been noted to contain the products of intellectual and artistic elites as a "best of ourselves"[37] motif (Schmitt, 2017). But the record also exemplifies how the creative disciplines in the arts and humanities have increasingly collaborated alongside the established efforts of engineering fields and the planetary sciences to deliver perceptibly legible statements for posterity or, plausibly, distant ETI. Perhaps, this growing influence of the arts in developing communication stratagems is best illustrated within the increasing number of professional artists, composers, designers, and other creative disciplines who contribute their diverse expertise towards interstellar message design workshops in order to help advance the development of future communication strategies for extraterrestrials using aesthetic, semiotic and acoustic principles (Vakoch, 2011a).

In addition to these communication applications, several contemporary projects seek to re-address a perceived imbalance in how a future recipient may interpret extant bodies of curated information by providing annotations to the corpus of documentation for subject matters which may otherwise be omitted from the archaeological record. The artist Joe Davis's infamous reactionary project *Poetica Vaginal* focused upon the deliberate censoring of vulvas (and, in general, misleading or misrepresentations of females' forms, prerogatives, and attributes) within both the Pioneer and VGR messages, serving as an artistic intervention; *however*, a U.S. Air Force colonel prematurely terminated the project upon learning of the proposed contents. Despite this intervention—*of an intervention*—several test transmissions were apparently pre-broadcast to nearby star systems. In a similar reactionary project to the lack of a Beatles composition on this record, a Beatles historian decided to correct this injustice by broadcasting the *Across the Universe* composition to Polaris, alongside a brief adduce written by Paul McCartney. Given the encoding strategy employed, it is likely that this bit stream will remain unintelligible; however, this has not deterred multiple media outlets from erroneously hailing this project as the first musical composition transmitted into space (this honor in fact belongs to the theremin concerts of *Teen Age Message*, broadcast in 2001). From assessing the chronology of messaging practices, such sensationalist claims and associated rhetoric are a common occurrence, with other glaring examples including: "the first person to think of something," "first to achieve a type of construct," or "first to send something, somewhere specific." Most claims correspond with the ideologies outlined in sections one and three above, bringing into sharp relief the dominating worldviews still shaping messaging practices.

There are also several "art for art's sake" projects which focus on using space as a "cultural laboratory" for expanding our cosmic consciousness by either enlarging the existing body of non-legible material available off-world, or by representing human cognitive and aesthetic sensibilities through highly stylized and symbolic depictions (such as the Planet Labs satellite art initiative). While these surrogate media will likely remain inaccessible for recipients who do not share these visual or cultural conventions, they serve primarily as a means of introspection and form of social critique for the creator's environments. Dispatching prosthetic extensions of our minds, mentality, and memories in techno-artistic forms, therefore, serves to project these aesthetic values and other heritage media into space as part of preservation or commemorative activities, safekeeping customs, conventions, knowledge, Western scientific imagination, and other cultural faculties, as per the prior legacy crafting category.

In particular, the wonderfully intricate *MoonArk*, compiled as a collaborative endeavor at Carnegie Mellon University, has been cultivated to showcase highly abstract, visual productions from contemporary arts and data visualization practices, in addition to tangible precious objects to serve as an "imperishable trace of humanity" on the lunar surface.[38] The project author's descriptions and use of language explicitly state that *MoonArk* serves as an aesthetic museum which reflects the spirit of the cultural histories contained, primarily for the benefit of contemporary populations, and secondly as an artistic message for posterity. Another, more unconventional artefact of material culture that fits this definition may be seen in the artist Katie Paterson's ephemeral work *Campo del Cielo, Field of the Sky*—a fragment of an iron-nickel meteorite from Argentina which had been melted and recast into a facsimile of itself, before being literally returned to space by ESA. It has since *re-re-entered* Earth's atmosphere.

The "Biodiversity" sapphire and platinum disk depicting a symbolic pentagonal map which represents symbiotic relationships in various terrestrial and aquatic biomes. The golden spiral emerges from Earth, with views of the five oceans orientating the emergence of different life. Latitude/Longitude of the deepest points of each ocean are demarcated at the points of the star (graphic and disk designed by Mark Baskinger with Steve Tonsor, Matt Zywica, Christie Chong, Deniz Sokullu, Maggie Banks, Natalie Harmon, Bettina Chou, Deborah Lee, Carolyn Zhou. ©2015 MoonArts Group, Carnegie Mellon University).

To briefly mention a related material-engagement practice, several temporary, passive reflector satellites[39] that conform to this subcategory have also been consistently launched as a means of ideologically interacting with contemporary surface populations across national borders, much to the frustration of global astronomy observation programs that campaign against such trivial gestures (as these establish precedents for "acceptable disruptions" with similar devices, as well as more intrusive Low Earth orbit constellations).

5.2 The Intersecting Crossroads of Belief, Custom, Spirituality, and Political Expressions

Belief and ritual custom serve as some of the most powerful and intimate social functions that are readily familiar to a majority of human cultural and social dispositions. While religious expression and the "spirit" of cosmism, in addition to similarly-held nuances within manifest destiny, expansionism, neoliberalism, and exceptionalism, alongside other constellations of political and cultural ideology, have tangibly influenced the creation of many of these messages, it is debatable how apparent spirituality and belief systems—as descriptive contents—have physically been in the corpus of messages and transmissions.[40] Certainly, much of this technological infrastructure is owed to the secular aegis of Western scientific tradition, but when looking at the exoatmospheric archaeological record (specifically objects of material culture which directly interact with human agents), and every layer of national space programs in general, we find that concepts of divinity, faith, and transcendence permeate much of human reasoning, enthusiasm, and exploration of this new frontier.

The early Apollo missions to reach lunar orbit were, perhaps, the first instance of religious customs manifesting in outer space and aspiring to bridge sacred and secular traditions together for the nascent Space Age exploration of this frontier, through ritual customs, readings, and symbolic gestures. The Apollo 8 crew reading of ten verses from the Book of Genesis is one recognizable example. However, these actions received much criticism on the grounds of fairness of representation (from the plurality of other faiths, in addition to criticism by atheists), and from advocates for the secular division of state affairs from religious principles. The VGR, while trying to reflect a broad philosophy of cultural heritage across the Earth, chose to purposefully abstain from including religious material on these familiar grounds (and likely also in context with NASA's implicit caution to astronauts against expressing future religious gestures during state-sponsored missions). While I do agree in principle with this decision, given the limited information container, the record contents and curation process—in addition to the written accounts supplied by the authors (Sagan *et al.*, 1978)—are deeply meditative, and evocative of spirituality and cosmological beliefs, framed under the "language of science."

Despite these principal difficulties, there have been several physical contributions to the exoatmospheric material record over the last six decades which ascribe to the mainstream theological mythos of Christianity, Islam, and Judaism faiths. As an unsanctioned activity during the Apollo 15 mission, the astronaut David Scott left his personal Bible on the dashboard of the Lunar Ranging Vehicle, with Apollo 14 astronaut Edgar Mitchell seemingly accomplishing a similar feat. In tandem with these lunar deposits, a series of microform King James Bibles flew to the Moon (and back to Earth again) aboard the Apollo 14 module for, as the authoring Apollo Prayer League (2020) members state in their organization's remit:

> prayerful support of the Apollo astronauts and the skill of NASA employees who built the rockets they flew ... but most notably—to land a Bible on the moon in honor of fallen Apollo 1 Astronaut Ed White who had planned to carry one.

Furthermore, script-based submissions for interstellar transmissions frequently evoke several deities and the praising of prophets, alongside (as the science fiction author

Andrew Chaikan notes) the projection of heavenly virtues onto extraterrestrial agents (White, 1990), and general appeal for ETI, as messianic figures, to intervene within our planetary (de)evolution.

5.3 Principles Governing the Activities of Commercial Enterprise in Outer Space

The history of commercial activities within outer space already spans several industries wide and many decades deep in facilitating terrestrial services—enterprise that will only continue to expand, given the ever-increasing market value of orbital data collection for a plethora of global economies, and the nature of spacecraft as highly expensive, elite commodities that enable the myriad kinds of modern lifestyles across our planet's surface. Technologists and entrepreneurs who advance such applications, in turn, are also seen as visionary pioneers or trailblazers who advance the distant social dreams of human futurity. But this rhetoric also serves to perpetuate tropes of transcendence, destiny, and salvation—familiar convictions, that are rooted in the mythos of religious faiths and ancient imaginings (Noble, 1997).

Without delving too extensively into this broad category, the first two decades of the 21st century have seen an exponential increase in commercial, for-profit entities contributing to the design, sponsorship, rhetoric, accumulation of contents, and launch of message artefacts into outer space as publicity stunts or to serve as a *rite of passage* for their technological capabilities, and also to foster public intrigue for profitable enterprise within space (including satellite-services, corporate claims to orbital ranges, asteroid mining, and laying a prospecting stake within the future colonization of astronomical bodies). Some salient and variegated examples of these lucrative marketing activities can be seen within the SpaceX launch of a Tesla Roadster with "Starman" as a dummy payload *and* technological benchmark stunt, Team Encounter facilitating development of the Cosmic Call interstellar radio messages to encourage crowd-sourced contributions for a solar sail propelled spacecraft (which unfortunately never materialized), the Japanese corporation Fuji-Xerox funding fabrication of the *A Portrait of Humanity* diamond wafer (which unfortunately never materialized), and Penguin UK hosting a transmission competition entitled *Break the Eerie Silence* to promote the launch of Paul Davies' SETI book (Davies, 2010) using the commercial organization *Sent Forever*.

The Deep Space Communications Network (DSCN)[41] is another transmitting organization specifically established to broadcast publicly contributed contents into space and have, to date, already realized an impressive back catalog of eclectic, distance-limited messages (which arguably capture the zeitgeist, and prevailing attitudes of human populations to a more accurate degree than curated, representational accounts). Readers may be familiar with the *Craigslist messages* which were broadcast by DSCN—a project that manifested when the Craigslist CEO Jim Buckmaster won an auction for transmission "airtime." Thereafter, he decided to broadcast several thousand public website adverts. While extraterrestrial recipients, if any, may be uncomfortable with this sudden bombardment of classified adverts taking up precious observation time, the prospective inhabitants of Alpha Centauri may be otherwise preoccupied with deciphering a binary data stream encoding a video format of the 2008 blockbuster *The Day the Earth Stood Still* featuring Keanu Reeves as Klaatu. Given the subject matter, I am unsure as to whether the intended recipients should be amused, dismayed, or insulted by our depictions of other intelligent life. Perhaps, the most visible oddity for advertisements within space, however, may be presently found in the weird Doritos broadcast of a tribal tortilla chips commercial to hungry extraterrestrials that may share in our gastro-preferences. I feel compelled to ask the reader whether they indeed feel comfortable in knowing that their outer space diplomacy interests are now being represented by the PepsiCo corporation. All these messages will likely remain as patterned noise, with some recurring semiotic significance.

Lomberg (2004) also highlighted a growing number of proposals for data storage in space as a lucrative (and novel) market scheme using lightweight magnesium plates of microfiche or other media to litter our celestial environment with birth and family history records, marriage certificates, legal briefs, prayers, secretive gestures, novelties, etc., in a similar manner to that advertised by the recently announced *AstroGrams* initiative. As national space programs have largely abandoned geocentric orbit, it is likely that we will continue to see the commercial sector expropriate this futurescape and, in doing so, fabricate a cacophony of local artefacts and EM messages for various applications—personalized gestures to eternity that will likely preserve conflicting, idiosyncratic, and illegible accounts for individuals and hegemonic populations that are capable of funding such endeavors. Separate to this, and as a testament to the significant impact that Space Age messaging activities have had on "re-shaping" terrestrial populations, several existing space messages have also been re-fabricated as part of the space memorabilia collectors' market. Replicas of the Apollo lunar plaques, *Pioneer Plaque* and VGR are now commercially available "to experience it [the VGR] for yourself the way it was meant to be played,"[42] providing yet another dimension to Jarvis's evocative time capsule experience category, emanating from the classic cultural history of human space exploration.

6. Promoting Outreach and Education in Outer Space Affairs

Several national space agencies and non-profit space-advocacy organizations, in addition to facilitating *names in space* initiatives, have also continued to develop artefacts and other crowd-sourced media projects to inspire public engagement with their missions, or foster educational opportunities within exploration and engineering activities for an assortment of audiences.

The Planetary Society is one of the more prolific organizations to successfully realize dual purpose objects aboard spacecraft that conform to this definition, including the *Marsdial*—an aesthetic-instrumental gnomon, devised specifically for Mars' climate, which aids in the color calibration of the Pancams on Martian vehicles, but also apparently serves as a highly aestheticized chronometric device for classrooms to learn more about this planet's environment. The Perseverance rover has kept up this tradition, in addition to also carrying an assortment of other cultural cargos. It can be argued that the VGR and *Juno Lego* statuettes also correlate with this mission outreach objective; however, more befitting examples of this category can be found within ESA's CHEOPS mission—an exoplanet space telescope—which contains two large plaques engraved with a rich tapestry of around 3,000 kids drawings from member states (see color insert), in addition to the Japan Aerospace Exploration Agency's (JAXA) regular inclusion of etched artwork styles on aluminum plates within their various missions.

Both of these latter examples, in addition to previously mentioned Planet Labs activities, present the physical body of a spacecraft as a blank canvas for public decal works in order to inspire mission engagement and intrigue, as well as to justify the benefits and financial costs associated with aerospace and interplanetary missions. Personally, I'm curious how these painted decal works may affect the thermal control properties and movements of spacecraft over time.

Furthermore, JAXA (and the agencies' precursor organization, NASDA) have held annual space camps for children that have occasionally produced simple bitmaps which would be subsequently transmitted towards local stars and constellations as symbolic METI activities. I am especially fond of the wonderfully absurd "rice dumpling and tea" bitmap, which was successfully transmitted by a NASDA facility to the binary star system Spica (in the Virgo constellation) in 1997. In addition to these projects, there

1997 NASDA message "Rice Dumpling and Tea" (bitmap courtesy Shin-ya Narusawa, redrawn from originals created by NASDA Space Camp attendees. Re-illustrated by Paul E. Quast).

have also been several symbolic radio transmissions broadcast to nearby space probes, intended as a gesture for promoting engagement within a mission, or supporting public opinion of space-exploration budgets. While these transmissions can be quite limited in scope, several of these projects have been established as competitions for multinational participation, such as ESA's playful *Wake Up Rosetta* video campaign, or sending phrase-constricted, Unicode messages, such as the recent NASA Voyager 1 (*#MessageTo-Voyager*) and New Horizons (*The Ultima Thule Flyby* greetings) campaigns. Regardless of how educational activities manifest within this material culture record, the benefits of such activities are proportionally tied to the hands and minds of those who immediately benefit from such projects, leaving us to consider how these initiatives may be viewed and interpreted under the lens of future spatiotemporal observers. Perhaps, they could be found as examples of "art for art's sake" projects similar to how we have historically perceived parietal artworks, or as presumptive archives for educating ETI about us? Maybe they will be interpreted as idiosyncratic legacy deposits secreted away from contemporary eyes, or as unique social documents preserving autobiographies from the nascent Anthropocene era?

A Taxonomy of Messages from Earth?

Clearly, much of these overarching motives and principles for crafting the broad variants of messaging artefacts are a very human affair—material signs and expressions indicative of the numerous ongoing pantomimes, frequent arguments, and perceptual boundaries that continue to disparately shape and divide our present conceptions of this terrestrial stage. While the above hierarchical listing of expositions underscore much of this social rationale and constructive dialogue behind the purported creation of messages as intentional, conscious extensions to the exoatmospheric archaeological record, it is incumbent of me to state that these observable conceptual trends in material culture depositing practices are non–Cartesian and, as such, cannot delineate clean, taxonomic demarcations. There is no neat divorce in the often-intersecting relationships between our mental faculties, and that of externalized material cultural practices (Malafouris, 2004). They are two inseparable sides of the same coin that are intertwined in the idealist creation of articulate, symbolic relics without verifiable recipients, retrieval dates, or other properties congruent with traditional time capsule practices.

While the intent of this discussion has been to provisionally establish a taxonomy for the cultural history of exoatmospheric messages as a basis for further context in the ensuing discussions, we should not solely restrict ourselves to thinking in such definitive coloring, on the grounds that purely reviewing these enigmatic puzzles under set mental templates may also limit our scope for understanding these messages as a crossroads for our cognition, material practices, social histories and cultural lifestyles (amongst other entwined characteristics). We should err on the side of caution when introducing such demarcations within a practice that is, to make a long story short, essentially crafting prostheses for an embodied cognitive-material relational process, seeped within a panoply of socio-historical traditions and species-specific perceptual responses to emersed environmental settings. The complexity of these ongoing archaeological discussions pertaining to material culture cannot be pursued here in great detail (for an extensive, robust review of some general practice arguments across archaeology fields, see Hicks and Beaudry, 2010). However, it is crucial to provisionally understand these relationships between the agency of our minds, and the "myriad kinds of things" ourselves, before expecting imagined, exotic actors to do so across vast spatiotemporal distances.

It is arguable whether such motives and inclinations, uniquely bound between the porous realms of idea, mind, memory and matter, may be feasibly externalized in the phenomenology of messaging artefacts for comprehension across periods of deep time and space, without the supportive crutches of cultural, ecological, biological contexts, and other vital referents that are unique to *living social settings*. Regardless of this uncertainty, these autobiographical surrogates, and the voices that they collectively immortalize, will hardly remain as the sole storehouse of information available for the archaeology of the future to base their inferences upon. Still, it is worth understanding how the entangled social, historical, cultural, and disciplinary relationships within these underlying motives filter to the mental foreground when we craft *metonymy* that intend to "speak for Earth." The succeeding texts in this volume (predominately based in the literature of the first four categories, but in some relevant context for the latter two segments), will provide some inkling into how we tend to think *with* and *through* this surrogate media; often revealing the range of frequently overlooked idiosyncrasies, values, narratives, and dispositions we transfer along with this material culture beyond our atmosphere. Such variegated *gestalt* may be more revealing of our minds, philosophies, cultural histories, lifestyles, social anxieties, attitudes, value codes, traditions, and conflicting psychological preferences than we may wish to portray.

NOTES

1. An internal archaeology quip to define the unknown nature of an obscure artefact or insignia, in the absence of also knowing what distant minds may have attempted to partially signify within their distinctive cultures.

2. This inter-generational bequeathing of knowledge can be viewed in the *information-preservation* category. Unfortunately, there is a distinct lack of inscribed evidence available for Ashurbanipal's heir, Ashur-etil-ilani.

3. These ritualistic practices are often associated with the death and rebirth cycles as raised by Jarvis (2003).

4. As a benchmark for segregating space-time capsules from votive message cornerstone deposits, a lowest threshold for information capacity and complexity should also be established for further reference. Considered implications for information theory, content structural complexity, and appropriate corpus samplings should also be incorporated into any such threshold.

5. A rather poignant example can be seen in the use of Adolf Hitler's 1936 Olympic Games address as a mimetic tool to enable human recognition of an extraterrestrial response message in the novel *Contact* (Sagan, 1985). This is known as a "crib" or a means of signifying the intended artificiality of a signal as a purposeful message.

6. Messages rarely belong to one set category or another, and likely embody a range of these objectives to varying degrees. The APOH catalog has listed this information according to stated project aims as ascertained through statements of intent by authors, interviews, use of plausible encoding stratagems, suitability of message contents and the intricacies explored within maintaining information commensurability.

7. *Weltanschauung* denotes our comprehensive conception of the universe, and of humanity's relation to it.

8. Such ideologies tend to be interpreted as linear advancements in social, biological, or cultural evolution of the entire human species, rather than representative of a specific population's era and their cultural backgrounds.

9. It is arguable that 20th century works in science fiction by Olav Stapledon, Ray Bradbury, Robert A. Heinlein, Issac Asimov, and Arthur C. Clarke, among others, established the ideologies of colonialist claims in outer space.

10. This is an oversimplification of the many competing ideas that led to the ideological foundations of the USSR.

11. I have chosen to focus on the two superpower protagonists of the era due to the sheer quantity of material culture launched by both nations in respect to other launch-capable nations. It is worth noting that France, Australia, and the United Kingdom were also main Western powers within this race. However, the competition is regularly narrowed to the "classic" U.S. versus USSR rivalry. The race also possesses several overarching connotations worthy of investigation including: the interlinking of competition with catalyzing technological advancements, the often romanticizing narrative of the superpower space race as an explorational program that presents Cold War victories as universal human achievements, the celestial expression of colonialist agendas during a period of time when the old national empires were fragmenting, and also how these material cultural legacies manifested across the fourth dimension as "layering" (Quast, 2021).

12. This is subject to the disputed claim of whether the inoperable Venera 3 probe automatically released its lander probe after losing contact, or whether the derelict spacecraft simply completed a flyby of this planet.

13. Henceforth *Outer Space Treaty* (OST). The OST also features several security, safety, and liability provisions.

14. There are various known configurations of Soviet pennants residing on the surface of, or in orbit around/nearby, the Moon (around ten deposits), Mars (around three deposits), and Venus (around 16 deposits).

15. The *Apollo Goodwill Messages* were also deposited on the lunar surface by this mission—after the U.S. flag was erected. Several other small memorabilia objects were also deposited, along with the Apollo 11 lunar plaque.

16. Given the initial military value of space exploration, it is perhaps not surprising to note that the U.S. Navy's TRAAC satellite was the first spacecraft to transport Western cultural property (the poem *Space Prober* by Thomas Bergin) into orbit in 1961, an object, that Gorman (2019) notes, preserves a conflicting vision of our early perceptions of space exploration. A similar conflict of manifested intent can be found in Trevor Paglen's *Orbital Reflector* which was due to inspire public intrigue in art and space, but ended up as inoperable debris when it was not activated due to political tensions forcing a U.S. government shutdown of infrastructure in 2018/2019. Dual meanings and symbolic conflict are a staple and often overlooked part of many messaging projects, in favor of consistent, candid narratives that favorably provide coherent stories about human achievements.

17. It can be argued that developing space technology is deemed as a perceptible stage in the social evolution of statehood—social Darwinian milestones which ascribed to the outdated linear trajectory of primitive to advanced societies that was initially proposed by 19th century anthropologists (and subsequently proven by the next generations of anthropologists to be wholly misrepresentative of how actual human societies develop).

18. It is likely that the most frequent national deposits in space are flags, and "made-in-country" stamped objects.

19. White (1990) accurately corrects SETI to SETILO—"Search for Extra-Terrestrial Intelligence Like Ourselves."

20. The current technological phase of scientific investigation and research began as a result of a seminal paper on search strategies written by the physicists Philip Morrison and Giuseppe Cocconi in 1959 while Frank Drake independently began *Project Ozma* observations of two nearby stars in these proposed ranges using specially designed equipment. Concurrent to this well-known history, early SETI pioneers in the Soviet Union also extensively contributed to this literature, searches, and international workshops (Gindilis and Gurvits, 2018).

21. At the same point in time as Cocconi and Morrison's publication (1959) on radio frequency messages, Robert Schwarts and Charles Townes (1961) recognized the use of optical masers (i.e., lasers of varying EM frequencies) as a suitable interstellar communication medium for OSETI (Optical SETI), and Ronald Bracewell (1960) also published a seminal manuscript on the search for extraterrestrial artefacts. Unfortunately, the acronym SETA has been reappropriated by pseudoscientists in their quest to discover alien precursor artefacts on Earth, rather than attribute early human technological ingenuity to the ongoing experiments of ancestral human civilizations.

22. The radio pioneer Guglielmo Marconi allegedly sent the first broadcast into space in 1909 as described within a 1919 edition of the *New York Times*, but there is a lack of discernible documentation for this likely weak premodern METI signal to meet present criteria (Zaitsev, 2006) or indeed exit the Earth's radiation belts.

23. It is worth noting that early METI attempts were still, in part, subject to the initial ideological vacuum of the Space Race (i.e., to be the first political power known by other minds, to garner ETI secrets for national security interests and advancements, or to determine whether advanced extraterrestrials embodied communist or capitalist ideologies). Messaging as a historical behavioral phenomenon emerged in both parts of the Cold War geopolitical arena, initially as a territory marking process for newly discovered places, but SETI conferences would become a rare stage for interactions between groups of Western and Soviet scientists.

24. Like the *Morse Message* using the Pluton-M planetary radar, the *Arecibo Message* was an outreach gesture as part of a technological demonstration of the newly upgraded Arecibo Observatory. As such, these projects are reminiscent of the prior Space Race national activity of transferring planetary pennants or mementos.

25. In this case, the probe crashes on an inhabited planet, and its power supply was repurposed to create weapons.

26. I find this to be a populist claim to attract publicity and sensationalist media headlines. It is worth noting that this consideration is often claimed as an afterthought, as opposed to serving as an intellectual exercise resulting from robust engagement with theoretical and practical discussions in SETI/METI literature.

27. The transmission was aimed at 1A 0620-00, one of the nearest known black holes—a fitting homage considering Hawking's lifelong work on the theoretical physics of these exotic, gravitational behemoths.

28. Examples of these posthumous memorials can be seen within the *Fallen Astronaut* plaque commemorating the deaths of astronauts up to 1971 (the year of the Apollo 15 launch), a dedication to the Colombia Space Shuttle crew (mounted on the back of the Martian Spirit rover's high gain antenna), the complete list of U.S. military Vietnam casualties on the Stardust-Next probe, and also the cremated remains of Eugene Shoemaker and Clyde Tombaugh included on board the Lunar Prospector and New Horizons probes respectively.

29. As a personal note on these legacies, this author's own biography and picture were apparently immortalized on the Beresheet lunar lander as analog inscriptions within the Arch Mission's archive, though I suspect this information will be quite inconsequential for a future discoverer, if it indeed survived the probes hard impact.

30. I would assume an astro-engineering project that is accessible with modern human spaceflight capabilities.

31. These proposals also likely stem from historical losses of knowledge from antiquity, such as the numerous fires at the Library of Alexandria and the sacking of the House of Wisdom in Baghdad, along with the contemporary loss of archaeological sites of world heritage significance in modern conflict zones.

32. It is worth noting that both these aforementioned science fiction examples possess a strong, psychological correlation with concepts of ancestral worship (Jarvis, 2003), and an envisioned destiny of needing to pass down our "superior" knowledge to information-deprived or socially destitute descendants. The language used to describe and promote such projects generally envisage the archive of information as a "moral missionary" designed to dictate to recipients our perspectives of who they are, what they should know and how they ought to behave (according to our civilizations' ethical customs and social standards). These same principles also partially underscore the granite-inscribed lingual tenets of the wonderfully obscure *Georgia Guidestones*.

33. Examples of these transformative "central project" experiences which are listed in White's *Overview Effect* (1987) include: Gothic cathedrals, Egyptian pyramids, the Apollo lunar missions, and U.S. space shuttle program. This is also known as "Cathedral thinking," a phrase used within establishing principles of long-term planning.

34. This satellite was intended to intentionally de-orbit and survive re-entry into Earth's atmosphere in 50,000 years' time—making this one of the few "space-time capsules" with both an intended audience and target date.

35. Quoted from the Arch Mission Foundation website, accessed on April 24, 2018.

36. Essential information herein is defined by the information-preservation community as contents that will need to be bequeathed for particular existential-risk applications (e.g., signposting hazardous waste deposits, documenting perishable planetary climate records, outlining rogue gene drives, and chronicling the proliferation of persistent organic pollutants, amongst other legacies for long-term monitoring of ecological stability etc.). The Beyond the Earth Foundation's *After the Horizon* program is presently documenting these enduring legacies under the thematic ethos of "do not go, and need to know."

37. In describing the "best of ourselves" within these messages, the choice of whose genomes to include within the *Immortality Drive* on the ISS possesses significance for what we categorize as "the best" of human qualities, characteristics, and traits worth preserving. In addition to the relevance of these choices in modern cinematic depictions of space colonization missions, one cannot also help but draw cynical comparisons with Arthur C. Clarke's (1964) sentiment "Space has room for many things, but not for 'your tired, your poor, your huddled masses yearning to breathe free.' Any statue of liberty on Martian soil will have inscribed upon its base 'Give me your nuclear physicists, your chemical engineers, your biologists and mathematicians.'"

38. The inclusion of these objects is reminiscent of the KEO project and several other initiatives that proposed to transport physical samples of soil, seawater, air, and human blood (or hair follicles) into outer space as "relics."

39. Examples of these launched reflector satellite projects include the failed *Mayak* (2017), the successfully deployed *Humanity Star* (2018), and the inactive *Orbital Reflector* (2019).

40. The role of social, political, and spiritual belief is, by and large, associated with language used in

advocacy campaigns for human spaceflight and colonization efforts (Billing, 2017); however, its influence within the creation of messages as technological prayers or harbinger devices for these customs should be the focus of a separate, devoted volume. A brief account of the role of theology in stimulating message construction under the rubric of Western scientific and (cultural) paradigms is detailed later (Traphagan, this volume).

41. Not to be confused with NASA's Deep Space Network of facilities that support interplanetary probe missions.

42. Quoted from Ozma Record's *VGR Kickstarter* campaign to redevelop the message as a replica product.

Part 2

WHAT WE SAY

The Tyranny of Assumption

*What We Should and Should Not Expect
ETI to Understand in Our Messaging*

JOHN W. TRAPHAGAN *and* KELLY SMITH

To date, there have been a variety of attempts of varying levels of seriousness to send messages in the direction of hoped-for intelligent, extraterrestrial life. There have been at least 31 intentional attempts to message alien worlds since 1974 (Zaitsev, 2011; Quast, 2018a). Most of these don't represent major risks as they involve one time, relatively low power, transmissions. Several dozen were just publicity stunts, as when Pepsi hired one of the EISCAT radar arrays in 2008 to share the good news concerning their breakthrough tortilla chip technology with any aliens living in the Ursa Majoris constellation (Capova, 2013b). But there are several projects underway to attempt communication in a much more serious and sustained way—for example, by using powerful lasers aimed directly at nearby star systems with known, potentially habitable, planets. The goals and motives of these new projects range widely. On the one hand, *METI International* has assembled a team of scientists and other experts to begin preparation for the kind of METI effort one would expect serious scientists to undertake—a carefully crafted message with a systematic approach to transmission. On the other hand, the amusement park millionaire William Kitchens has created the *Interstellar Beacon Project*, which proposes to "backup humanity" by beaming the entire contents of Wikipedia to the stars (The Interstellar Beacon, 2020; Kitchens, personal communication, 2014), without concern that some broad consensus—or even discussion—among the population of Earth should occur before sending such a message.[1]

As a result of these developments, the debate about whether this is a good idea has blossomed into something of a cottage industry (Vakoch, 2016; Brin, 2014; Pichalakkattu, 2018; Haqq-Misra *et al.*, 2013; Peters, 2017; 2019). Given these recent developments, it's worth considering more carefully exactly what sorts of implicit assumptions are built into the messages we send, whether these are actually reasonable, and how they impact what we can expect ETI to glean from such messages (and subsequently, how ETI may choose to react or respond to such constructs). To be fair, we will focus here on what is, in many ways, frequently considered one of the most serious messages thus far in terms of its construction—the original *Arecibo Message*—since it's far too easy to discount the silliness of a Doritos commercial.[2]

The Russian radioastronomer, Alexander Zaitsev, argues that for an attempt to send an interstellar radio message to be considered credible, it must satisfy three criteria: (1) the

Frank Drake's original 1962 pictograph, which was blind tested via an electronics and cryptology magazine in the same year while also being submitted to numerous Nobel laureates for decipherment. The scientist Barney Oliver was the only respondent to figure out the messages' structure, but not its encoded contents, before responding to Drake with his own pictograph message. Note: the bitmap grid has been arbitrarily assigned to frame these information "bits" in an image format as per our visual preferences (bitmap originally compiled by Frank Drake [1962]. Re-illustrated from data by Paul E. Quast).

choice of the target star should be reasonable, (2) the energy per bit of the information should be sufficient to make the message detectable against the background noise of cosmic sources, and (3) the message should include a key that would allow aliens to decode it (Zaitsev, 2012). Among the five attempts to message ETI that Zaitsev argues are credible is the first such effort, astronomer Frank Drake's broadcast from the Arecibo radio telescope of binary information towards M13, a closely packed cluster of stars roughly 25,000 light years from Earth. The distance, of course, means that the *Arecibo Message* does not satisfy the first of these requirements very well, but it does represent a serious attempt in that the message's primary significance lies in demonstrating the ability of terrestrial radio astronomers to construct and send a message, rather than to generate interstellar discourse (SNACI, 1975). Indeed, the message was designed to "describe some characteristics of life on Earth" that staff members at the National Astronomy and Ionosphere Center believed would be of interest to alien civilizations (SNACI, 1975). As Vakoch notes, Drake was concerned about how we might communicate without having direct physical contact with ETI, given the enormous distances involved and, thus, the impossibility of providing sufficient context to make translation possible (Vakoch, 1998a). The *Arecibo Message* therefore carried more interpretive significance as a literary device for discussions amongst present-day terrestrial populations than it did as contact media for foreign ETI cultures.

Drake's 1962 approach involves the creation of a two-dimensional pictograph that consisted of 551 bits of information, the only factors of 551 being the prime numbers of 19 and 29, which represent the bi-dimensional lengths of the sides of the bitmap message when properly formatted (Vakoch, 1998a). Assuming that the recipients of the message were able to decode the message, orient it properly, and translate it into a symbolic system that makes sense to them, the content of

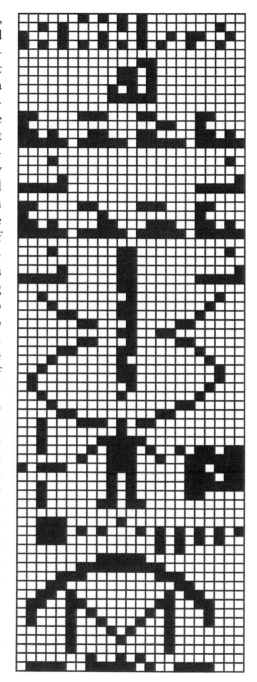

The Arecibo Message. Also featured in *A Simple Response to an Elemental Message* as repeated image #1, and repeated throughout Cosmic Call Series, along with *Lone Signal* crowd sourced media (bitmap originally compiled by Frank Drake [1974]. Re-illustrated from data by Paul E. Quast).

the message includes information that is intended to show human knowledge of nature, the technology used to send the message, and information about our species and its local Solar System. This approach later informed the *Arecibo Message* that displays the numbers from 1 through 10—atomic numbers of elements hydrogen, carbon, nitrogen, oxygen, and phosphorus, which together make up DNA on Earth, along with formulas for sugars and bases found in the nucleotides for DNA. There is a graphic representation of a human figure with the average height of a male (why male, as opposed to female?) as well as the world population of humans. And below these segments are graphics of the Solar System indicating which planet is Earth and identifying the Arecibo radio telescope with the physical diameter of the transmitting dish.

Implicit Assumptions in the Arecibo Message

Of course, it is not possible to send *any* signal without making at least some assumptions, but it is important to be clear about precisely what these assumptions are so they can be examined critically to ensure that they are reasonable. The assumptions related to successful communication associated with the *Arecibo Message* seem to fall into two broad categories related to the ways in which ETI will think, as compared to humans:

1. Rational (scientific) convergence: Any ETI capable of receiving our signal will understand basic aspects of science, mathematics, and logic in more or less the same way humans do.
2. Cultural assumptions: Any ETI who deciphers our signal will be positioned to accurately interpret information about humans (or at least not form a deeply misleading picture of human culture).

It should be noted that there are other implicit assumptions in operation within the message, including the idea that our message will not harm aliens in any way, either in terms of their civilization as a whole, or in relation to the individuals (if they have individuals) that happen to run across and interpret our message (Traphagan, 2017). However, even if we ignore the complex issues associated with assumptions about our potential impact on extraterrestrials, assumptions about the ways in which intelligent aliens might think present problems that deserve a closer examination.

Rational Convergence

Obviously, there is little point in sending a signal at all unless ETI can be relied upon to at least: (1) recognize it is a signal from another ETI and (2) decode its basic meaning, at least in terms of the evidence we provide of our own status as ETI. Would ETI recognize this as a signal? Assuming that ETI were actually monitoring *for* a signal, it seems highly likely that they would, given that any species able to receive the signal would have radio technology and be able to differentiate such signals from natural EM phenomena. It is thus hard to imagine (though not impossible) that such a species would fail to immediately recognize this as a transmission from another intelligent species.

However, would alien ETI be able to reconstruct the pictograph? This is less clear, but probably still a good bet. The second figure in this essay shows the message as it was

Arecibo Message alternative arrangement as 23 rows and 73 columns (re-illustrated from data by Paul E. Quast).

Arecibo Message arranged as concentric circles. There is no reason to believe—or disbelieve—that this information will be formatted in this way but note how our conventions of left and right spatial organization in two-dimensional imagery breaks down, while also interfering with content acquisition. Human figure shown in the south of the image (near center) (illustrated from data by Paul E. Quast).

intended to be viewed by ETI, but it is worth noting that if the message is decoded into 23 rows and 73 columns, it is unintelligible. Certainly, proper decoding might be too much to expect of a single researcher or someone who only spent a small amount of time on the project, but if we assume that alien ETI would assign a high priority to deciphering such a signal (see cultural assumptions below), a team of researchers with the assistance of supercomputers would likely be able to figure out the small number of ways to decode the signal so that it made some sense. However, we need to be careful with this assumption. Amusingly, the technical challenge here is not always apparent to non-scientists who comment on these matters. One author recalls a lengthy conversation at a conference with a colleague who assumed that if we sent something visual, all the aliens had to do was "look at the video," without any appreciation of the fact that video is not transmitted as it appears on their computer, but rather as a series of binary bits that must be assembled in a particular fashion.

Assuming that alien ETI managed to correctly decode the message, would they be able to glean the basic scientific information (e.g., physical constants, chemical formulae) from the signal? Again, it seems hard to imagine (though again, not impossible) that an alien civilization technologically advanced enough to receive and decode the signal would fail to recognize ordinal numbers or basic physical constants such as the atomic number of hydrogen. It could be that they use a very different system for *representing* these truths, of course, in which case determining that this is what they are looking at (to use a visual metaphor) might be a real challenge. As linguist H. Paul Grace notes, we need to keep in mind the fact that the customary way that we think about language and how it is used is a product of the reality construction in which that language itself works and is used to construct reality (Grace, 1987). Our particular reality construction shapes what we expect (and what would be unexpected) when it comes to information being presented to us—and that which is unexpected can be difficult to understand even when it is simple. As Sheri Wells-Jensen has noted (personal communication, 2019), aliens might, for example, see the content of the message as overly simplistic—the work of a child—and, thus, decide it is not worth their attention. Nonetheless, this may not be an insurmountable problem if aliens use reason anything remotely like our own. They must have *some* concept of numbers, for example, and thus can reasonably be expected to pick up on a simple symbolic representation of ordinal counting, etc.

Would alien ETI be able to glean other scientific information from the pictograph? There are actually two separate problems here: First, would aliens be able to make sense of the *visual* representations used? There are several critiques of pictorial representations as an approach to communicating with ETI. Perhaps most notable is the problem that the emphasis on pictorial representations is a product of a strong human reliance on vision as the primary means of making sense of the world (Vakoch, 1998a). Such a representation presents problems if the recipient at the other end evolved in such a way that a different sense was its primary way of dealing with the world. An example of this might be an intelligent bat-like species that relies on sound and echolocation, as opposed to light, as its main way of interacting with the world. Such a species might have a very difficult time in making sense of the message, even if they succeed in decoding its basic structure (Traphagan, 2015). Foucault notes the emphasis on the visible that arises in Western art and the sciences during the 17th century and, in relation to observation, privileges the visible over other things (Foucault, 2001). Vision is so second nature to humans that we typically fail to even consider these factors—to the point where a chapter such as this would be exceedingly difficult to write if we were under an obligation to avoid all visual references.

An additional problem lies in the fact that the message received is symbolically represented as 1s and 0s. How would one know to translate the numbers as the colors black and white? The message is "visual" because we chose to interpret the signal that way. It was sent using frequency shifting in which one frequency somehow represents black and the other white, as a means of generating something that can be interpreted visually—but how would ETI know that the frequencies represent colors? What happens to the "visual" representation if ETI decides that the frequencies represent tones (perhaps their language is tonal) rather than colors? Maybe that wouldn't pose an obstacle to interpretation, but it's very difficult to be sure. The important point here is that if a different sense is second nature to ETI, or they use a way of communicating unlike ours, it is possible that ETI would miss significant aspects of a visually-oriented message, not due simply to an inability to see them, but because their entire conceptual framework is based on a different metaphorical "picture" of the world around them. In fact, ETI might not even recognize that the message *was* an image.

This problem is especially acute when it comes to understanding some aspects of the message that seem obvious even to scientifically illiterate humans, such as the image of a human. Vakoch notes that semiotics (the study of signs) provides a way to think about the communication of ideas and that we must be able to differentiate a sign from the thing being signified. For example, the term "automobile" represents a particular type of object that people use for transportation; the word is a sign that represents the physical object being signified and, therefore, is not something we should confuse with the thing itself (Eco, 1976). This may seem obvious, but it is important in relation to sending any sort of pictorial message, like the *Arecibo Message*, to ETI. Without an understanding of what a human looks like (signifier), it might well be impossible for ETI to understand what the image of a human stick-figure is signifying. In other words, without knowledge of an actual object as a referent, the sign is potentially meaningless (or, worse, misleading) because the relationship between the object and its sign is arbitrary. There is nothing that necessarily relates the sound or word "automobile" with a picture that represents it (which is not even the automobile itself, but an image *of* an automobile). It might be argued that the relationship between a line drawing of a

1995 and 1996 NASDA messages (left to right) (bitmaps courtesy of Shin-ya Narusawa, redrawn from originals created by NASDA Space Camp attendees. Re-illustrated by Paul E. Quast).

car and a car itself is less arbitrary than the linkage between the word for the car and the object, but there remains an underlying cultural component in that one has to have some socio-historical concept of a wheeled vehicle, in order to make the connection without having both the car and its drawing available to compare.

This problem exists for all of the information in the *Arecibo Message*; even the binary representation of the numbers from 1 to 10 are signs that represent those numbers, which are, of course, signs that point to a particular kind of ordering of objects or things. An image like the stick-figure may work fine if ETI were to have bodies shaped like ours, because they might be able to draw a comparison and conclude that this is a—albeit very strange—representation of the people who sent it. However, if they were octopi residing on a world where all organisms are radially symmetric, there might be no reason to make this correlation. In other words, although it certainly seems likely that any ETI will

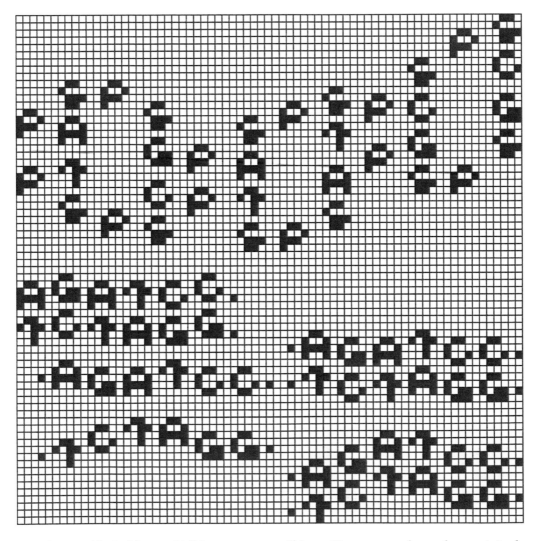

Greetings to Altair, bitmap #4 (bitmap courtesy Shin-ya Narusawa, redrawn from originals created by Professors Hisashi Hirabayashi and Masaki Morimoto. Re-illustrated by Paul E. Quast).

understand counting and cardinal numbers, we need to be careful not to assume that something even as simple as cardinal numbers will be symbolically *represented* in a commensurable way between humans and aliens, even if the basic information contained in that representation is the same. Although this may not necessarily represent an insurmountable obstacle to interpretation, it may still be quite difficult to overcome.

What we can be confident of is this: some aspects of our message would be clear and others mysterious. Even if we assume that aliens would have little trouble interpreting the numbers one to ten, this alone doesn't convey much other than what they already knew from the mere existence of the signal—that we must also be technological and thus mathematical. Perhaps, given that ETI received the message, the image of the parabolic mirror would be at least suggestive of a radio telescope, especially given that the message itself seems likely to have been transmitted from something similar to the technology with which it was received. However, the "M" underneath the mirror will likely

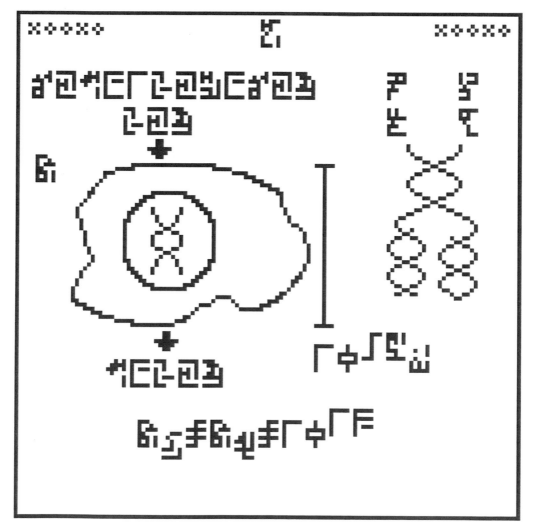

Cosmic Call 1, bitmap #18 (bitmap courtesy Yvan Dutil and Stéphane Dumas. Re-illustrated by Paul E. Quast).

confound them. Indeed, one of the authors initially took this to represent the support structure underneath the mirror.

The problems continue. One might think the picture showing the double helix structure of DNA just above the figure of the human would be easily interpreted, since it is chemical in nature and aliens must surely understand chemistry. But, in fact, it would likely be extremely difficult to interpret. After all, it took *human* researchers many years to hit upon the idea of the double helix as the correct solution to DNA's structure, and we knew the entire relevant context in great detail. So, it's not at all clear if alien chemists would be able to do the same given only a two-dimensional picture and cliff notes of the chemical formula. Even if they did figure it out, they would likely have to guess at why we chose to convey the structure of this particular molecule, since it's hardly obvious from the chemistry alone that this is the primary means of transmitting heritable information on Earth. Amusingly, in classes on SETI, when asked to interpret the section with the DNA structure and human stick figure, students have sometimes asked why there are antlers coming out of the head of the human. Thus, the mere proximity of the two images could actually be confusing rather than informative. And in the case of the figure of the human, it's anyone's guess what aliens would make of this, especially given the very low resolution. It certainly could be a biological organism, but it also could be a number of other things—including something that exists on the ETI home world, but not on Earth. It is important to keep in mind that ETI will not be interpreting the message in a vacuum and, since their interpretation will involve some sort of comparison with what they already know, it is possible that one of our images will invoke a connection with something on their world that is nothing like what we intended to communicate.

The ability of ETI to decipher the representation of the population of humans also seems dubious, not only because there is no necessary relationship between the number depicted and the number of humans on Earth, but also because it assumes ETI is a collection of individual biological organisms, like humans. If ETI were to be post biological, which as Dick has noted is a reasonable possibility given that ETI's "civilization" may be millennia older than ours, the notion of equating intelligence with a collection of individual biological organisms may not make a great deal of sense to them (Dick, 2003). In short, even with respect to the most "straightforward" aspects of the message, it is likely the aliens would be able to easily interpret some parts of it, have good guesses about other parts, and be clueless (or dead wrong) about the rest.

Cultural Assumptions

The preceding analysis of the *Arecibo Message* suggests some important assumptions contained in the construction of the message that fall into two categories:

   1. Ideas we have about the nature of aliens.
   2. Ideas we have about "meanings" that aliens are likely to assign to elements of the message or the entire message.

We tend to assume a great deal about our unknown alien interlocutors. For example, we take it for granted that aliens will be sufficiently interested in communication with whomever sent the message to put their top scientists on the task of deciphering it—and this may well be necessary for even minimal success in understanding it. But this need not be the case. If our message happens to arrive at a world of xenophobes, they may not wish to put their resources into figuring out what outsiders wish to say to them. Those who intercepted the message, or their governments, might even suppress knowledge of its existence. On the other hand, they might assign it a very high priority for reasons we do not anticipate. Xenophobes might well interpret our message as an existential threat and devote significant resources to deciphering it in order to take commensurate action, such as sending something back that tells humans to (putting it mildly) "buzz off," or, worse, encoding a virus in their response designed to destroy our own computational infrastructure (Langston, 2014). Perhaps, ETI would interpret the message as a call for help or an annoying version of an interstellar chain letter. Without any knowledge of the society and values of ETI, we should be cautious about assuming that any message we send would be greeted positively or even neutrally. In other words, sending a message does not necessarily have to be interpreted as indicating a desire for a warm, friendly relationship.

Another problem that arises with messages to ETI is that, although the intended meaning of the message may be difficult to interpret, there are also subtexts carried with any message that convey unintended information about the sender, and it is hard to know what aliens might make of this information (Traphagan, 2016).[3] Perhaps most problematically, if the message were interpreted as having been sent by an alien *civilization* to make contact, that assumption would be entirely wrong. The *Arecibo Message* was sent out by an extremely small group of astronomers who do not even speak for the government they live under, much less the planet as a whole, nor does their way of thinking necessarily represent that of humanity more generally, as humans are generally far more interested in Doritos than biochemistry. ETI might reasonably conclude that Earth is a politically unified place where reason and science reign supreme, particularly if their world has been politically unified for a long time and they no longer think in terms of political diversity and conflict (if they ever did). In other words, not only is the content of the message difficult, and potentially impossible, to interpret; the nature of the origin of the message may be vague or even directly misleading, depending upon the cultural context of the beings who intercept it.

Conclusion

By raising the problem of lurking assumptions in the creation of a message for ETI, we are not suggesting that the task of messaging ETI is likely to be fruitless, nor that meaningful communication is impossible. Rather, the important point to take away from this is that the nature and potential influences of these types of implicit assumptions need to be carefully considered and systematically addressed in any attempt to craft a message if we truly hope for some sort of coherent interstellar dialogue to emerge. Meaning is dependent upon context. The good news is that any potential ETI we might contact will in some sense share an important context with us. We both occupy the Milky Way galaxy. We are both subject to universal physical laws and natural forces, and even if we do not necessarily symbolically represent those laws and forces in similar ways, it is reasonable to assume that we understand many of the same things about how our shared physical context works. It may even be reasonable to assume that evolution would operate in largely the same way throughout the universe; therefore, an ETI living in an ecosystem similar to ours might even be much like us from a biological perspective as a result of evolutionary convergence (Martinez, 2014).

The impediments associated with human-based assumptions tend to become most robust when we think about the social conditions and symbolic structures that humans and ETI use to *represent* the world around them and communicate with others. It seems highly unlikely that we will share much in terms of context when it comes to cultural values and social variables, even if we are very similar from a biological perspective, because our societies will have developed independently and will almost certainly be at significantly different levels of technological capability or social complexity—they may, for example, have economic and political systems that we have never even *imagined* on Earth. Our own experience has shown repeatedly that the influence of different cultural contexts deeply shapes the ways in which we see the world and interpret the meanings of actions and objects in the world—and assumptions associated with social context have the potential to not only limit understanding, but to generate significant *misunderstanding*. Careful consideration of assumptions in the crafting of messages intended for extraterrestrial consumption needs to take into account not only these types of problems related to mutual understanding, but also the potential for misunderstandings generated by tacit assumptions that have the potential to influence what can be understood and how a message may be interpreted.

NOTES

1. It should be noted that both authors serve on *METI International's* Advisory Council and Smith serves on *The Interstellar Beacon's* Advisory Board, so both organizations are open, at least in theory, to dissenting opinion.

2. One can question the seriousness of the *Arecibo Message* in terms of its likelihood of ever being received. It was pointed at a target in the Hercules constellation about 25,000 light years from Earth and the signal will be extremely weak by the time it gets there in the far future. However, it is reasonable to view the message itself as a serious attempt to construct content that might be interpretable by "smart" aliens, and the issue of detectability is a matter of technological abilities of the receiving end of the transmission, which is not possible for humans to predict.

3. A decent catalogue documenting *some* of the overarching assumptions baked into SETI/METI practices, along with challenges that arise from these varying human subtexts, was created by the analyst Anthony Judge (2000).

SETI and Terrestrial Visual Perception

What Can We Learn About Communicating with Other Species from Human Visual Perception and Visual Ecology?

CHRISTOPHER GILLESPIE

The deep history of our planet, and that of most planets orbiting a star, is one of unimaginable and relentless flows of energy—sunlight descending and covering the surface, geothermal heat rising up from the interior of the world. On Earth, wedged between these vast sources of power, billions of generations of living organisms leveraged the physical rules of nature for their own survival in an unending process of evolution. There have been many biological responses to the energy that surrounds and maintains life. Some responses have involved the manipulation of this energy, whether it is the vocalization of sounds to communicate, or even the direct generation of light in the form of bioluminescence. Others evolved responses, have involved detecting and responding to this energy through sensory systems. For humans alone, our sensory systems can respond to the intensity and wavelengths of light (sight), the orientation of the body (balance), the complex modulation of patterns of energy transmitted through the air (hearing), the texture of a surface (touch), cellular distress (pain), odors in the atmosphere (smell), and the ability to taste.

Sensory science is vast, and often disparate.[1] This becomes even more apparent when we consider differences between species for a single sensory system, or those senses that humans do not have. It is well known that other species can see EM wavelengths outside of those we can see[2] (infrared and ultraviolet); and we have also observed species making use of magnetism and electricity, as well as being sensitive to other properties of light such as polarization (Kirschvink, 1982; Heiligenberg, 1973; Dacke *et al.*, 2003). Each one of these senses has its own evolutionary history and implications for communication within and between species.[3] Though for several reasons, this essay will primarily discuss the perception of light and the eye.[4] Firstly, it is a sense we use to communicate most, but not all, non-transitory information, whether as a "store" of language in the form of text, representations of slowly changing environments, or symbolically charged cultural positions. Secondly, it makes use of the only force that is adapted to long range communication over interstellar distances. What is particularly interesting about light is that, alongside the sheer density of energy that arrives from the sun, the interaction with the surfaces modulates the spectral energy of that light. It contains information about the surfaces which makes it suitable for information acquisition about those surfaces. As well as these changes, the combination of reflection and absorption means that interface information is mixed with geometric and spatial organization. In this way, light then becomes a vessel of signs regarding not only substance, but also of space. Both of the particular and of the holistic. It should be no surprise then that light proved so useful to animals that even early in the development of life, the physical form of the eye and color perception appears to have come into existence.[5] From the fossil record of early creatures (approximately half a billion years ago) we have identified that eyes, structurally similar to those that exist now, were present on the Trilobites (Levi-Setti, 1993). Evidence for color vision has been derived from observations on "living fossils." These are species that have been shown to have not physically evolved for millions of years, such as Lungfish, Lampreys and Coelacanth, which all contain evidence concerning the evolution of the perception of color in the eye. In the Lungfish for example, we see that they can respond to four different ranges of wavelengths of light (for comparison, most humans are trichromatic and have three) (McCall, 2006; Hart *et al.*, 2008; Collin *et al.*, 2009).

As SETI lies at the intersection between human sensory perception and other species perception, such information can help us identify the hidden or implicit assumptions underlying our assumptions and expectations in our search. After all, we are implicitly drawing from our own experience to try and create solutions to future problems we may encounter. Ideally, by understanding the scientific explanations for differences between terrestrial species, we may be better primed to resolve the conceptual difficulties underpinning SETI. The purpose of this essay is to consider some limited but key observations related to the study of human visual perception, and the broader field of visual ecology. The first few sections will primarily consider the basics of how our visual system organizes the light it receives on the retina, and the final sections will broaden out by considering the perception of color, and the connection between the environment and perceptual experience.

SETI and Terrestrial Perception

On October 10, 2016, an interstellar radio message was transmitted called *A Simple Response to an Elemental Message* (henceforth ASREM) from the ESA Cebreros ground station in Ávila, Spain. This consisted, in part, of 70 photographs encoding various ecological themes about our world. They ranged from pictures of animals, geographical scenes, and human legacies in the environment, all the way to a time-lapse image of a night-time scene. Pictorial communication, and those images that capture ecological information, is a complex process of viewing and is subject to the interpretation and perspective of the individual who is looking at the picture. No matter how one chooses an image, they bring with them unintended meaning that can be picked up when one chooses a different framework from which to study them in greater depth. This interpretation is often assumed to start at the cognitive level as we experience the image, and bring to it memories, expectations, language assumptions and creative thought. However, even before we experience the image, our sensory system is processing and transforming the light that is encoding the region of space we are looking at into an object, within an image. For the purposes of this essay, I will describe this as a process of *translation*. Here, the visual system has prioritized or learnt local aspects (*features*) of the optical flow of light that are important in making sense of the scene. It then translates the information that is present on the retina, in the context of higher cognitive functions such as memory, expectation, and attentional direction, into our experience of a single, coherent reality.

In the case of ecological images, we are rarely aware of the underlying information we are cognitively processing to arrive at our conclusions about what an image is showing us. For instance, in one of the ASREM images we can see an orangutan swinging from trees. We can readily identify objects and environmental context from the presence of the trees and may infer that this is all the information we are reading. But there are more subtle cues present that are no less salient than the simple presence of an object. To take one, the orientation of the animals' body, is bound to our understanding of *uprightness*. Another environmentally relevant source of sensory information would also be changes to things such as the relative angle of hair to the direction of motion, as this is related to the density of the air or medium of action affecting the cues that enable us to perceive the orangutan as being in motion.

The difficulty in reading from this perspective and understanding how complex information from the sensory system is driving and presenting to us, is due to the fact that it is the basis for our own comprehension. To take an example from the image, there are spectral features that indicate daytime (such as white highlights), at the same time our nervous system represents the day-night cycle in the circadian rhythm, and takes such features in the light as meaningful to that process[6] (Fisk *et al.*, 2018). Therefore, there are implicit connections between the visual content of images and the referential experience of the time of day. There have been several contemporary artists who have used our ability to divorce these two processes and enable us to directly question

A Simple Response to an Elemental Message (illustration ©Paul E. Quast and ASREM project).

Orangutan (*Pongo abelii*) Suma reunited with a baby before moving through trees, Gunung Leuser NP Sumatra Indonesia in 2006. Featured in *A Simple Response to an Elemental Message* as image #48 (photograph ©Nature PL /Anup Shah/WWF [Image Number: 01046159]).

our phenomenology. The experienced tone of light itself is more readily perceived in the work of Scottish artist Katie Patterson, who has directly replicated the spectral content of moonlight, enabling us to experience an interior room illuminated by a type of light we only normally experience in the outside environment. The phenomenology of the experience of night is also questioned by the work of the English artist Darren Almond. Using the long-exposure times of cameras, he generates photographic images using light from the Moon. This results in images that have the definition of daytime but have the tone of the light we experience at night. The eye cannot physically respond to light over such long time periods, so the experience of the night in the images is "felt" and experienced in a way that our sensory system could never directly produce. In other words, the light has been translated into a different phenomenological experience than we would have had if relying on our sense of vision alone.

Our scientific understanding of the day-night cycle has also been bound to our understanding of light and heat. One of the things we came to understand is that our perception of light is but a slim fraction of the continuum of energy light can have. With heat, there is a type of light we refer to as infrared (IR henceforth), and this can be visually presented to us using "false" coloring. We can see this in the second image in which the global trends of the temperatures across the surface of the planet are illuminated (see color insert). However, this involves the visualization of raw physical data, and ecological data in the real world may have unique features that can only be drawn out in other contexts. For instance, the artist Richard Mosse uses specially designed photographic film[7] that allows a camera to capture IR. Through this process, we begin to see how the apparent experience of the world would begin to vary if we ourselves would be able to see IR. But this use of IR is again a treatment of the environment, mediated through technology. For other species, whether IR or ultraviolet light, perception is formulated around its natural occurrence and potential survival benefits. True perceptions of IR have been shown to be comparatively rare. Two known species, pit vipers and a beetle known as *Melanophila acuminate* (Goris, 2011; Schmitz and Bleckmann, 1998; Bennett and Cuthill, 1994), are known to have evolved the use of IR for radically different purposes; for the former, it is the basis for a method of nocturnal predation, and for the latter, a way of avoiding forest-fires.

Warm-blooded animals, as well as fire and sunlight, emit this type of light. Because it is emitted rather than reflected, it would not be efficient for discerning objects as we normally experience them, but it would be good for detecting sources of this kind of light. In addition, there are other factors to consider; objects tend to absorb heat and passively emit IR light and, environmentally, sunlight is warmer and more IR rich than the night-time. This phenomenologically unseen light cannot be said to be conceptually identical, in terms of its behavior in the environment, to the spectral light we normally perceive. On Earth, IR is associated with warmth production, so could be said to be process-associated-light, while spectral light is more readily reflective to bodies, and could be said to be object-associated-light. There is another aspect of this story, after all; we find the use of IR useful for night vision goggles. But, it must be remembered that we ourselves emit such radiation, so any attempt for a warm-blooded animal's eye to see IR would be instantly swamped by the IR light from the blood that feeds the nutrients to the cells. As it is also readily absorbed by substances such as water, it also becomes unviable for environments in which heat is readily conducted away. Here we see something important; IR light (much like the reflecting light we perceive) has its own story, its own information about its potential use. That story can inform us about the kinds of limitations and exploitations that are possible for the species that make use of it. In other words, whether intentional or not, the medium of a message potentially contains information about the entity that sends it. This meta-information is far richer and more generally nuanced than may be fully understood, until you study the sensory ecology in which it is exploited.

The "Fundamental" Sources of Visual Experience

Our reality is experienced as a vista of objects and identities, present in space, and changing within time. Through our reality, we have come to understand the dynamics and substances that are the source of those experiences. In turn, we now know that this reality is divorced from the physics that underpins its appearances. Our reality is stitched together by the interplay of pulsing patterns of neural signals. Transmitted from the retina, they are woven together with other aspects of the nervous system. The idea of "seeing," while synonymous with light, is not entirely accurate. Even light is complex as a wave of energy. Like waves on the sea, they can be more or less powerful, or frequent or infrequent. These are the equivalent of intensity (or brightness) or frequency (loosely color, but this shall be discussed later.) In the Bridget Riley painting *Blaze 1*, we can see some of the complexity of this process in action. In her painting, she has used both black and white, alternating between these colors to create circular patterns within one another. This generates a feeling of motion and movement that should not really be present.

The raw light pattern may appear to us as the sole "fundamental" source of visual sensory information, but it is not. Our experience of the world is not a simple mapping onto light, nor is it solely modulated by things we have learnt—it is partially a product of physiology and cross-sensory activity. The visual system makes use of physiological facts, such as having two eyes producing a "binocular disparity" which acts as a signal to the brain (Wheatstone, 1838), changes in attention (Carrasco *et al.*, 2004), the position and direction of associated sounds (Jack and Thurlow, 1973), and biological motion (Neri *et al.*, 1998). These directly contribute to our experience of the world. Three-dimensional depth (or solidity) is derived from binocular vision—but not solely as monocular vision can produce three-dimensional scenes[8] (Ponce and Born, 2008; Vishwanath and Hibbard, 2013). Patterns such as the "rotating snakes" illusion can induce motion as one finds in optical illusions (Kitaoka and Ashida, 2003). In more phenomenologically dramatic examples, auditory stimuli can even be seen to generate entirely fictitious flashes of light[9] (Shams *et al.*, 2002). These physiological and stimuli-based sources of information produce further convergence and divergences with other species. Animals with eyes on the side of their head will not have access to two near-identical, overlapping images, while the statistics of a texture can also be picked up from a distance by echo location as much as it can be in sight (Zagaeski and Moss, 1994). To take the latter example to see how this leads to further, unexpected overlaps with vision, a recent experiment identified that bats (*Micronycteris microtis*) determined the presence of an insect on a surface by comparing the surface texture with the changes in those statistics caused by the acoustic shadow, due to the insect blocking the sound reaching the surface. Shadows occur in sound too, but unlike shadows in visual sight, they are based on the perspective of the bat that emits the sound (Clare and Holderied, 2015).

Although the sensory system is a composite and a complex biological process, it is worth examining what happens to "raw" information derived from the patterns of intensity of light as this information lies at the heart of the chain of events that are translated to experience. It will also become clear that the process of interpreting and translating light into experience is far more sophisticated and structured by the ecological structure of the world than is obvious.

Light and Early Vision

Virtually all human cultures have developed ways to manipulate materials so as to create sensory sensations that we can comprehend—what we have referred to as "material culture" throughout this volume. For the visual system, we manipulate the material surface of objects to modify the projected image of that region of space to produce a huge variety of pictorial devices, from drawn pictures of the world, to letters representing sounds, line drawings, or photographs. It is often the case that we can make assumptions about how pictures can be used due to its unending versatility for communication—it can almost seem lawless in its seeming infinite variability. It can seem so natural and yet, at its heart, it is a cultural technology that leverages the functioning of the visual system in visual ecology to be able to present whatever we wish to present.

Let us briefly consider what implicit perceptual assumptions underlie the *Arecibo Message* pictograph (See Traphagan and Smith, this volume) and think about what it is we have manipulated to generate this simple, but highly abstract image. For the purposes of our analysis to draw out our own perceptual assumptions, let us naively consider our own interpretative image. In the simple human-inclined reconstruction, we have patterns of varying levels of abstraction which are embedded in spatial regions that are brighter or darker than each other. Such binary patterns are useful to us as ways of pixelating data and visualizing it. Without any spatial information that relates relative instances between these values, *our* natural default assumption is to present them at equal and regular spacing (for a discussion of this, see Traphagan and Smith, this volume). But even here, the use of symbols is difficult for our visual system to process in a pictorial way, and so to address this legibility issue, a common solution has been to translate this information into arrangements composed of squares, whose tessellation permits the creation of a picture. Here, a simple description leads to a set of concepts— contrast differences, edges/corners, geometric extent/regularity and contrasting colors. We will go through each of these concepts and look at how each level is represented within the visual system. However nearly all stages, from the initial retinal processing to the visual system of the brain, are surprisingly complex and it would not be possible to go into the fullness of the entire response.[10]

The cells of the retina do not just catch a singular region of light, rather they respond to an extended catchment area along the retina. Such sensory regions were initially described by Sherrington (1906) in observations that regions of the skin of a dog elicited a trigger for the scratching impulse. Later, the regions were identified, and the concept generalized such that receptive fields were also shown to be present on single neurons by Hartline (1938). However, these regions are not always homogeneous; in the case of the retina, the relative position of where light lands on this receptive field of a retinal cell can potentially affect the activity of the cell (Kuffler, 1953). In the interior of the cell, it will activate whenever light is present on it, while the region surrounding this area will act antagonistically and only produce activity when no light is present. The overall activity of the cell is the combination of both responses.

Some of the differences can be unexpected, given our experience of the world. One may think that we pick up light equally across the retina but, in fact, the retina can have blind spots where no light is received, and the visual system must actively "fill-in" for us to experience a single coherent viewpoint (Kolb *et al.*, 2001; Sterling and Demb, 2004; Ramachandran and Gregory, 1991). The retina then sends an already complex, processed signal along the optic nerve to the brain. Here the early visual system responds in a highly structured way to specific features preserved within this signal. Hubel and Wiesel (1959; 1962; 1968) discovered several important aspects of the response of these neurons in the early visual system. One example was the response of certain cells to the directionality of regions of contrast. In a classic set of experiments,[11] they discovered that individual cells within the visual system of cats and monkeys responded to a rotating black line, with the cell responding more energetically when the bar becomes increasingly vertically orientated. There are numerous other features that the visual system has been observed to respond to. This has included things as specific as the presence of corners, bilateral symmetry (Wagemans, 1995), or the appearance of a closed boundary (Gerhardstein *et al.*, 2012). Bilateral symmetry,[12] in particular, is associated with biological organisms. As such, symmetry is useful for attending to signals connected with faces or animals (i.e., pareidolia) and, due to the strong set of associations and preferences for humans, can easily trigger the sense of seeing some meaningful interpretation such as those that occur with a Rorschach inkblot.

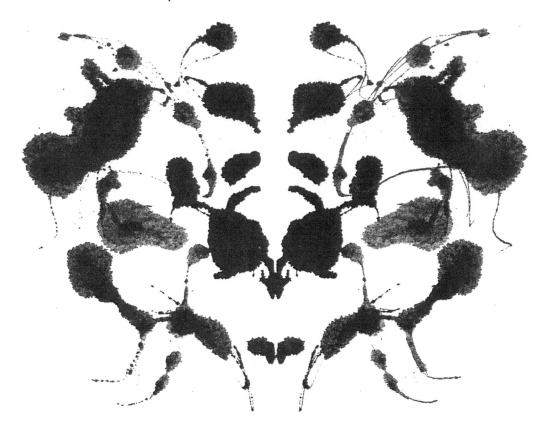

Rorschach Inkblot Test. Featured in *The Last Pictures* as image #73 (photograph ©Trevor Paglen. Courtesy of the Artist and Creative Time).

So far, we have only discussed what features of the optical environment the individual cells are responding to. The question for us then becomes, what is the visual system doing with this information, and how does it start to become woven together dynamically to allow us to perceive objects in space, changing with time?

Perceptual Organization

An image was engraved on a plaque adorning the antenna shielding on both the Pioneer 10 and Pioneer 11 spacecraft as they journey beyond the Solar System. The plaque consisted of a simple line-drawing that presented a diagram of our Solar System, aside two unclothed humans that appear to be floating giants set against the tiny, unscaled planets. When a human visual system looks at this image, it appears straightforward. But let us consider what we experience; our eyes see the short black marks and we perceive those black marks, in one context, to belong to the frontal view of a human body, and in another context black marks appear as part of a frame that sits behind those same humans. The overall effect is that we perceive familiar objects with identities,[13] such as the perception of images intended to be perceived as planets. But we can easily disabuse ourselves of this identity and understand that they are simply circles. An abstract spaceship that is, by scale, the size of the Solar System. Two bodies that float, upright as if being pulled towards a source of gravity, in front of a rigid frame of

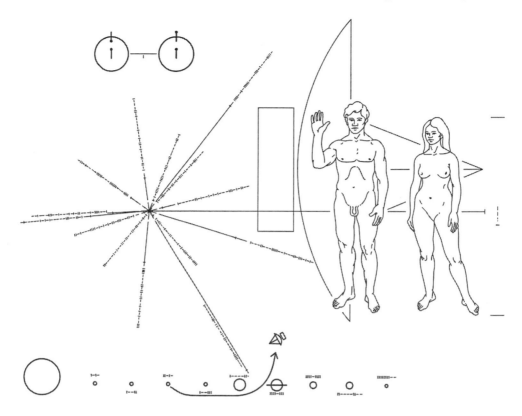

The Pioneer Plaque, also featured in the Voyager Record, along with *Lone Signal* crowd sourced media (graphic from NASA/ARC, vectorized by Mysid. Plaque originally compiled by Carl Sagan/Cornell University).

lines. The second we analyze the image, with respect to itself, we are pulled through various levels of abstraction and information, allowing us, with some common sense, to understand what it is that was intended by the human designer—to present humanity and its home planet amongst the Solar System from which the Pioneer spacecraft has travelled from.

This illustrates an interesting fact about the visual system of humans; it is a dynamic system in which we experience and can consciously parse "meaning" and associative visual structure. In other words, the cells perform tasks to lend our perception a structure (known as perceptual organization) that we, as viewers, can sometimes modulate by attending to different aspects of our experience[14] (Koffka, 1935; Loffler, 2008). A fundamental process in the dynamics of these cells is the appearance of discrete lines or marks across the visual field to produce the sensation of a single, coherent object boundary. This process is known as contour integration. In this process, the neurons are constantly accessing the likelihood that individual regions of contrasted luminance are likely to be associated with other regions, and if it is likely, treat them as if they form a singular contour (Field *et al.*, 1993).

We can perceive this process in action easily when we see the overall shapes implied by a connect-the-dots picture before we sit down and draw the lines as they are meant to be. Kin to this scenario, where the picture book is a puzzle in which we must create the underlying images, the visual system is translating discrete and separate information in light into an object or boundary. The visual system uses an incredibly diverse set of features in the environment to identify a probable object boundary, and this strongly overlaps with those features that the visual system is sensitive to,[15] such as curvature (Bertamini and Wagemans, 2013), symmetry (Machilsen *et al.*, 2009; Sassi *et al.*, 2012), closure (Kovacs and Julesz, 1993), familiarity/predictability (Sassi *et al.*, 2010; 2014; Sassi *et al.*, 2012), and sharing a shape profile with other objects in the visual field (Gillespie and Vishwanath, 2019).

In the processing of drawing or painting, artists are implicitly using the tendency of the visual system to determine implied contours and boundaries. Here, we encounter another factor that is part of this perceptual organization where the visual system interprets simple shapes to be foreground or background (Grossberg, 1994; Cutting and Vishton, 1995; Baylis and Driver, 1995). We can see these effects in action in the *Pioneer Plaque*. The details of the bodies appear to us as the frontal surface of the external boundary of the male and female body. Likewise, the curves of the lines appear as surfaces that lie behind the humans. It also appears that the planets and the bodies all belong to a single space—they are all background to a foreground.

Even more complex effects can be discerned when semantic information is salient to the organization of the objects. In the young/old woman illusion for example, the lines and shapes have two possible interpretations, and our visual system will alternate between each possibility (Kornmeier and Bach, 2005; 2012). This all occurs before our higher-level cognitive processes interpret the meaning of the image, and illusions such as those mentioned above demonstrate that the visual system is constantly attempting to develop an interpretation of the retinal information.[16] As an extreme use of such processes, we see transformations in shape and identity play out in M.C. Escher images where these processes are manipulated to produce absurdly complex tessellation changes across an imaginary transition plane of time.

There are two things to draw from this. Firstly, our vision is interpretive and a process of translation. We are familiar with seeing objects and shapes in clouds or textures, yet generally our culture is not aware of the implications of this, and its suggestion of a dynamic, translating, or transformative form of perception. Secondly, the simplest of our pictorial presentations—line drawings—represents the type of picture that is most adequately designed to make use of the rules our visual system has developed in interacting with a world full of light, and eyes that can easily respond and distinguish subtle changes in contrast. This is not always the case, for instance, in water-based environments, whether misty atmosphere or murky sea, this assumption may be a lot weaker.[17] We could equally receive a communication that is based on changes in gradients of changing contrasts, rather than abrupt marks associated with drawing.[18] We know, for example, that our own visual system extracts shape information from shading (Pentland, 1989). However, we also see other responses such as a jellyfish which can respond to changes in light overhead. In the dramatic Jellyfish Lake located in Palou, a type of golden jellyfish has evolved to migrate across the lake on a schedule based on the time of day.[19] Indeed, sophisticated responses to light have been seen in other types of jellyfish. On some families of box jellyfish, we have seen eyes that are unnervingly similar in structure to our own[20] (Coates, 2003). In context, then, we can speculate that a species whose evolutionary development was based around a general response of lighting and shading may find the most comprehensible pictorial mode to be that of shading, rather than the line.

Color Perception

When we reflect on aspects of our experience of reality, other than the overarching themes of space and time, we often have a sense of the importance of color to our perception of the world. It can seem as if color is some fundamental aspect of the world itself. This centrality has a long and distinguished intellectual history of its own, but one of the central arguments was whether light itself was made of three or more discrete "types" or, as Newton (1704) proposed, that all colors lie along a single continuum. This argument itself was resolved when it was determined that certain cells within the retina of the eye respond to differences in the frequency of light (Wald and Brown, 1965). In effect, the discrete appearance of primary colors is a property of our biology, whereas the continuous appearance of color comes about due to the continuous nature of visible light. However, even here, the story of how our sense of color is structured is neither explained fully, nor clarified entirely. The response of the cells does not correspond to our notions of primary colors. Rather, there are three individual photo-pigments; these are pigments that chemically respond to a range of wavelengths allowing the retinal cells to detect energy across those wavelengths.[21] These loosely correspond to short, medium, and long wavelengths, but this can vary quite significantly across species (Jacobs, 2018). In truth then, our experience of color is not identical to the frequency of light; it is merely synonymous with it. The fundamental informational source of color is relational; it is a process of combining three overlapping ranged cellular responses. Essentially, the endpoint of our experience begins with a ratio of how much of a certain frequency of light is absorbed by photo-pigments (De Valois *et al.*, 1958). There are some specific consequences of this physiological fact. Firstly, if the three photo-pigments absorb equal amounts of certain frequencies of light, multiple different patterns of wavelengths can, in principle, trigger the same ratios. Therefore, two or more essentially different lights can trigger the same perception of a color[22]—these are known as "metamers" (Cohen and Kappauf, 1982).

Information important to the visual system is different from the methods we use to encode information with light. As a physical energy, light can be used to produce information by manipulating and layering individual wavelengths together. Equally though, the bundles of frequencies that trigger certain *colors* are informative about the physiology of an animal and, as we will see, that physiology maps onto the kind of visual ecology they develop into (Cronin *et al.*, 2014). In relation to SETI, the duality of the informational uses of color can occur together. As we have fine control over individual wavelengths in creating artefacts of material culture, light can be used to determine the functioning of visual equipment. To explicate my point, let us take the case of the *Beagle 2 calibration plate*; a "test card," created by the artist Damien Hirst, which was due to be used as a calibration target plate for the stereo cameras and spectrometer instrumentation on the British *Beagle 2* Mars lander (see color insert). Let us imagine a situation in which such a plate was found by another species.[23] When analyzing a plate of this sort, they are presented with two possibilities. On one hand, frequencies of light are core tools in encoding information. On the other, clusters of close frequencies and the technology used to create such colors, such as those chemicals that make up the spots of colors, can contain information about the physiology or general bodily scale of our species. If one sends information, whether a physical object or image, one needs to appreciate that it may be both information *from* us, and information *about* us.[24]

Visual Ecology

Humans are only one subset of the broad variety of color-perceiving animals, and it is worth asking ourselves why such specialization came about in the first place. One theory concerning the evolution of color vision is that it arose as a response to the complexity of detecting an object in shallow water, where complex movements across the surface, column and base of the water caused distortions such as flickering that made simple "detection" unviable (Maximov, 2000). In other words, if there is some parameter that can be calculated that is independent of the flicker and changes to the variability in the illumination, it would unlock the ability to determine the presence of an object despite the changes. The simplest solution to this would be to compare the ratio between two frequencies of light, which would not change because any distortion to both frequencies would remain stable. In other words, dichromacy solves the problem of variability in shallow water; allowing the organism to attribute changes caused by a flicker to environmental variability, or phenomena that relate to another organism. There are potential connections then to be found and derived between species and their environments. As it is shown, one of the images presented in the "simple message" depicts how light flowing down through the sea, silhouettes the nearby fish. It is thought that dichromatic aquatic color vision is more efficient at detecting and silhouetting fish in this scenario than trichromacy (Sawicka *et al.*, 2012).

Evolution has permitted species to develop biological strategies to counteract the ability of predators to make use of features of the environment such as these. In nearly

Giant Kelp (*Macrocystis pyrifera*) forms dark, lush canopy forest in cold waters, southern California, c. 2005. Featured in *A Simple Response to an Elemental Message* as image #70 (photograph ©Norbert Wu/Minden/Nature PL [Image Number: 00977115]).

all countries, we are aware of the way in which wasps and bees—animals able to sting—tend to possess tones of yellow. Correspondingly, there are insects like hoverflies that use their markings and color to hide their inability to sting or attack. When we combine color and the environment, these developments with biological strategies are tied to the environment.[25] The results of this often make intuitive sense,[26] where animals in snowy environments are likely to be white, or the dragonfly is reflective and lives around the surface of water (Forbes, 2009; Troscianko *et al.*, 2009; Stevens and Merilaita, 2011). An interesting question to contemplate is the relative lack of transparent skin on land or air-borne organisms, but we do rarely find it in certain organisms, such as glass-frogs or the beautiful glass-wing butterflies.[27] Animals can also be observed to have an underbelly that is lighter or a different shade from its body. This is known as countershading and may be explained as an adaptation to the fact that luminance is brighter from above, and so the body pattern is intended to disguise the animal as seen from below. This general problem is also more evident in the sea, where the sunlight entering the water is sufficiently bright to silhouette the physical body of a fish. In this case, a biological response of bioluminescence is used to diminish the silhouetting, which is known as counter-illumination (Ruxton and Sherratt, 2004). These types of evolutionary strategies can vary dramatically, with diverse effects producing colors and patterns matched to environments such as forests or coral reefs (Cott, 1940).

It is important to recognize that the combined visual ecology of species across the Earth have produced a successful strategy that occurs in many different environments and employs many different biological mechanisms. In other words, there are types of visual patterns which reflect or inform us about the nature of the environment that the patterns arise in.

Between Information and Experience—The Present

As a species, our historical inclination is to attempt to devise new methods to communicate. We create new styles of painting or music, we devise new ways of writing mathematical insights, we attempt to create ways of communicating with people that do not yet exist, and we even attempt to communicate with other species.[28] Fundamentally though, this process is a feedback loop between the visual environment and the human visual system. When we are exposed to novel methods of communication, our nervous system needs to learn how to interpret and make sense of the new structural information before we can begin to perceive the intended communication. We tend to not be overly aware of the degree to which this translation occurs because perceptual organization has evolved to occur on extremely short timescales.[29] However, sometimes we become aware of this by accident, such as when we are in darkness and mistake a coat hanger for an intruder, or when attempting to perceive three-dimensional information in a "magic eye" picture. Even for communications to human posterity, we may assume too much about what they will naturally be able to discern from our images, given their perceptual experiences will be a product of their experienced environments.

In the case of sending messages from our planet into space, we find a huge variety of pictorial devices; whether they are of our natural world, such as those we find in ASREM, obscure bitmaps, or the pioneer probes in which we have the social signals of two figures and a "roadmap" of our home system. In all these examples, we are making use of our sensory system that translates a world consisting of objects and parts, background and foreground, and light and dark. However, if we look outside of our own species and consider ETI in general, we have additional problems to the ones that arise in visual communication between humans. For instance, while all species need to navigate an environment containing physical bodies, even if a species can see light and color as we do while also having similarly structured eyes, it does not follow that other species will *translate* the global signals of space and time in the same way we do. Even if they perceive their relationships to the world in a similar way to us, it may be that *their visual systems* find it easier to process different information to create *their versions of pictures*. We need only think of the difference between a line-based etching such as Edward Hopper's *Night in the Park*, or a post-impressionist painting such as Van Gogh's *Starry Night over the Rhone*, to gain an intuition how this could lead to different consequences, and broadly how diverse the perspectives available across our own species social and cultural histories are.

As a counterpoint to this problem, a second strategy employed to communicate with ETI has been to encode physical properties of our world into technological or scientific signals. By relying on information alone, we may suppose that an ETI may be able to form a conclusion about what we intended to message, by working out what chemical, mathematical, or physical properties are encoded in our message. In other words, by stripping out social or naturalistic cues and leaving only information relating to the physics of the world or some arbitrary subset of perceptual features and distilling them into a picture, it may seem intuitive to us that we reduce the difficulty for any ETI in interpreting and devising a response. But, if we take our own sensory system as an example, our perceptual process is translating features based on how we evolved; the clearest examples that demonstrate this process of perception through translation-in-action, are the transitory figures and animals in clouds and in cliff-faces. It may be that any informational signal we send may be distorted and perceptually re-organized by another species into something that makes sense to them anyway and, therefore, the information we intended to convey will be lost, or in the very least, rendered ambiguous under the evolved cognitive filters that led to an independent extraterrestrial culture.

When we begin to parse the implications of communication and SETI, it is common, and understandable, that the mysteries of intelligence, sentience, and experience tend to dominate our thinking. After all, doesn't SETI, by implication, simply mean communication with another intelligent species? It should be equally clear that experience and intelligence are evolutionary developments that have built upon the foundations established by sensory perceptions in resident environments. The possibilities inherent in these foundations, and their diversity, are still being investigated on Earth, and it is therefore useful in understanding the difficulties and possible avenues related directly to inter-species sensory perception, and hence, prospects of communication.[30]

We are bound, no matter how we translate our communicable materials, by our experience. In our case, we trace out our sense of the world with phenomenological concepts such as objects, time, and space; such experiences appear self-contained and fully formed to us but are intangibly bound to the fragile interplay of external flows of energy, and the biological strata whose consequence is the intelligence and social interactions that have refined, recomposed, and decomposed these very concepts. It is equally problematic that we wish to create "information" to send to another world, without any knowledge of the possible paths of evolution open to other intelligences. There may be two strategies that can be employed— that of encoding information we wish to communicate which is discussed later in this volume, but also ways of encoding information about our perceptual foundations. After all, we see a super-abundance of phenomena like color or camouflage, which can be informative about the environment, and hence speak to our physiology. Despite the general problems inherent in speculations about communications of this sort, what we do have is an immense potential body of knowledge in sensory perception and sensory ecology. A vast *dictionary of connections* between environment, and the nervous systems' translation of the physical world into experience. Should we ever receive a message from another species, it is likely we will need to know that dictionary by heart to be able to understand what, and who, it is we are attempting to communicate with.

Between Information and Experience—The Future

The usefulness of sensory ecology, and its potential for greater insight, can be considered by thinking about how perceptual psychology and sensory ecology have been used for this purpose before. In the latter case, perceptual psychology has been related to the semiotics of communications with a focus on the importance of a Gibsonian ecological perspective[31] to how communications with ETI receivers may proceed (Sonesson, 2013). In the case of sensory ecology, a general attempt to use observations of the natural world to strengthen the speculations of what may be *out there*, was attempted in the 1970s by Doris and David Jonas (1976). Here, we see the possibility of understanding how our sensory perception works—it provides a skeletal frame over which we can make assumptions that can be examined and explored thoroughly, without succumbing to the over-reach that arises from un-anchored speculation. There is also a sense about how we can address the phenomenological difficulties inherent in species-to-species communication, as well as exploratory scientific investigations. These are worth addressing separately.

One of the major issues that arises when we consider the scale of the difference in the experiences between species is how this could disrupt our attempts to develop methods of direct communicating. While it is true that there is a layer of intractability about this problem, in the future it may be possible to investigate aspects of changes of experience. One of the most recent developments that speaks to this core issue is the recent development of glasses that correct color-blindness in certain individuals, such that they are newly exposed to both the vividness of color, and the potentially specific aspects of our general experience of color (such as purple). A brief investigation of videos online for such reactions lead us to a rabbit hole of emotional responses that diagnose the shock of an *unveiling of experiences* that, until recently, was simply not possible. In the future, studies may document how we as a species cope with the transition from an understood absence to a shared common experience, and how it shapes our relationship to communication. In such an example, we may be able to discern more subtle issues in communications with other people or species with whom we cannot expect a one-to-one correspondence of phenomenological experience.

In conjunction with this, the notion of a sensory ecology as a dictionary will allow us to work with base level sensory processes that enables us to define plausible categories. The examples I have visited previously are the connections between a migratory bird's sense of time and their perception of a magnetic field, and the navigational use of the Milky Way for dung beetles (see endnotes). Both lead us to clear notions that the experience of waking/sleeping, time, and seasons, will be more firmly grounded in the directional space traced out by having a fixed horizon, or orientating field of stars. In this way, while we cannot conceive of what it may be like, it is not beyond us to see that the attractiveness of the distant horizon, or the association with confusion and darkness, would be linked in much the same way that coldness and darkness are associated with night and winter. However, we can go further than this due to the technological and scientific tools and knowledge we have developed since the 1970s. In principle, we can model hypothetical sensory systems, environments, and read data with neural networks, to investigate whether sensible structural information can be extracted from hypothetical sensory apparatus. To put this in more concrete terms, we should consider hearing and music—we can model and describe the acoustics of the space, the musical instruments, the human vocal cords, the inner ear, and the translation of the signal from the physical, through the sensory and into the nervous system. In other words, there is an Environment-Apparatus-Network chain that can be examined experimentally for hypothetical astrobiological perception. In addition, the use of virtual reality, and Deep Learning algorithms, could also allow us to further identify the viability of such physical strategies to perform tasks that we observe in nature.

It may now be that a field of astrobiological perception is viable, both as a hypothetical spur to understand terrestrial perception and also to develop new technologies based on other species' sensory capacities. As a whole, this leads us to a somewhat optimistic proposition. While in our worst-case scenario there is no way for us to have truly satisfying communication with the alien (whether terrestrial or astrobiological), it may be that we have sufficient tools and understanding to build a grounded, and substantive, idea of the vast richness of experience that may be beyond the horizon of the Earth's sky.

NOTES

1. Due to its centrality to human behavior, sensory science dovetails with insights in fields as diverse as physics, philosophy, biology, social psychology, and the arts. The sensory system is further influenced by factors such as environment, evolution, logic, and expectation. A third related issue is that the frameworks used to interpret the sensory data have been influenced by novel technologies being used as metaphorical concepts to think about the nervous system. It is worth looking at the volume *Theories of Visual Perception* (Gordon, 2005) for a consideration of different theoretical frameworks and their relative strengths and weaknesses.

2. Certain humans have been known to detect some ultraviolet light under specific conditions when the cornea has been removed, with the protein's sensitivity to the wavelengths of light dropping from 380 to around 300 nanometers.

3. The most intuitive case is that of migratory birds which use vision to perceive magnetic fields. The relevant proteins are also associated with regulating circadian rhythms. Alongside this, birds produce vocalizations and have sophisticated behaviors such as flocking and intra-species social interactions.

4. It is true to say that the importance of the phenomenology of sight, is often inflated in Western cultures for social and historical reasons (as described by Foucault). However, for the notion of astro-biological perception, a broader definition of "seeing" is warranted as we may expect a greater degree of communication if two distinct perceptual systems have a common functionality. We could broadly consider sight as being the active, intentionally controlled process of perceiving a contiguous, temporally stable percept. Under this rubric, sonar would fall into a common category with sight. Passively hearing, on the other hand, has a greater degree of receiver characteristics with precepts that are unstable (as in, you cannot hold the whole utterance in a single moment). The visual counterpart, bioluminescence can also be thought of in this way, and the closest analogy may be in the cuttlefish that generates beautiful and sophisticated color schemes for other cuttlefish.

5. Evidence suggests that there have been over 40 instances of eyes developing rapidly for Metazoan development. For an overview of the evolution of the eyes, see *The Evolution of the Eyes* (Schwab, 2017).

6. In certain birds, the proteins responsible for magnetic perception are also responsible for the perception of blue and the circadian cycle (Pizon-Rodriguez *et al.*, 2018). We can speculate then that their sense of time has a strongly spatialized component associated with the magnetic horizon. As a counterpoint, we can consider how the dung beetle responds to lighting experiments designed to emulate starry nights (Dacke *et al.*, 2013).

7. The technology was derived from Cold War, anti-camouflage surveillance technology. It specifically used an IR color film called *Kodak Aerochrome* that was initially developed for aerial bombing. To aid in human perception, it picked up wavelengths outside of the normal register of the retina and rendered them in brighter hues. This resulted in green foliage appearing as lavender, crimson, and hot pink, thus highlighting camouflaged individuals.

8. To investigate further why monocular stereopsis occurs, Vishwanath (2016) presented an image to a participant and placed a pointer between the image and their eyes. They asked the participant to primarily focus on the pointer but maintain their awareness of the image (covert attention). Eventually, the pointer was taken out of view, while the participant continued to look straight-ahead. This was sufficient for the participants to report qualities associated with stereopsis (such as a sense of tangibility, or object protruding into space), and tangential qualities such as the feeling that the scene had been miniaturized.

9. The sensory modes are not ring-fenced from each other, and processes such as ventriloquism arise because of spatial signals cross and are resolved between interactions between (say) vision and hearing (Teichert and Bolx, 2018).

10. Gollisch and Meister (2010) discuss the processing that occurs as early as the first retinal response.

11. The observations were made by measuring the direct statistical activity of neurons of anesthetized, paralyzed animals (initially cats, then monkeys). A simple two-dimensional bar was presented to the animals and rotated while the measurements of neural activity were recorded.

12. Bilateral symmetry is a ubiquitous feature of a huge number of organisms across nature. However, there are a small number of exceptions associated with defenses, a fidler crab with one larger claw for example. It's informative to consider the trade-off between bilateral symmetry and camouflage. The former makes an organism more detectable, and the latter hides it. The co-existence of both suggests that the combined benefits are worth the evolutionary compensation required to develop camouflage.

13. In the essays titled "The Tyranny of Assumption" and "How We Think We Know What We Think We Know…," familiarity is discussed as a social function. However, for the purposes of this text, it suffices to say that regardless of the framework which describes the concept of familiarity, it has perceptual consequences. One could ascribe the relationship between the social and perceptual as indicating that the social phenomenon of familiarity is treated as an evolutionarily relevant factor, embedded in the biological processes that underpin the perceptual response to the physical world and, therefore, it is allowed to directly influence those very-same processes. For example, perception is ring-fenced from some social or cognitive functions, while open to interference from others.

14. We can get a stronger sense of this process by thinking about painting. We can attend to a single gestural mark, groups of marks that delineate (say) a hand or a tree, or the whole content and style of the scene. We can alternate between composing and decomposing such structures by changing what sets of features we look at (looking at the color blue, looking at some gestural marks, looking at the boundaries of an object). This, in turn, leads to higher-level perceptions appearing and disappearing accordingly, and leading to the re-formulation of interpretation.

15. To return to the example of the dung beetle, the perceptual encoding of the Milky Way is a hyper-specific feature of the night-sky. In comparison, humans see a dynamic process that creates linkages leading to constellations and disparate clusters of stars during night-time. It may be that the dung beetle may perceive the Milky Way as a stable, interlinked cluster of points more akin to our perception of a tree canopy where leaves appear interlinked with a whole, directional motion against which the empty sky can be seen through.

16. Psychophysical experiments often make use of the fact that such interpretations dynamically change and break down depending on context. An example is priming, where showing a person a tangential stimulus will make them more efficient at detecting a target, when the target and tangential stimuli have common features.

17. It is easy to find ecologies on Earth that have, or produce, air-borne molecules that disrupt vision such as water droplets surrounding waterfalls, mist around swamplands, chemical vapors around volcanoes, etc.

18. In an interesting volume *Other Senses, Other Worlds* (Jonas and Jonas, 1976), the authors speculate about extraterrestrials and the possible information that they could encode within other sensory modes, such as smell, and elaborate on how society and art could develop with a different dominant mode, while relating their speculation to aspects of terrestrial sensory ecology.

19. In this case, the behavior relates to the process of photosynthesis and the creation of nutrients. Interestingly, we have seen examples of radiosynthesis using gamma rays in fungi around the Chernobyl Nuclear

Plant. This leads to the idea that if life could survive around neutron stars, such as pulsars with emissions of rotating bands of gamma radiation, then migratory behaviors could arise based around radiosynthesis.

20. While eyes from multiple organisms may be comparable, it does not imply that their experiences are akin. The closest proxy we may have is whether the behavior of the organism is complex. Whereas we see intelligent behaviors associated with octopi, there is a much lower level of behavioral sophistication with jellyfish. This may reflect a relative diminishment in the sensory environment presented from the visual senses, regardless of how "advanced" the eye may be.

21. The three cone-cells of the human retina correspond to longer wavelengths (L cones, peaking on 560 nm), medium wavelengths (M cones, peaking on 530 nm), and short wavelengths (S cones, peaking on 420 nm). This varies as they carry different opsins, named OPN1LW, OPN1MW, and OPN1SW respectively.

22. Metamerism is a diverse topic in its own right, and includes metameric matching, as well as metameric failure. Matching is particularly common when there is a low saturation value for light, so beige or dark colors tend to have more metamerism than highly saturated colors. Outside of producing identical patterns in the eye, metamerism can occur whenever two materials are presented under different light sources. The well-known version of metamerism is called "Observer Metamerism," which refers to color blindness.

23. Their initial response could involve a complicated assessment of the generalities of an object; whether it appears safe, the scale and structural organization of the object in comparison with their own bodies or technology, and the context they found it in. Such initial framing devices could arguably disrupt an analytical interpretation of the functions for the parts of the object.

24. It's worth framing an analog of this scenario: we have a human and an AI. The human must solve a puzzle designed by an AI, and an AI must solve a puzzle designed by a human. The solution to both puzzles depends on clues in the other puzzle. To solve the problem, they must share clues, but to do so they also must be able to decode this tangential information. In effect, information *about* how the other makes sense of the world, is the key to integrating the clues in their own individual puzzles. If the AI or person can decode a possible interpretation of how the other entity creates puzzles, they then have a chance of reducing the entire set of possible solutions.

25. We can extend our notion of crypsis to include other senses, such as organisms that modulate their vocalizations to resemble the vocalizations of other organisms (Ruxton, 2008).

26. The connection between environment and camouflaging strategies is well observed to begin to develop over decadal intervals. The classic example is the effect of the industrial revolution on moths where, over the course of several generations, there were significant coloration differences between moths present in polluted environments (which became darker), and those that maintained their original coloration.

27. An interesting thought experiment is to consider how transparency could develop in environments where shadows indicate to organisms the presence of predators. Above the sea, the more common strategy appears to stream-line the shape of the body profile to minimize the scale of the shadow.

28. Dolphins have been suggested as a test-case for investigating communication with other intelligent species.

29. Perceptual organization is tied to the motion of the eyes. When something unexpected appears in view, it can take about 200 microseconds to initiate an eye-movement, which would then linger in that position between 20-200 microseconds.

30. It is important to note that this refers to truncated communication without social cues (e.g., gestures, tone of voice). Invariably, the limitations are even greater in this context when we have, arguably, removed the larger part of our own standard communications. We see the consequences of this in our own social behavior when we create abbreviations such as "LOL," and emojis to compensate for text-based communication.

31. It is beyond the scope of this footnote to elaborate on the full extent of the terms of the historical, philosophical, and methodological arguments about the interpretation of sensory signals. However, it is worth noting two important conceptual "schools." In James Gibson's 1979 *Ecological Perception* (This should be distinguished from sensory ecology, which is a broader term that I use in the main text.), the energy of the optic-flow of light is sufficiently rich with invariant patterns and details that an understanding of the stimuli in this context, can lead to an understanding of the phenomenology of vision. In comparison, David Marr (1982) proposed a computational understanding of perception, whereby the nervous system extracts and processes hierarchical information, thus phenomenology is a product of such computational processes. These are, by no means, the only two viable theoretical frameworks.

Approaches to Communicating with Extraterrestrial Intelligence

John W. Traphagan

The process of searching for extraterrestrial intelligence (ETI) implies that some scientists, at least, hope for the possibility of communication between humans and imagined interstellar interlocutors. This seems a rather trivial observation, but it is important to recognize that contact may not involve communication in any interactive sense of the word, particularly if it happens via radio telescope, or other formats of protracted transmission technologies. The distances involved are vast, making the time between contact events likely to be exceptionally long. Even if we happen to run across intelligence in our local galactic neighborhood, we could be looking at 50, 100, 200 years or more between an initial "Hi, we are here" and a response, despite the fact that our transmissions will travel at the speed of light. The question of contact itself is largely a technical one involving specific types of devices, such as radio telescopes or lasers, to send or receive messages to other intelligent beings, or simply to alert ETI to the fact that there are others who are interested in contact and, if possible, some form of interactive communication. The nature and meaning of contact with beings from another world, of course, involves a complex of technical, cultural, linguistic, and other variables that will shape the manner in which any form of communication might unfold (see Traphagan and Smith, this volume).

Note that I refrain here from using "other civilizations" when writing about contact scenarios because, although this is a common trope in SETI literature, communication will not be between civilizations. Rather, at least on our end, communication will involve individuals who have made a decision to send out a message or have been fortunate to receive a message (as a *de novo* transmission or response to human signals). The event of contact will, most certainly, become quickly embedded in larger institutional and cultural structures such as government bureaucracies and social media discussions, but at no time will a communicative event involve interactions between ETI and the civilization of Earth. Why? The answer is simple: Earth *does not have a civilization*. It has many different cultures, governments, religions, etc., but we lack unification and, thus, if we were to make contact, it would not involve representation of our world, but of some segment of our world that happened to be directly involved in the contact scenario (Peters, 2011; Traphagan, 2017). We have no way of knowing what the other end of the cosmic phone line would be like, but if they are anything like us, it is possible that contact with ETI would actually be contact with certain representatives—official or not—of the people who live on that world.

This is an important point often overlooked or ignored in the literature on SETI, but that will profoundly shape how a contact event unfolds, and how approaches to communication need to be conceptualized. What language or *languages* do we use to communicate with ETI? Although there are basic patterns among human languages, such as the use and ordering of subjects, objects, and verbs, those languages vary considerably in terms of how they are syntactically constructed as well as how they are represented in visual form (Cornips and Corrigan, 2005). Language, of course, shapes how we experience and interpret the world because it influences how we process information about the world.

Although anthropologists and linguists have debated, at times intensely, the validity of the Sapir-Whorf Hypothesis mentioned in the Hollywood blockbuster *Arrival*, the concept of linguistic relativity, as it is more generally known, raises important questions about the relationship between language and cognition and the extent to which language shapes, or even determines, cognitive categories and, thus, plays a primary role in constructing our reality (Kay and Kempton, 1984).[1] This issue is important when we think about communication with ETI, because we will be, by definition, dealing with beings whose cognitive capacities have independently evolved in physical and cultural environments different from those of humans.

I want to begin this essay with a brief example to underscore the complex problem that communication presents—simply among humans of different cultures, let alone between humans and possible aliens who evolved on a planet many light years distant from our own.

An Example from Earth

Most people who have studied a second language quickly learn that certain phrases and terms are difficult to translate from their native language to the one being studied (and *vice versa*). For example, the term *boke* ボケ in Japanese translates roughly as senility in English, but the meaning of the term is significantly more complex than the English translation because it not only indexes a condition of cognitive decline, but also a moral condition related to cognitive decline that becomes associated with the person experiencing dementia in old age. To add complexity to this, the term sometimes can be used jokingly, despite the fact that it has rather serious meanings related to late life, and it even can be used along with other terms to indicate conditions of mental disorientation, such as *jisaboke* which literally means difference in time disorientation and refers to jet lag (Traphagan, 2000). It seems that when translating from one language to another, the full meaning is somehow not conveyed or requires a great deal of description for the meaning to come through. Space does not allow me to enter into a detailed discussion of the complex scholarly discourse that has existed among philosophers and linguists concerning the extent to which literal translation is possible (Wilson, 2015); however, a simple illustration will provide a window into the nature of the problem.

The counting of objects would seem to be something fairly easy to translate across different languages given how significant and common enumeration systems are across all cultural backgrounds. Humans universally can count the number of things that they see in the world, and the number of objects one happens to be observing usually is not open to a great deal of interpretive debate. If you and I are standing in front of two

Alfa Romeos, it is unlikely that we will disagree on the number of Alfa Romeos we are observing. This becomes a bit more complex if our vision is obstructed in some way, but that is not a matter of counting, but one of perception, and whether we can observe the same material things in the same way (see Traphagan and Smith, this volume). Counting the number of cars in a driveway, is just counting the number of cars in a driveway. In the same way, the counting of the number of pencils in your pencil box is a straightforward exercise in identifying how many pencils you have. Three cars are three cars; five pencils are five pencils. Or are they?

Table 2 shows the manner in which the counting of cars and pencils is represented in written English and Japanese (compiled by John W. Traphagan).

Table 2: Counting in Japanese and English

English	Japanese
One car	車一台
Two cars	車二台
Three cars	車三台
Four cars	車四台
One pencil	鉛筆一本
Two pencils	鉛筆二本
Three pencils	鉛筆三本
Four pencils	鉛筆四本

Obviously, these are very different written languages, and they convey meaning in different ways. English uses letters to represent phonemes, or units of sound that have no meaning on their own. When one strings these phonemes together to form a word, they also form a meaningful idea. Japanese works quite differently, using ideograms to represent complete ideas. Thus, the character 車 can be pronounced as *kuruma* and means car, vehicle, or wheel. It can also be pronounced as *shiya*, such as when it is attached to another character to form the word *okadensha* 赤電車, which refers to the last train on a rail schedule. There is no direct relationship between the sound used for a kanji character in Japanese, and the phonemes that combine to make a word. In fact, many years ago while living in Japan, I received a phone call in which the person asked for someone named Yūko (友子). I told them that they had the wrong number, apologized, and ended the call. My wife, who is Japanese, then asked who was on the phone and I said, "Wrong number, someone asking for Yūko." My wife's name is Tomoko (友子), but she responded, "Oh, that was probably for me." The first character in her name can be pronounced in more than one way, but there is no way to tell by looking at it which is the correct way, and it means "friend" regardless of which of these ways it is pronounced.

If you look at Table 2, you probably can figure out which kanji characters refer to numbers. From one to three, it is fairly obvious, although when one gets to four the character seems less directly representational of the number of things being counted. If you look at the characters to the right of the numbers, it will be much more difficult to interpret in terms of meaning. These are counters, which identify characteristics such as the size and shape of an object being counted. Thus, for cars, it reads *kuruma ichidai, kuruma nidai, kuruma sandai, kuruma yondai,* and for pencils it reads *enpitsu ippon, enpitsu nihon, enpitsu sanbon, enpitsu yonhon.* The meanings for these phrases are "one large machine called a car," "two large machines called a car," and so on, and "one cylindrical writing instrument," "two cylindrical writing instruments," and so on. There are more than a hundred different counters that are used to identify the types of objects being counted. This may seem strange to an English speaker, but it is no more arbitrary than the need in English to differentiate between one and more than one thing being counted by sticking an "s" at the end of the word to make it plural. If I count, one car, two car, three car, you will understand what I mean despite the lack of pluralization. One might ask why it is necessary to indicate the difference between one and many versus the size and shape of an object, and the answer would be that it is a cultural variation that is basically arbitrary but so deeply embodied that we think of these as being *natural* ways of counting. However, the Japanese approach is no more, nor less, natural or accurate than the English approach.

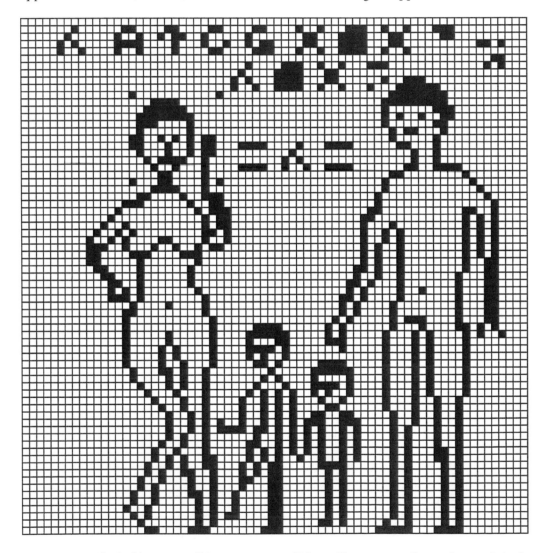

Greetings to Altair, bitmap #11 (bitmap courtesy Shin-ya Narusawa, redrawn from originals created by Professors Hisashi Hirabayashi and Masaki Morimoto. Re-illustrated by Paul E. Quast).

The point here is that when we translate between English and Japanese, information does not transfer on a one-to-one basis. The information about size and shape is lost when we translate counted things into English, and the information about plurality versus singularity is lost when we translate from English into Japanese (plurals are not usually used when counting in Japanese).[2] This difference does not generally impede communication between English and Japanese speakers, because we both have a concept of counting and represent that concept in similar ways. But when it comes to more complex meanings, communication between speakers of different languages on Earth can become quite difficult. There are some terms in Japanese that simply do not map well into English without a great deal of explanation, and varying cultural backgrounds can make this problem more acute. For example, when my wife meets with friends in Japan and starts talking about television shows from their childhood, I quickly drop out of the conversation. I understand most of the words, but the context is so remote from my own experiences that I really don't catch much of the meaning in the conversation. In other words, when one factors in cultural knowledge—whether it is related to historical events, or linguistic patterns viewed as natural—it becomes quite difficult to convey and interpret meaning accurately and effectively.

Some SETI researchers have recognized the problems that arise in trying to communicate using *human* language and have raised questions about how we might find other approaches to communicate with ETI.[3] Beyond the technical question of how best to send a message that can be received and identified as a communique, there is a basic problem of how to encode a message in such a way that the communication of ideas may actually be possible. And, of course, there is always the problem of which ideas should be communicated. Do we try to "say something" about ourselves? If so, what do we say? Do we simply try to point out the fact that we are here, and at least some of us are interested in contact? If there is a way we could communicate without language *per se*, which would still convey something about humans, and would that be relatively easy to interpret by ETI?

Music, Mathematics, and Communication with ETI

Recognition of the fact that human languages are highly varied and represent meaning in complex ways, has led some scholars to explore the idea of universal forms of representing ideas among humans as a way of possibly communicating with ETI. Psychologist Douglas Vakoch argues that music might represent a productive way to convey meaning to extraterrestrial civilizations, in part because music involves linked phenomena of acoustics and mathematics. It seems likely that an advanced extraterrestrial civilization will have a concept of mathematics and the tones in musical scales (although there are different ways of constructing scales,[4] and ETI that does not rely on the sense of hearing may have no concept of music at all) that can be mathematically represented as frequencies. Musical tones, like phonemes, do not carry any particular meanings, but the structure and organization of a musical work might suggest the presence of cognitive organizational abilities and modular intelligence to an extraterrestrial that intercepts a musical message sent from Earth. Vakoch argues that we could create a message with a "musical primer" to reflect basic aspects of music perception, such as interval size and pitch direction, as a means of introducing a "basic vocabulary that reflects both components of music and parameters of human auditory perception" (Vakoch, 2010a). This would not, of course, convey meaning in the sense that words do, but it would convey information about humans as intelligent beings and would also, perhaps, tell ETI something about our auditory capacities (see color insert).

There are, however, problems with this approach. While we may recognize unfamiliar music as music, this is because humans universally create musical forms. We have no way of knowing if beings on other worlds do the same thing. Therefore, although music seems universal to humans, there is no reason to necessarily think it is a universal behavior that intelligent beings use to express themselves. Indeed, humans do not even necessarily agree on what constitutes music as opposed to noise, and some musical forms, such as free jazz or scat singing, seem to challenge the boundaries of how to define a collection of sounds as music. In fact, many contemporary musical forms, such as noise music, have challenged the definition of music and some composers—for example, John Cage—have focused some of their work on exploring the nature of silence in music (Hegarty, 2006; Kahn, 1997).

When meaning is associated with musical forms, like with language, the association of particular sounds with emotions or particular images may well be arbitrary. For example, what is it about Beethoven's 6th Symphony that makes it "pastoral"? The problem here, other than varying personal preferences, is that the listener brings to any musical piece or genre a set of subjective experiences, and cultural knowledge structures that are mapped onto the music in the process of interpretation (Traphagan and Traphagan, 1987). It is not unreasonable to think beings that grew up on a world different from ours—and with possibly different sense organs—rather than interpreting Beethoven's 6th as pastoral, or even as *music*, might simply interpret the work as noise. Keep in mind that although today we view Stravinsky's *Rite of Spring* as a masterpiece, the debut performance in Paris did not go well (to put it mildly), and both music and choreography were abhorred to the point that it incited rioting among the audience (Chua, 2007).[5] In short, our capacity to interpret music—or even to recognize something *as* music—is embedded in cultural context. While music may be a universal human competence, it is clearly not constructed, nor interpreted, in universal ways even among humans (Cross, 2001).

Another approach to communication with extraterrestrials has been the idea that mathematics might represent a basis for finding common ground with ETI. Mathematician Carl DeVito notes that many scholars have argued that our mathematics should be understandable to any technologically advanced society (DeVito, 2013). On the surface, this makes sense. For ETI to build the kind of equipment necessary to receive a message sent from Earth, a knowledge of mathematical principles is inevitable. Simple ideas like numbers and counting should be obvious to ETI, but even that, as noted above, brings difficulties because we may conceptualize the nature of counting in different ways,[6] or at least think about what is meaningful in terms of counting in different ways—as is the case between English and Japanese speakers. Indeed, throughout human history there have been various ways to symbolize numbers and the current, largely universally used system for symbolizing numbers is not the only way to do so. It is important to keep in mind that the system by which we symbolically represent numbers is based on the number ten, which we then use as a convenient way to represent very large quantities, such as the number of stars in our galaxy (4×10^{11}) (DeVito, 2013). The fact that this is convenient for us may be a product of our biology—we have ten digits on our hands and ten on our feet—which may have led us to model the universe mathematically on the basis of tens. There are, no doubt, many other ways one could model the universe mathematically that might not appear obvious to us,[7] just as our approach might not appear obvious to ETI. Therefore, although it is reasonable to assume that ETI will have an understanding of mathematical principles that overlaps with our own, it is *not* reasonable to assume that said understanding will be symbolically represented in a way that is mutually understandable between ETI and humans.

Visual Images?

The preceding discussion of music and mathematics, as frameworks for communication with ETI, raises the problematic issue that simply because something is a universal competence among humans, and as with mathematics at least, is likely to be understood by aliens, the manner in which that understanding is symbolically represented is not necessarily common or even commensurate. Another idea that has been frequently used as a means to convey information to ETI is visual images. The Pioneer and Voyager spacecraft all have messages that were designed to, at least, communicate our existence should those craft be found by aliens at some point in the distant

Computers on Parade, East Berlin (original title: Berlin, 750th anniversary celebration, parade, computer). How would post-biological beings begin to interpret this photograph? Does humanity celebrate technological advancements, or do computer overlords organize elaborate parades? Featured in *The Last Pictures* as image #82 (Bundesarchive [Bild 183–1987–0704–077/CC-BY-SA]).

future. Launched in the early 1970s, both the Pioneers 10 and 11 spacecraft have plaques that display symbolic representations of the people who sent them, our location in our star system and the galaxy, a silhouette of the spacecraft, and schematic representation of the hyperfine transition of hydrogen (see Gillespie, this volume). Unfortunately, if either of the spacecraft were to be found, it would likely be quite difficult to interpret the plaques. If ETI were able to conclude that the drawings of the man and woman were beings, would they think that they are the same species, or two different species? How would ETI interpret this if their form of reproduction were asexual, or if they were post-biological beings (Dick, 2003)?

The problem that arises here is that specific meanings, as well as meaningfulness itself, are not immediately evident without prior knowledge of the beings that created the message. Note the use of an arrow to point out the location of Earth within our solar system. Would a species that had never developed the bow and arrow have any ability to interpret the meaning of the arrow as visual symbolism to signify *pointing* to something? Or indeed that the triangular end was the *pointing* indicator, and not the flat end? And when we look at the figure of the humans, it is important to keep in mind that it might appear that the man is strangely asymmetrical, with some sort of odd appendage sticking out of his right side. What does that appendage mean? Hello? Stop? For Americans at least, the same hand signal can mean both hello and stop. Humans can correct for the lack of knowledge about something because we tend to share basic understandings of our own world, and our various cultures have evolved in close proximity within similar environments. Without *any knowledge* of human history, biology, and culture, it would be quite difficult to interpret the visual content of the *Pioneer Plaques* (Trphagan, 2010).[8]

Both Voyager spacecraft carry messages containing significantly more information than could be placed on the *Pioneer Plaques*. The VGR included on both Voyagers incorporate not only visual images on the records, but also a combination of images, sounds, music, and greetings in numerous languages recorded as frequency-modulated acoustic data. The spectrographic images include things like a woman nursing a child, men running in a track meet, various forms of architecture, an X-ray of a hand with a woman holding her hand next to it, images of planets, a man on a spacewalk, and so on. Assuming that ETI could figure out how to access the data (there are instructions provided on how to create and use a phonograph player), it is difficult to know the extent to which these images and sounds could be understood. One very clear point, however, is that the images are highly biased. There are no images of war or pollution of the Earth. In other words, the data on the records do not reflect a particularly balanced representation of humanity, assuming it can even be interpreted. This raises an unfortunate possibility that if ETI were able to interpret the data we selected to represent a concise autobiography of Earth, a wrong or deeply misleading understanding of our world might be a result of their efforts (see Capova and Quast, this volume).

Challenges to Communication with ETI

Over the past twenty years or so, SETI scholars have devoted increasing intellectual attention to working out how we might best go about constructing messages that could be interpretable by ETI. These efforts are not only important when it comes to creating

Photograph of astronaut Edward Higgins White II on a spacewalk during the Gemini 4 mission (photograph from NASA/James A. McDivitt [S65–30433]).

messages aimed at aliens. In fact, thinking about the problems that arise related to communication with intelligent beings on another world may help us in interpreting any message we receive in the future. Douglas Vakoch's important edited volume *Communication with Extraterrestrial Intelligence* (2011b) explores the wide range of problems, from the technical issues of how to go about sending or receiving messages, to questions of how we might make ourselves understood in any message we decide to send (Vakoch, 2011b; 2014). There has also emerged an, at times, heated debate within the SETI community about the ethical concerns associated with METI, or the construction and sending of messages from Earth with the intention of making contact with ETI. METI, also known as Active SETI, the idea of sending messages toward imagined alien civilizations, presents a variety of problems. The simple fact is that if we broadcast and our messages are received, we will have, at the very least, indicated that humans exist and have developed technology that allows us to send information into interstellar space. There would be no going back. How either a message or the knowledge of our existence would be interpreted by beings from another world is entirely a matter of speculation. And there are fundamental ethical issues associated with METI—such as whether or not one person, or group of people, have a right to "speak for Earth," and potentially place all of humanity at risk should ETI prove to be belligerent—that remain unresolved among humans, despite the fact that we have already started sending messages and have been doing so for quite some time (Smith, 2017).

The moral issues that surround contact are elements in a wider array of challenges that humans face should we make contact with alien interlocutors. Communication is difficult without something shared, because those things shared provide a basis for interpreting information through careful comparison. What might we unwittingly— and therefore unknowingly—give away? Perhaps the most significant problem to be addressed in developing approaches to communication with ETI is generating a sophisticated awareness that those things we think of as natural—whether cultural, behavioral, ethical, or even physical—have the potential to impede our capacity to interpret the intent or meaning of anything we receive. The same problems should exist at the other end of the contact scenario. In other words, at the core of the problem of communication with others, whether alien or human posterity, is the capacity to empathize. As scholars have noted in several fields, human empathy is based on our ability to share emotions and to recognize and understand the thoughts, interests, and feelings of those with whom we interact. The emotional and cognitive systems through which we interpret our world and construct reality, are closely linked to empathic response and the ways in which those empathic responses function or evoke not only our internal

Making friends with the sea—These orphan children at Marathon, Greece, were brought from the interior of Asia Minor by Near East Relief and had never seen the sea before, c. 1915 or 1916. We often cite space as a metaphorical shoreline without realizing that not every earthling has access to these basic terrestrial environments, let alone one requiring aerospace technologies. A cropped version of this photograph featured in *The Last Pictures* as image #4 (photograph from the Library of Congress/Near East Relief, George Grantham Bain Collection [Reproduction Number: LC-USZ62–130734]).

cognitive and emotional systems but do so in ways that are closely related to the social contexts in which those systems were formed, and in which our experiences are constantly embedded (Shamay-Tsoory, 2011).

We need to be careful to recognize that even the basic knowledge of things like physical laws, that we will no doubt share with aliens who can send and receive interstellar messages, are not neutral, but are constructs generated through, and limited by, who and what we are as physical and social beings. In other words, as physicist-philosopher Henry Margenau argued in the middle part of the 20th century, "physical reality is that which is known at a given stage of history," which implies that physical reality itself changes with discovery and the generation of knowledge—we do not know or experience a reality that is permanent, nor is a permanent reality directly reflected in our mathematical and other representations of experience (Margenau, 1952; 1977). In other words, we live in a contingent reality that is both a product of our natures as physical and mental beings, and also shaped by our experiences of the surrounding world. The same will hold true for aliens and generates a significant limiting factor in how communication may unfold, and whether or not communication or understanding is even possible between beings who developed on different worlds and, at least to some extent, in different realities. This is not intended to suggest that communication is impossible with intelligent aliens, but should that moment of contact arise at some point in the near or distant future, humans (and our interstellar interlocutors) will need to work hard to overcome the inherent limitations generated by the fact that what we know about the universe is not the universe itself, but an abstraction from the universe itself. The greatest challenge in developing approaches to communication with ETI lies in finding ways to transcend the symbolic and cultural structures that shape our experiences, and the ways in which we express the meanings associated with our encounters with the universe in which we both live.

NOTES

1. We can think in terms of cognitive domains that people use to organize and categorize their worlds. For example, if I were to ask the reader to list ice cream flavors, and the reader happened to be an American, then vanilla and chocolate would likely be among the first flavors to come to mind. However, if I were to ask a Japanese person the same question, it is possible that macha (green tea) would replace chocolate, which is less common. There would also be different flavors on the two lists. An American is unlikely to have black sesame or squid ink flavored ice cream on the list, but this would not be surprising for a Japanese menu. On the other hand, a Japanese person is less likely to have a flavor like rocky road. In other words, we carry in our heads cognitive frameworks that we use to organize and categorize the world, and these frameworks are significantly shaped by the cultural and experiential contexts in which we live.

2. In the rare cases in which plurals are necessary, it is possible to add an ending to some words. Thus, for example, the term *watashi* means "I" but *watashtachi* means "we" or "us." One can also use the term *hitotachi* to refer to a group of people, although it is not absolutely necessary. The term *hito* can itself refer to one or more than one person. How these are used is largely depending upon context.

3. I have raised the idea that visual/physical modes of communication such as ritual might prove potentially useful (Traphagan, 2010).

4. At a recent TED talk (2012), Vakoch discussed the prospect of musical scales in reference to message design, including the arbitrariness of traditional Western (12-tone) and Eastern (5-tone) scale divisions, alongside the use of a 31-tone scale, as proposed by Christiaan Huygens, and indirectly used in the 17th Century Netherlands.

5. This very same concerto was included in the Voyager compendium. It is anyone's guess how this avant-garde composition may be received by an ETI should the Voyager spacecraft be intercepted at some point in the future; however, Timothy Ferris (who co-selected this masterpiece for the VGR) whimsically

quipped, "The composition built no bridges between cultures. It was more like a shout across the river" (Sagan *et al.*, 1978).

6. If we consider Naomi Mitchison's "brainy starfish" example (1962), if the organism resembles the sensory channels of an actual starfish, they may perceive their surroundings as a series of chemical patterns, saturations, and directional concentrations. Sensing quantity values this way could therefore inform the developmental history of extraterrestrial mathematics, and similar branches of scientific advancement.

7. DeVito (2013) briefly discusses some vestigial remnants from previous base numerical units, employed by past human societies, that still remain in use within our dominant base-ten system. Examples include the measurement of time (60 seconds per 60 minute), and angle measures (one degree contains 60 minutes).

8. This lineage of depicting visual information influences message designs today, for instance in *A Beacon in the Galaxy* (Jiang et al., 2022) which updated the Cosmic Call 1 bitmaps (not, as stated, the Arecibo bitmap).

How We Think We Know What We Think We Know and How the Unknown Others May Think They Know What We Think We Know (and Show)

Paul E. Quast *and* Klara Anna Capova

> This is a message, a message addressed to you constructed in a code-system called English marked with Roman letters in ink on sheets of cellulose, or as liquid crystals on the screen. It is here and now, perceived by the senses, interpreted by a being with a brain, body, and history, living in the world…. By using this code-system I evidently hope to make myself understood, to awaken in the mind of the receiver similar thoughts and ideas I have when I formulate this message
>
> —David Dunér (2011)

Dunér's remark draws into sharp focus the logical premise of this chapter: how do we endow a mental thought that has been rationalized by current human minds (illustrated as material signs within inscribed media, or streams of photons), to distant observers who may not share similar cognitive faculties, sensory modalities, cultural and social histories, symbolic forms of expression, a phylogenetic lineage, or environmental commonalities with us? These surrogates of present-day material culture are also required to function across the fourth dimension, often far removed from their creators by decades, centuries, or millennia, before they may be inadvertently found and contemplated by foreign cultures which may be profoundly different from the minds that manufactured them. There is a lot to unpack here about "making matter mean," and inevitably any such approach requires us to exercise intersubjectivity in addition to also employing a healthy skepticism and due diligence in hopefully finding any singular, verifiable approach forward.

The crux of this human-epistemic problem of envisaging how to extend our cognitive faculties outwards, stems from knowledge of how we imagine societies that reside far away from us (in space or time) in a similar manner to the way we imagine human societies from antiquity. Communication—or more likely in this case, uni-directional monologues conveyed by disembodied surrogate messages—can be a technically challenging feat, given the fact that we do not possess extensive experience within even articulating our own ancestors' linguistic, pictorial, or cultural habits, despite sharing many convergences in biological, environmental, and cognitive properties (alongside storehouses of contextual knowledge about these civilizations). Formulating an answer to this semiotics challenge, or the complexity of related, follow-up questions, is by no means a simple task and has been the stable subject of multidisciplinary SETI literature, information transfer systems, and communication theories (as well as popular science fiction inferences) for much of the last century, if not longer. Most of the theoretical work and practical experimentation within this "incommensurability problem," in turn, has been conducted under the particular auspices of Western scientific traditions. However, these paradigms and aesthetic principles, inherited from ancient and modern European civilizations, are extensively intertwined across our interactions with myriads of technological objects, and supplemented by complementary cultural histories with (in some cases even personal) worldviews. We may notice that these relationships with material artefacts are not even held universally by all contemporary human societies. How we utilize *this knowledge* in approaching our analogies[1] can also be subjective, and dependent upon how various disciplines in the sciences differ in rationally gathering and inducting reason or intent from common data (Denning, 2011c).

It is within (and in comparison to) this Western cultural inheritance that we describe other distant minds (ETI, or terrestrial) under value-instilled labels such as "advanced," "technological," "sentient," and "intelligent." These definitions, in turn, also function upon analogies and prepositions rooted in the nuance of "progress" from the Age of Enlightenment (17th-19th century), and its affiliated cultural worldviews and scientific histories—"what we know and learned from experience, *they* should also understand in the same way." But not everyone reading the same "manual of reality" will necessarily come to the same conclusions. For example, just because a subgroup of humans who share similar cultural histories and customs apprehended their surrounds using higher mathematics for scientific research and technological strides,[2] does not mean that other forms of intelligence will, or are necessarily inclined to, follow these exact models of phenomenology (Barker, 1982; Sefler, 1982). The nature of reality is far more complex and diversified than we originally imagined, and we may never recognize a *fundamental reality*, but only relativist versions dependent on the unique faculties of biological species—in our case, humans read generational *anthropomorphists editions*, but even these readings do vary.

Often when theorizing about prospective ETI that we do not share any kinship with, or distant human posterity, we are seduced into believing that current human analogies are the hallmark signs of any "advanced civilization." Therefore, we tend to calibrate our minds and communication approaches around organizing such efforts to fit our assumptions about how societies generally operate. We tend to imagine "the other" as a mirror of our own minds and achievements, however vague, baseless, or partial this may be. Our choice of analogies to employ within *our* theories (and the messages we choose to craft from these speculations) is significant, as those facile decisions will become operational assumptions. As we have seen in the formative years of SETI, they may then serve as a lectotype for guiding the parameters of our search *for decades*; "The problem with analogies is that they are highly persuasive, inherently limited, and easily overextended" (Denning, 2014).[3]

As previously noted, we inevitably began undertaking SETILO based upon a very truncated perception of ourselves, extrapolated into the futurescape of the Earth system, using phenomena we could readily measure at the time. But the end goal(s) of ETI, or even human posterity, may be fundamentally different than our own.[4] To borrow from the Vakoch (2011c) mountain climbing metaphor, we generally tend to perceive a distant observer as inhabiting a higher vantage point along our linear pathway uphill, but we seldom stop to consider whether there may be alternate branches, or separate thoroughfares, that others might embark upon. Perhaps, they may be even traversing hidden routes that lead them towards alternative summits, not the forecast advanced, technological peak we have envisaged. We also seldom look back downhill, except to pluck out suitable episodes of human society to support our assertions about unobserved minds.[5] Our cognition, cultural histories, and rationale forethought rendered within communication concepts, may serve as the metaphorical climbers, each possessing their own novel prerogatives, histories, sensory communication apparatus and embodied understanding of their location—properties that reinvent as they ascend or descend *Mount Vakoch*.

It is within this struggle of inference and projection that we condense our "meaning" of human thought into a collection of communication faculties and mental schemes encoded within messages that, until very recently, largely aimed to induct an observer (or distant climber) into *our* stark means of abstract exchanges and expressions, curtailed by *our* dominant senses and sensitivity ranges—methods frequently employed to interact with close cultural cousins of *our own* biological species. In context with discussions advanced across previous essays, and to explicate our point, let us briefly explore some of the communicative schemes already bound within exoatmospheric messages from a contrarian approach. We will refrain from contributing to the speculative literature associated with prospective recipients, and re-mapping terrestrial experiences onto other minds. Instead, over the duration of our discussion, we simply intend to raise questions as to whether we could reasonably expect the particularities of present linguistic and semiotic media to remain legible for distant spatiotemporal agents, basing our argument in context with our own extensive investigations into past human societies.

Chomsky, Contact, and the Death of Tongues

Dunér's "message" example specifically highlights the perplexity of whether natural languages[6]—regularly regarded as the most influential, erudite invention of the human mind for conversing with fellow speakers—may adequately convey the signified "meaning" in thought to a distant intellect that may, or may not, be familiar with our arbitrary symbolism, or associated lexical, syntactic, and semantic conventions. The linguist Noam Chomsky has surmised that natural language faculties are a means of conveying information to other inducted speakers, but the primary function of these lingual systems is internal. They enable us to self-order our idiosyncratic thoughts, beliefs, and personal philosophies (what is pointedly referred to as an internal or *I-language*), before ostensibly expressing these mental ideas into the world across multiple communication channels (external or *E-language*). Natural E-languages are an adequate means of expressing our thoughts and infrequently exchanging information with contemporary audiences through signs they are contextually familiar with. But, in this sense, it should be acknowledged that these particular subsystems of communication are also contingent upon a much broader array of behavioral patterns, social etiquettes, gestures and customs, not to mention shared cultural contexts and histories for mutual comprehension.

These are only some complementary E-language factors that do not readily translate over spatiotemporal distances in removed material prostheses of our minds.

This synopsis invites us to contemplate whether the coded structure of natural E-languages, as the products of human minds (and associated semiotic, phonetic and grapheme systems), may be unique, embodied epiphenomenon that are contingent upon living usage, alongside specific cognitive, cultural, technological, environmental, and temporal contexts to interpret. Alternatively, could these code-systems remain legible outside these mental circumstances in messages, as prospective artefacts of material culture-exchange? Lingual constructs have been a steady fixture of crowd-sourced interstellar transmission projects as simple outreach and public engagement activities but, in recent years, information rendered within native terrestrial tongues has also come to frequently populate space-time capsules and some scientific interstellar radio messages as representations for these language structures. It is therefore worth provisionally unpacking some of the mental baggage underpinning our linguistic systems to contemplate whether they may retain their abstract, interpretive context, or serve as indecipherable, unreadable messages beyond our own terrestrial context.

Understanding a language or a plain sentence of alpha-numeric script, juxtaposed into a wide variety of punctuated arrangements, and subject to syntactic, lexical, and grammatical rules that endow it with meaning, is simply not a process of extracting definitions from each arrangement of "words" and tallying these results together as a direct translation (Saint-Gelais, 2014; see also Traphagan, this volume).[7] Moreover, words themselves are arbitrary carriers of semantic meaning, reliant upon cultural connections, contexts, and established familiarity with object associations, a disparity that can be plainly seen within the multilingual words used to define commonly experienced referential objects.[8] Interpreting lingual expressions therefore, entails accessing the meaning hidden within arbitrary assemblages of signs, while understanding the correlation between how these symbolic units integrate with the often unspoken, vernacular assumptions, hierarchical rules, graphemic systems, and cultural histories that lie embedded within the formation of such conjugations. In turn, readers beyond our temporal horizon will need to grasp meaning from what will likely be limited sample corpora for alphabetic, syllabic, or ideographic script systems, whereby "Languages can vary wildly when matching words to meanings, and there doesn't have to be any kind of pattern" (Wells-Jensen, 2018). Words and associated objects also fall out of use over time. Without knowing what is signified (or if we even share what is signified[9]), we *can* and often *have* assigned arbitrary meanings to construct false narratives. Clearly, such erroneous "translation efforts" defeat the purpose of our intention to simply understand what others wish to tell us, but they may also stymie actual deciphering progression.

In addition to these principal difficulties, the informatician Geoffrey Bowker (2008) contends, "acts of committing to record (such as writing a scientific paper) do not occur in isolation; they are embedded within a range of practices (technical, formal, social)." This is true for the inscribing party and their embodied mental faculties, but it is also valid for the translating agent who will never approach such a cryptographic exercise as a neutral observer. To paraphrase the theorist Mieke Bal (1997), when there is very little informational context supplied for a location and described processes or concepts (such as the structure of DNA outlined in Traphagan and Smith, this volume), the end-reader may inevitably imagine one to fit their presumptions, alongside approaching such a message under the lenses of their own social histories, knowledge bases, cultural dispositions, and un/familiarity with our abstract symbolism. Stories under this *social construction of knowledge* from recent history, such as Ishi "the last Yahi"[10] (Denning, 2011a), are regularly evoked as a sobering reminder of our difficulty in directly grasping unfamiliar anecdotes, lore, social customs, and linguistic expressions from an unacquainted culture—even if the dialogist is readily available for direct exchanges over years.

Any reader who has had trouble in extracting meaning from the Old English epic poem *Beowulf* will be familiar with the archaic difficulties of how languages simply change over time (Tolkien, 1940). David Crouch and Katarina Damjanov (2015) assert:

> [The] crisis in language has at its core a problem of ontology: a failure to recognise that the words we use to describe ourselves are always, and have always been, bound up in our technological modes of being.

As Crouch and Damjanov describe, language can be extensively altered by how we integrate various and frequently changing technologies[11] as a means of augmenting our memory, gestures, knowledge, and interpersonal communications with our fellow present-day Terrans. An intriguing, contemporary example of this, can be plainly seen within the increasing adoption of script abbreviations, acronyms, memes, and a corpus of evocative "emojis" as convenient substitutes for lingual expressions on mobile devices. Often, such expressions are one of the diverging characteristics of distinctive subcultures or movements. The epistemic and logical versatility of our languages to invoke "all that matters" to a contemporary populace is clearly a strength, but ironically also one of this cultural technologies' greatest weaknesses, given this lingual evolution comes at the cost of slowly annihilating what these code-system previously signified for prior generations, or how they interacted through them.

Ø = 16cm

The A and B sides of the disk of Phaistos, with small, identified field numbering created by archaeologist Louis Godart. Many of the signs resemble those from Linear A and Linear B scripts; however, the language depicted on the disk (if any) remains undeciphered (graphic from authors D. Herdemerten, Derivat-Arbeit Perhelion/Wikimedia Commons, CC BY-SA 3.0 DE).

Beowulf is a relatively modern work when compared to the decipherment of inscriptions written in extinct human languages—Linear A, Etruscan funerary inscriptions, and the Khitan, Byblos, and Indus scripts,[12] being some of the more well-known examples of marks that survive, but at the expense of their meaning and social history (Pope, 1999). These languages were certainly meaningful to their native speakers, but this cultural significance may not necessarily remain consistent, or mutually translatable, for our divergent global civilizations. Inter-cultural contexts do vary dramatically and can also radically change for any number of reasons. Even with a large corpus of redundancy documentation available for some of these languages, brute computational force or exhaustive cryptographic exercises have not been capable of elucidating precisely what ancient minds thought, did, and said within obscure symbolic units (Hauer and Kondrak, 2016; Zhang, 2017). These imperative contexts are usually the first interpretive threads to erode when a language falls out of everyday use, despite the thoughts remaining "preserved" within enigmatic script on disembodied surrogate medias—objects that still conserve archaeological clues for their users' social and cultural histories, outside of any stored informational contents.[13]

Many of the admirable translation precedents set by the decipherment of a few ancient languages such as Linear B, Demotic Egyptian and hieroglyphs, Babylonian cuneiform inscriptions, and the majority of Mayan glyph studies, also reveal a rather discouraging prospect. Successful translation campaigns for obscure human languages have historically relied on the investigating agent possessing some evidentiary resources gained from ancient-to-modern knowledge transfer systems (such as bilingual documents with recognizable translation keys,[14] common nouns in cartouches, or an innate foreknowledge of other neighboring cultures' phonetic-syllabic signs, morphemes, and scripts)—factors which eventually led to working decipherments for further scientific examinations (Daniels and Bright, 1996; Chadwick, 2008). The saga of learning from Mayan glyphs presents an excellent example of such hardship, despite the surrounding regions still containing many living descendants from these cultures that could readily provide context for allegorical stories as part of formal archaeological investigations (Finney and Bentley, 2014).[15] The words of the Mayan scholar Michael Coes (2011) summarize this point further: "no script has ever been broken, that is, actually translated, unless the language itself is known and understood."[16]

It is likely that the meaning signified within arbitrary semantic units of ancient scripts will remain inaccessible, in lieu of any familiarity with how these cultures linked properties of their languages with displaced (and questionably shared) referents in the real-world, or the social histories of these minds that continually crafted and reinvented *their material signs* (Quast, 2021). Such is the nature of information loss over intervals of time. Perhaps there is wisdom in recognizing this complexity and simply understanding these material signs as a representation of dynamic exchanges between the inducted, fluent users of the code-system. Despite our difficulty in accessing these meanings, however, it is plausible that the hierarchy of information structuring within these code-systems, may prove insightful when examining the dialects of the fluent sample language users.

Universal grammar, as theorized by Chomsky during the defined *cognitive revolution* in linguistics, is a tacit argument that suggests our lingual faculties, underpinning the capacity for early-age language acquisition in infants, is made possible by an innate (and seemingly species-specific) genetic endowment which exists independently of life-world experiences. In contrast to John Locke's *tabula rasa* hypothesis of infants as "blank slates," that effectively acquire language through a laborious mental process of deducting from their resident dialect (which would likely take far longer), Chomsky's theory asserts that children are born with a hereditary "hardware" with established syntactic rules—neural properties that assist with initially learning the features of a native language through exposure (Chomsky, 1999). Children (under normal conditions) have been observed to be equally capable of learning any dialect of any human language that they are exposed to over early formative years, in addition to also developing a broad range of grammatically correct utterances outside of those they directly perceive during this time. (This is Chomsky's transformational generative grammar theory.) These observations lend credence to Chomsky's thesis that there is already some inherent, syntactic faculty present as a biological endowment, allowing infants to gradually acquire an initial language "state" through interfacing with their living socio-cultural surroundings, while also peer-learning from other material resources.

This neurobiological pre-programming of lingual subsystems, and how mechanisms are seemingly tailored to learn from resident societies, presents some salient implications for the prospective acquisition of natural languages by foreign minds—agents attempting to grasp and learn solely from disembodied thoughts, represented as symbols across displaced artefacts of material culture. Such minds may disparately exhibit alternative language faculties, socio-cultural histories, sensorimotor systems, or prescribed minimum cognitive design requirements that may further impede this acquisition (Chomsky, 2000). But this same bio-cognitive theory also strongly implies that there are common syntactic principles underscoring the structure of all natural human languages apprehended by children—a constraining numbers of ways, governed by hierarchical mental paradigms, in which natural languages can be organized in human brains, which Chomsky refers to as the "language organ." In the very least, this "organ" could signify how restricted our lingual-cognitive capabilities in fact are, as apparent in how natural languages developed within the confines of this intracranial, syntactic infrastructure over time. Think of this as *common roots for language development*, with ensuing branches then subject to variation. If substantiated, then learning about the structuring principles of this "language organ" by cross-comparing these assemblages of contextless symbols from multiple lingual systems, may prove insightful for those studying our social faculties and intelligible interactions from afar, even if these investigators never interpret our conveyed meanings.

It is uncertain whether prolonged exposure and a laborious, generational approach towards gaining an understanding of unfamiliar language expressions and conventions could, in theory, enable a truly external scholar to clench such intricacies embedded within different lingual constructs—assuming of course, these strings of symbols are accurately interpreted as such. Our brief philological history of learning from the past through one-way fragments of dead tongues is one avenue we have questionably been successful within—languages separated by *mere* centuries or millennia, with little to no evolutionary changes in our biological "hardware" or other common contexts. This may indicate that cultural and social immersion are fundamental factors for understanding simple lingual contexts within these arbitrary graphemic systems, much less the overarching syntactic, lexical, and grammatical rules that cooperatively endow it with signified meaning, intent, and structure.

Insights such as Chomsky's theories on bio-mental endowment, however, should be cautiously recognized as analytical procedures for studying disembodied intellects, or a "brain in a vat" approach (Putnam, 1982)—isolated methodologies, which readily benefit from our capability to experimentally cross-analyze neighboring lingual systems, while enabling us to crucially proof check our inferences, and correct ensuing errors if (or when) found. There are, of course, many counter arguments to Chomskian universal grammar, such as the contested case presented by the linguist Daniel Everett (2005) for the non-conformity of the Pirahã language to established lingual and mathematical archetypes. Chomsky's theories of language acquisition, while certainly influential in the cognitive revolution, have since been contested by several branches of the cognitive sciences conducted under the general disciplinary banner of "psycholinguistics." At the risk of overtly generalizing these extensive sub-disciplines, this area of study broadly examines interrelationships between the structure of the human brain and how our cognitive faculties are mapped and organized at the neurobiological level, while shedding light on how memory, its mental rules, and categorization mechanisms (arising from cognitive and sensory precepts from an environment) then shape linguistic acquisition.

In contrast to universal grammar, psycholinguistics seeks to unravel the neurobiological plasticity of the "non-vat" brain in living, experiential environments, conducting perceptual experiments helpful for understanding how webs of memory and meaning in the human mind interplay with, and support, language acquisition. A common strand of these studies looks at how parts of the brain psychologically respond when exposed to perceptual stimuli and discrepancies that conflict with an individual's established syntax, amongst other factors— behavioral responses to recognized "errors" that activate different parts of the brain depending upon the individual's memories, enabling us to study how we may mentally map lingual responses. Perception, reflection, and living memory—*the behavioral and psychological properties*—are deemed as the mechanisms behind language acquisition, rather than a special language organ. Certainly, these studies tell us much about how human brains organize and map patterns of lingual expressions, but these lessons have only begun to steadily filter into mainstream SETI literature as contextual indicators when contemplating ETI's capabilities. Regardless of these language acquisition theories, for an observer separated by vast expanses of space or time, they may not have access to the social, cultural, ecological, and psychological contexts necessary to extract such astute lingual patterns, from likely impoverished fragments of arbitrary, symbolic systems; extracranial assets, representing displaced referential objects.[17]

In a recent METI International workshop *Language in the Cosmos*, several papers (including one co-authored by Chomsky) have argued that it would be perhaps possible for humans to understand an extraterrestrials version of universal grammar, if we think they are humanoid (Patton, 2018) and, much like our own crowd-sourced messages, if they choose to utilize more than one sample of their xenolinguistic systems (assuming these transmissions contain bilingual documents much like the Long Now Foundation's *Rosetta Disk*, this may open up possibilities of finding one alien language that may parse lingual concepts and processes in a relatable manner).[18] The cryptology expert Cipher Deavours (1985) articulated a similar, optimistic account, suggesting that a cryptological approach could allow us to learn from a message without understanding the contents meaning, provided we are supplied with plentiful language samples, logical puzzles, and other data for comparison to gain clues for decipherment. The success of such approaches towards understanding ETI messages, however, can be deemed anthropocentric as it rests on the premise that such humanoids will also share our cognitive faculties or sense modalities, amongst other prerequisites, and also choose to articulate their thoughts using similar cultural technologies (i.e., languages, aesthetic principles, grapheme systems, phonetic conjugations, bodily gestures, and other unlikely cosmic coincidences) that correlate with our systems of universal grammar. As Chomsky and Gliedman (1983) amusingly speculated about such an impossible encounter:

> If a Martian landed from outer space and spoke a language that violates the principles of [human] universal grammar, we should simply not be able to learn that language the way we learn a human language.

Given the fact that we do not possess any empirical evidence or experimentally verifiable data for such civilizations to begin with, these speculations will inherently remain as such. Insights from the approaches of psycholinguistics will continue to shed light upon the peculiar connections between our own language acquisition devices and the intimacies of the human brain, but it is reasonable to assume that such innate conditions would have also independently evolved in ETI's own linguistic frameworks—inherent capabilities that will be governed by *their own* biological, cognitive, and social domains. Simply put, these disadvantages will likely impede our capabilities to "learn" ETI languages, leaving much space for speculation. In the event we do receive an ETI linguistic (or any) signal, Traphagan (2014) opines that, prior to our translation efforts, "the invention of an alien culture will begin almost the moment that contact is made." Similarly, one of these authors has also poignantly outlined that, given the distances and timescales, contact may unfold as an accumulative, durational event with intermittent "layers" of overlapping messages arriving between first detection, and a sanctioned human response (Quast, 2021), in contrast to the "classic" depictions of *first contact* as a single message, single response affair. Given our likely slow progress of deciphering the limited, unrecognizable data, in addition to elongated reply waiting time, it is likely our window into these foreign cultures will always remain presuppositions. We unfortunately do not have any interstellar exo-missionaries[19] to supplement our impoverished ethnographic assertions about ETI. Xenolinguistics is a fascinating, interdisciplinary field and extensive school of thought (for a detailed exploration of theories, see Wells-Jensen, 2001). However, when viewed in context with our competency in interpreting ancient human languages and other ethological communicative models (Lestel, 2014), it remains difficult to ascertain whether we may in fact possess the cognitive dexterity for interspecies communication with extraterrestrials with whom we will likely have no common, verifiable ground. As Richard Saint-Gelais summarized (2014):

> The consequences these considerations have for interstellar communication are quite obvious … where sender and recipient do not share a common language; the latter cannot rely on an already established language competence with which to work out the meaning of the message but must instead start with the message itself and try to infer from it, conjecturally, the lexical and syntactic rules that endow it with meaning.

Natural languages have historically been a subject of contention for use within messaging applications, given their cultural dependence on established arbitrary symbolism and other syntactic conventions designed to dictate concepts and meaning to the initiated. However, the pedigree of artificial or "meta-languages" such as the early *AO Language* (Gordin, 1924), *Astraglossa* (Hogben, 1952; 1961), *Lincos* (Freudenthal, 1960), the

D-O Language (DeVito and Oehrle, 1990), and prepositional logic based upon lambda calculus (Turner, 2006), have also been posited as alternatives, using the principles of mathematics, alongside logical exercises. Without straying too far into the philosophies of astrolinguistics[20] and whether defined elements of platonic mathematics may be considered a universal *lingua franca* (Samuels, 2006), it is broadly accepted that we can describe *our* perceived, underlying reality in ever-increasing computational fluency. These descriptions corroborate *our* principles of scientific ontology and *our* sensory perceptions of *our* surrounding spatial environments. Despite our reality according well with these descriptors, it is likely our understanding of mathematics is also culture-species specific, created by embodied human minds, to resolve peculiar terrestrial problems, arising at particular points in our experienced history.

Russell Smith (1963) has asserted extensively that both languages and mathematics are sets of symbols that only have conventional connections with the objects or concepts they signify, while Traphagan (see Traphagan, this volume) has noted how even some modern human cultures also structure basic counting systems differently (within lingual expressions). Further, the mathematician Carl DeVito (2011; 2013) has extensively discussed the formalist logic of mathematics as a human invention; designed to suit our particular cognitive faculties, scientific, social, philosophical, metric, and technological applications, all based upon very human historical needs and values, "God made the positive integers, all the rest is man's work" (Dantzig, 2007).[21] Nevertheless, given this proposition of a shared bedrock of natural numbers, it seems plausible to still consider exploring basic constituents of the anthropic "operating manual for reality," and whether it may be accessible *from the ground-up* through artificial languages by another intelligent reader residing within our common reality.

This was the approach that was initially adopted within Freudanthal's Lincos[22]—a means of using simple arithmetic, mathematical operators, and ostensive syllogism to build up more complex equations for demonstrating the principles governing natural phenomena, and thereafter, messier topics such as human behaviors and reasoning through disembodied interactions. Freudenthal's Lincos was principally built upon conveying these examples as phonemes through the punctuation of a signal's properties (though how these units were intended to be encoded was not explicitly outlined), but there have been several periodic generations of these logical inferences formulated (Ollongren, 1999; 2001; 2004; 2010; 2011; 2013; Vakoch and Ollongren, 2003; 2011; 2013). It remains to be seen whether increased complexity will remain equitable with increased understanding of "human manual" operators, processes, terrestrial properties, and our signified "meaning" in thought. However, Ollongren's revised edition of Lincos has been purposefully developed as a metalanguage to begin bridging this interpretive gap for natural language systems with higher-order logical constructs, alongside widely availing of lessons gleaned from early computational languages (for a detailed discussion about these artificial languages, see Oberhaus, 2019).

Regardless of this theoretical discourse concerning how to properly develop and punctuate such artificial languages for ETI, in the meantime, there have been some practical experiments which have brought these constructs from the pages to the cosmos. The ideographic 5x7 bitmap "alphabet," produced as a propaedeutic guide by the

Opposite: **Cosmic Call 1, bitmaps #1, 2, 4, 5, 6, 8 (bitmap courtesy Yvan Dutil and Stéphane Dumas. Re-illustrated by Paul E. Quast).**

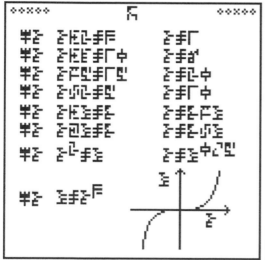

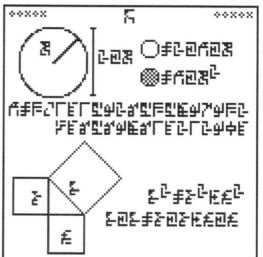

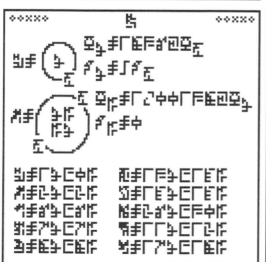

astrophysicists Yvan Dutil and Stéphane Dumas for the initial Cosmic Call 1999 transmissions, functioned upon the principles of Lincos to teach simple concepts such as arithmetic, chemistry, and physical constants across 127×127 bi-dimensional bitmap "pages." The later series of Cosmic Call transmissions in 2003 would further revise this guide, reducing these ideograms to 4×7 bitmap characters on a single, long 127-pixel wide bitmap page. These alphabetical characters are notable as they instill the phoneme principles, envisioned by Freudenthal's original Lincos, into unique visual ideograms that can be juxtaposed into a wide variety of logical arrangements, while incorporating punctuation as the negative visual space within these noise-resistant images—properties that accord well with the prevalence of human perceptual faculties, and aesthetic conventions, embedded in pictorial message designs for ETI.

Outside of these visually dependent formats, the hailing *"Rosetta Stone"* component of the Lone Signal beacon was designed by the planetary scientist Michael Busch to build a complex encyclopedia of 8-bit[23] concepts in order to detail more extensive operations such as orbital periods, proper motions, and chemical compositions for the Solar System (amongst other concepts, though these formats are based upon lessons from Lincos) within long strings of shifting EM frequencies. Similarly, the initial "tutorial file" for the recent *Sónar Calling GJ 273b* transmission series provided a smaller index of 8-bit mathematical concepts, useful for further lessons and subsequent comprehension of the enclosed musical scores. Furthermore, other popular media depictions have featured Lincos-styled encoded signals from ETI, such as Sagan's classic novel *Contact* (1985) in which humanity receives a three-dimensional matrix of blueprints for constructing an interstellar teleporter device. We will leave it to the reader to contemplate whether these content-laden signals may convey our meanings to distant minds.

Representing or Reflecting Meaning?

Within our preceding discussion about natural E-languages, the sensory information we wish to endow within our messages remains as an abstract mental thought, packaged within an extraneous format (i.e., script), that does not possess any tangible relationship to, or interpretive benefits for, decoding the information. The difficulty arises when we expect an essentially arbitrary, symbolic medium format (for example, a punctuated EM signal, or inscribed metal plaque of text) to lend an interpretive context, or allude to decipherment clues, for perceptibly random contents *we choose to transfer* as information to others.[24]

As a classic example of a natural science approach towards addressing this issue, we would hope that the use of recognizable numerical sequences (as familiar, shared information), punctuated within specifically chosen EM frequencies at the beginning of a human transmission to ETI, may directly relate to the ensuing arithmetic guide for an Ollongren-inspired astrolinguistics message. This strategy attempts to establish a mental bridge *between* the chosen medium and opening informational format in order to support provisional recognition of the signal as an artificial construct—and thereafter, foster decipherment work through information theory analytics (and ideally, accessibility to the related main contents we endow with a specific meaning). Reconciling this relationship *between* the recognizable sign and signified meaning, may be our strongest meta-semiotic approach for communicating what we wish to say (and how we can state it)—context that can operate outside of the message contents which, as famously coined by the philosopher Marshall McLuhan, ensures that "the medium is the message." But what if the medium affects, or constrains, the measure of meaning we can convey as a message? What if the message we desire to convey is just too cumbersome?

In respect to unidirectional communique for posterity, future human societies could run the gamut from Palaeolithic Age to Space Age, or to prospective Silicon Age cultures, and therefore remain unpredictable for even our broadest forecasts of posterity, in addition to the respective commonalities we may infer as shared information. Moreover, they may inherit a storehouse of knowledge or "digital amnesia" from the 21st century, which will inevitably influence how our descendants can approach a prospective messaging device from their antiquity (assuming these constructs are recognized as artificial messages, and not as ancestral debitage). Thus far, the majority of our cultural practices that craft messages for unscheduled retrieval by a target human audience have focused on the transfer of random, experiential information *we would like to impart* for unknown minds (i.e., tailored encyclopedia and autobiographical materials about humanity, our cultures, and Earth), without actively engaging with the finer underlying meta-semiotic obstacles of this communicative challenge—as visible within the range of arbitrary mediums and data-encoding techniques employed in such time-binding tasks. Whereas we can cautiously rely upon some biological commonalities with the immediate futurity of human audiences to extrapolate upon, "we are running twenty-first century software on hardware last upgraded 50,000 years ago" (Wright, 2004), in addition to limited ecological properties and other inherited social contexts, we cannot trust such coincidences at all when supporting interpretive conditions for ETI civilizations.[25]

There are already several excellent volumes (Vakoch, 2011b; Vakoch, 2014) dedicated to theorizing about sufficient approaches towards establishing contact (preferably, a mutual back and forth exchange) with ETI using EM signal strategies that "step out of our brains" (Cabrol, 2016). In a related, serious scientific effort, numerous studies have also discussed successive attempts to safeguard deep geological transuranic depositories from future human intrusion, using material "markers" intended for unknown recipients[26] as living memory-retention schemes—many of which are variably reliant on the persistence of some inherited aesthetic and cultural conventions. Some of these approaches have been discussed within this volume, and elsewhere, as alternatives within our proto-lingual and aesthetic channels. However, when we examine some of the stated timeframes required for projecting information across intervals of geological time, we inevitably end up considering how to *future-proof* those essential communique for human posterity by adopting measures that treat the futurity as "aliens" or Neanderthals—that is, distinctive intelligent societies that may not be comparable with *our* own cultural lineages, scientific ontologies, or social customs.

As once articulated by the late sociologist, activist, and cultural theorist Stuart McPhail Hall:

> Representation is a very different notion from that of reflection. It implies the active work of selecting and presenting, of structuring and shaping; not merely the transmitting of an already-existing meaning but the more active labor of making things mean [Croteau and Hoynes, 2003].

This line of inquiry gradually transitions us towards the multidisciplinary fields of semiotics, the study of mental signs as something that stands for, or "re-presents," something else to someone else, "not in all respects, but in reference to a sort of idea … sometimes called the ground of the representation" (Peirce, 1931–1935). Peirce's use of the term "ground" refers to how the contextual form of an expression relates to the signified contents or functional concepts thereof between communicating parties. Traphagan (this volume) has already touched upon semiotics earlier, but it is worthwhile expanding upon his discussion for further context.

Generally, we do not think about "real" objects themselves, but rather the surrogate memory value, and types of signs, that re-present them to our minds in some capacity.[27] An example of how signs are frequently interchanged with, and substituted for, experienced objects, concepts, and even other signs, can be plainly seen within our endnote about the "Marsdial" artefact in highlighting the arbitrary "word" unit as a referential symbol. We do not need to physically point at the fourth planet from the Sun as a representational cue, rather we can rely upon merely uttering a shorthand syllable or inscribe a four-letter, Roman-script word (i.e., the sign) to convey what we intend to signify to an English speaker (though deducing the correct, interpretive context between the planet and confectionary good requires further social context for referents between sender and receiver). Use of this shorthand symbol, however, is restricted to those who understand the English language, or recognize the semiotic context.

In Peirce's logical model of semiotics theory,[28] there are three distinctive categories of signs: the index,[29] icon, and symbol. All of these signs (to varying degrees) function on a recipient's convention of familiarity with the signified object or concept to attribute meaning through the observers' elaboration from the given sign (Saint-Gelais, 2014; Denning, 2014). Taking Joseph Kosuth's *One and Three Chairs* as a visual exemplar for our Peircean categories, this artwork provides a notable, comparative perspective for this relationship between mental signs and their signified object through the use of symbols (the use of the arbitrary wording in the title, alongside the dictionary definition for a chair), and an icon (the chair photograph).[30] The word "chair" clearly does not evoke in the mind of a person unfamiliar with the English language (or related members of this language family) a description capable of illustrating the signified common object. The word certainly does not resemble the object. Our word is dependent upon the virtue of its relationship with other signs, cultural contexts, and familiar properties that rely on predispositions. Though some of these properties may be shared between communicating parties, the word itself remains opaque as it clearly does not draw to the mental foreground a notion of the actual object for a non–English speaker—they require a work-around medium such as a native language dictionary, or picture, to unlock the symbolism of our word, to then "awaken the similar thoughts and ideas" we wish to communicate as chosen content.

The second of these signs, the "icon," possesses a relationship of similarity to an object as it shares directly observable properties that should form the basis for any common inception of the artefact.[31] In our Kosuth example, a photograph completes this semiotic circuit, as a chair is a commonly experienced object already known to both parties, and therefore a recognizable property of the chair (conveniently depicted in a recognizable medium for both parties) should convey visual clues necessary for the observers to deduce context for what is being evoked. Icons possess this advantage of providing a common, tangible reference point for an observer already familiar with a chair structure and the communicator to hinge their assumptions of meaning upon.[32] If we point to an image of a chair, while speaking gibberish, the other party is at least able to recognize our conversation is related to the object, with body language perhaps supplementing our lack of a common language. Icons are, perhaps, the most reliable signs to use for distant communications practices in order to signify contextual meanings for other minds. But, as outlined above, their efficacy to do so is contingent upon pre-existing triadic relationships between the sign-signifier-signified elements, i.e., the icon, the interpreters, and intended meanings. As Vakoch (1998a) notes, these icons need not solely consist of pictorial or visual representations, with Batesian mimicry (a form of camouflage system within the animal kingdom) relying on a direct relationship of similarity between a known property of an ecosystem, or another organism's characteristics, in order to conceal or deceive other creatures.

Icon development is a proactive and generative field of semiotics research which continues to question the invisible assumptions that underlie the cognitive production of meaning and reception of signs within information-exchange systems. There have been several messaging projects whose usage of inscribed visual icons have aimed to convey semiotical information to distant spatiotemporal observers beyond Earth—cognitive experiments that practically stretch the mental bounds of human meaning-making material practices, i.e., the Goddard approach using interplanetary spacecraft with inscribed metal plates (Goddard, 1920). Readers may be familiar with the seemingly alien inscriptions included on the cover of the Voyager records—ideograms which aim to plausibly correlate a mutual measure of time, counting, spin speed, and conversion of audio frequencies into spectrographic displays. This use of etched insignia, that are both tactile and visual, provides a multisensory channel to access the encoded audio and pictorial media. However, the strength of some of these icons is bolstered by the fact that the record itself (and spacecraft), serves as an *accessible common object* in establishing a rudimentary convention of familiarity for extrapolation. Voyager spacecraft, essentially, function as our "cosmic chairs." This, of course, relates to Marvin Minsky's amusing proposal (Minsky, 1985) to dispatch a cat to ETI as a familiar common conversation starter for further mutual deductions. Similarly, the *Companion Guide to Earth*, intended to be placed in Geosynchronous orbit, relies upon establishing this convention of similarity, initially through the visual and tactile rendering of Earth's tectonic configuration across the archive's spherical shell. This rendering may then be compared with the nearby planetary surface, an iconic representation that hopes to correlate with two-dimensional global projections and future tectonic configurations "re-presented" inside for contemplation.

The Voyager icons, of course, stem from the earlier *Pioneer Plaque* inscriptions, both of which manifested from subsets of the same minds. However, these signs would subsequently come to be uncritically employed in other messages as a proven concept.[33] While this lineage of consistency may be beneficial for interrelating message deposits should several renditions be recovered, depictions such as Pioneer's Bohr hydrogen atom diagram may not correlate with ETI's—or future humans—graphic depictions for the lowest energy state of this particular element (if ETI even visually process information and represent it in a similar manner). By this, we are not implying that this symbol is unrecognizable *as is*, but rather to raise caution against our tendency to overly rely on and reuse standardized signs, now recognized by millions of humans as a matter of convention, without critically reassessing whether we believe they may still objectively reflect the signified property of reality they are designed *to represent*.[34] The very act of relying on historical conventionality, or the cultural familiarity of established icons, may unintentionally convey the meaning instilled in such signs as an idea, much like the maligned archaeological *ideograph myth*,[35] "a piece of knowledge, a piece of wisdom, a piece of reality, immediately present" for others to easily understand (Pope, 1999).

Discussions about communicating with ETI using EM signals have principally

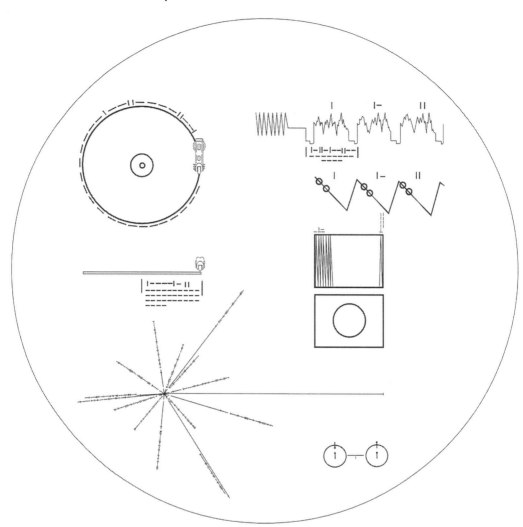

The Voyager Golden Record cover inscriptions. From bottom-right, clockwise: A diagram of the Hydrogen atom illustrating the change between the lowest energy states (spin moments of proton and electron establishes a common measure for time); a map of 14 pulsars and the frequency of their pulses (using prior time measure and converted into binary) which also points to our Solar System; instructions on how to play the physical record with provided stylus (guide denotes proper spin speed, and record duration); finally, depictions of the wave form of the video signals, including details on how to decode the images, and also how the first frame appears (a circle) (Graphic from NASA/JPL, re-illustrated by Nicolás González Montofré).

been shaped by the two generally opposing stances arising between the professions of the natural sciences (researchers who varyingly stipulate that establishing a two-way dialogue with extraterrestrials will be straightforward using mathematics and terrestrial science), and scholars of the humanities fields arguing the opposite position (Denning, 2006). These variable stances centralize around the premise of how we may signpost our signal, which initially remains as an unusual cosmic phenomena until it is recognized as an artificial message, using a peculiar pool of overlapping knowledge. This challenge is inherently semiotical. However, the jury is still out over the benefits, or

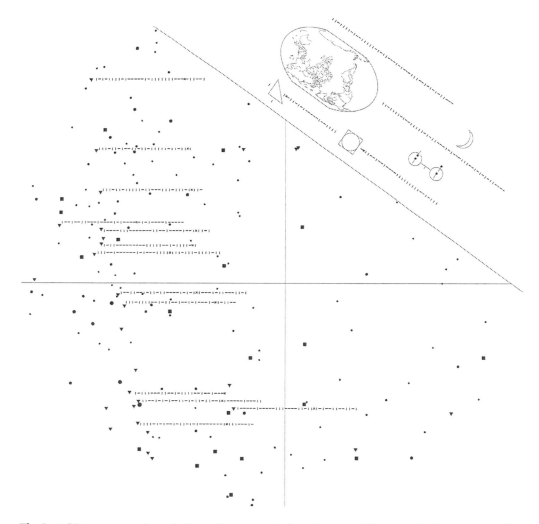

The Last Pictures cover inscriptions. From top-right to bottom: The same Hydrogen atom diagram depicting the transition between lowest energy states from the Pioneer Plaque (as a time measure using the "hyperfine transition" period between wave crest in these fundamental states); A projection of Earth's geopolitical map; Binary notation at the Southern hemisphere denotes Earth's spin axis rate; Crescent Moon notation denotes its orbit rate using to Earth-axis spin rates; Squared-Circle and Triangle are a verification system for universal geometric ratios to aid in interpreting math and symbols used. Finally, the larger portion of the cover illustrated a map of millisecond pulsars (triangles) with periods measured using Hydrogen notation, and locations for active galactic nuclei that emit radio emissions for tens or hundreds of millions of years (squares). Note the usage of black squares to signify bright sources of radio emissions in the cosmos (e.g., the active galactic nuclei), symbols which were initially proposed as a new international indicator to denote "radioactive waste buried here" in graphics for long-term communication proposals, as part of the 1993 Sandia Report for the WIPP depository (graphics ©Trevor Paglen. Courtesy of the Artist and Creative Time).

adequacies, of either proposed approach for message construction. In lieu of a "cosmic chair," or our capability to dispatch Minsky's cat to parts unknown ahead of a conversation with ETI, our principal mediums of choice (microwave and optical EM frequencies) present a challenge in terms of ensuring that the configuration of a self-contained

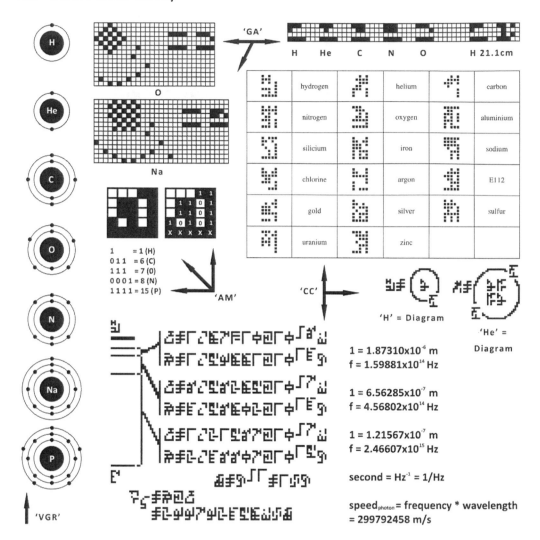

Redrawn graphical representations for atomic elements already contained in several physical and electromagnetic messages. Featuring numerical representations of elements (depicted in the Arecibo Message—AM), bitmaps as encoded symbols for elements (Cosmic Call's "alphabet"—CC), Bohr's diagrams (of the type used on the Voyager Record—VGR, and Greetings to Altair message—GA), a model which could be either a Rutherford or Bohr's diagram in the mid-right of mosaic (Cosmic Call), along with a bitmap at the center-bottom for the spectral signature of Hydrogen (Cosmic Call, previously illustrated in image #41) (Mosaic compiled by Paul E. Quast).

message reduces the risk of it being misinterpreted as a natural phenomenon. Freudenthal (1960), when developing his syllabus of Lincos, noted that certain characteristics of a transmitting signal could, in theory, be harnessed to represent formatting concepts—in his exercise, our concept of time through the signal's physical duration and regular punctuation thereof. Similarly, the physicist Philip Morrison would also later surmise:

> there is always one thing in common whenever there is communication; namely, the signal … by denoting in the signal itself we can make communication; we can invent a language, so to speak, by pointing. What we point at is not some other object. We point at and with the signal itself [Morrison, 1963].

This line of inquiry inevitably transitions us back into the contested territory between the natural and social sciences in how we can successfully punctuate—or modulate—this medium to reflect meta-characteristics that may, thereafter, serve as common referential points. Successful communicative approaches may one day oscillate between both camps but, at the risk of recursively dwelling on the principles governing the commonality of mathematics as a basis for artificial languages, we shall instead elect to focus on the use of semiotics within "pointing" in EM message construction.

In assuming that the underlying physical properties of chemistry are the same throughout the galaxy, as opposed to local celestial bylaws contemplated by the physicist Andrei Linde (Michaud, 2007), we can infer that the constrained array of elements, and their spectral emission signatures, might be a good place to start for expressing iconic messages (though, how foreign cultures represent these systems would likely be relative). Vakoch's studies, whose theories we have followed here, have extensively commented on the need for a signal to reflect the message concept it represents, rather than simply serving as a conduit for conveying more arbitrary cultural media. In his iconic approach for communicating chemical relationships and other baseline concepts (Vakoch, 1996; 1998a; 2008; 2010a), we are presented with a tangible correlation between the use of precise frequencies to mimic corresponding chemical emission lines, and how these constituents combine to create molecules in increasing complexity along a set chronology. The chemical equation for creating water is discussed as an example of this process ($H^{+}+OH^{-} \rightarrow H_2O$ or, in terms of radio frequencies, the hydrogen line of frequency 1420 MHz merges with a later appearing hydroxyl ion represented at 1640 MHz, to produce water at a new frequency line of 22 GHz).

If we could extrapolate from this approach, in context with the known common organic chemistry of interstellar space (Paterson, 2014), it may be feasible to begin building a particular compendium of chemical compounds, or an intersubjective "alphabet," as a system that possesses some semiotic relevance for terrestrial biology, human sensations, or synthetic applications. For lack of a better term, this would be a form of "multi-spectral Lincos" that is directly reflective of the building-block information we wish to convey, "content shown through form." Vakoch's manuscripts explore this format to express simple nuances and selective functions of terrestrial music that also reflect some of the perceptual characteristics of our species, in addition to "family trees" (limited models for transmitting genetic information from parents to child), predator-prey relationships, altruism, and other crude behavioral characteristics of terrestrial species.[36] It can be argued that the notion of a rational intellect at the other end of the cosmic phone call may serve as another common "object" of interest (Reijnen, 1990), but a rapport must be built before discussing abstract, detailed characteristics about unfamiliar human agents. As argued earlier in this volume, how would ETI understand the human form if they are not already familiar with it or DNA's structure? Bitmap human figures, such as those placed in the Arecibo message, Cosmic Call, and other public transmissions, may likely remain as opaque binary bits without a reliable means of articulating such abstract notions from signals (see Traphagan and Smith, this volume).

Building a significant syllabus of logical, step-by-step chemical tutorials, with Paterson's application in mind, may eventually allow us to physically *point* elsewhere beyond the signal to mutually-observable celestial objects of common interest (e.g., nearby interstellar clouds of dust, supernovae remnants, and ionized gases such as

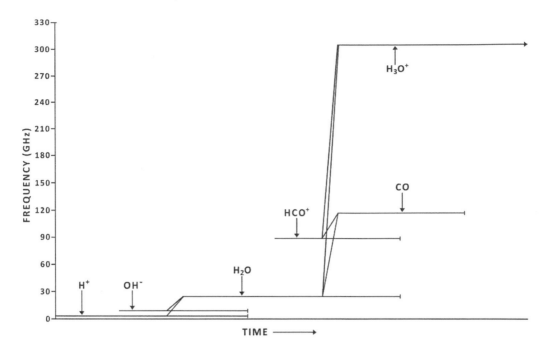

Along the transmission time vector (bottom axis L-R), the Hydrogen Line (H^+) of special frequency 1420.406 MHz, merges with a later appearing Hydroxyl ion (OH^-) represented at 1612.231 MHz, to produce compound dihydrogen oxide (water) at a new frequency 22.235 GHz ($H^+ + OH^- \rightarrow H_2O$). Expanding on this model, the Hydrogen Line and Hydroxyl ion frequencies soon end, placing focus on H2O. The new molecule Formylium (HCO^+), represented at the special frequency of 89.189 GHz is introduced (or may be built in a similar manner to the water model), before merging with the H_2O frequency to produce the ion Hydronium (H_3O^+) and also Carbon Monoxide, at 307.19241 GHz and 115.271 GHz respectively (H2O + $HCO^+ \rightarrow H_3O^+ + CO$). The CO frequency then discontinues, to place emphasis on the molecule of interest (H_3O^+) for further expansion. Besides Hydronium, all these denoted frequencies are identified as important astrophysical spectral lines by the SETI League and several international astronomical authorities for further scientific observation. Formylium has been found in the remnants of supernovae (for instance, Supernova 1987A), while Hydronium is often found in diffuse and dense molecular clouds (for example, Sgr B2 and the Orion-KL nebula) and the halos of comets like Hale–Bopp. The iconic message, essentially, attempts to establish a familiar semiotical foundation using signs for chemical phenomena that should be universally comprehendible, while moving the recipient's attention towards common referential objects for mutual extrapolation of the lesson (in this case, perhaps examining common astronomical objects containing these molecules together). Syllogism of these, or other similar molecular modelling, may be offered to reinforce the lessons (for instance, with an additional electron, H_3O^+ can break down into constituents—H_2O and H^+), in addition to repeating the leaps between chosen frequencies to ensure adequate observational windows are available to re/collect the entire signal (diagram illustrated by Paul E. Quast).

nebulae), incrementally building up common molecules that emphasize particular compounds, while also increasing the complexity of our examples to serve as an exposition of ideas. This could, theoretically, enable us to collaboratively foster competent semiotic exchanges, and allow us to peer into how other minds may structure their thoughts around the same reference points differently, rather than simply using such channels to communicate arbitrary information of meaningless or abstract intent. Perhaps the inability of our species to resolve the Fermi Paradox is due to how we initially approach communication with extraterrestrials as a form of dictation process for idiosyncratic meaning, rather than as opportunities for interspecies collaboration?[37]

It is likely that this content-poor approach of "multi-spectral Lincos" has been suggested before by both disciplinary camps and possesses several drawbacks unique to understanding the medium. Some of the obvious drawbacks include: the lack of clear information complexity for signal-processing algorithms, the prerequisite for ETI to simultaneously monitor a large cache of individual frequencies for select channel usage and, as a more human-behavioral issue, this slow approach may also be insufficient to meet the competing agendas of a number of disciplinary orientations that advocate for expedient communication (Denning, 2011b). However, it possesses a didactic capability to, in the very least, consider how to communicate signs for core metadata details about terrestrial biology (not just that of humans), simple epistemic frameworks for how we model our understanding of phenomenology and, as Vakoch's approaches explain, basic analogous processes using simple forms of narratorial operations—as intelligible contents for intelligent foreign cultures. Given the scope of representing everyone and no one in particular, this intelligible approach is certainly worth exploring as part of our extended preparatory toolkit in communication strategies for ETI.

Learning Lessons from "Contact" with the Past

To conclude, we'd like to break faith with the focus of our study and end with a cautionary tale—the story of Altamira cave in northern Spain (David, 2017). Around the year 1868, a local hunter encountered a narrow entrance along the side of a limestone hill leading into a cavity that had been revealed when a nearby tree toppled during a storm. In 1875, he would show this cavern to a local amateur archaeologist, Marcelino Sanz de Sautuola from the nearby town of Reocín, knowing that Sautuola was interested in such unusual sites. Sautuola's initial investigation of the cavity found unassuming debitage and strange black markings scattered across the cavern's walls and floor, but nothing distinctively grasped his attention at the time. While visiting the 1878 *Exposition Universelle* world's fair in Paris, Sautuola met the famous French prehistorian Édouard Piette and would observe several portable lithic artefacts from prehistoric antiquity that were recently unearthed within identified Palaeolithic settlements in French caves. Upon returning to his hometown, Sautuola logically applied what he had learned to study the nearby Altamira cavern. During the process of excavating and uncovering archaeological objects such as bone and tool fragments at the cavern entrance, his young daughter María began venturing deeper into the cavern. As the fable goes, she called out "Daddy, look oxen" to which her father then investigated and began to recognize a rich, polychrome ceiling of pigment, illuminated by the torchlight, to be magnificent zoomorphic depictions of bison and human handprints—a rich tapestry of parietal art dating back to the Upper Palaeolithic (Pike *et al.*, 2012), sealed away by rockfall for over 13,000 years.[38] (See color insert.)

After publishing an initial account of the findings in 1880,[39] Sautuola's interpretation of these paintings as products of Palaeolithic artisans was met with open hostility, ridicule, and derision by the French archaeological establishment. This was mainly due to the dominant viewpoint at the time that prehistoric humans lacked abstract thinking as nothing remotely similar to these wondrous depictions had been encountered before. Therefore, Sautuola's theory and findings were considered ludicrous. There were several openly disparaging, intellectual attacks on Sautuola's reputation, character, and work; one notable account by the eminent French Prehistorian Émile Cartailhac denounced the paintings as a forgery or hoax perpetrated on the cavern walls between Sautuola's visits. Only by 1895, when new, irrefutable evidence at the Grotte de la Mouthe cave in France literally resurfaced and corroborated the parietal art at Altamira, did scholars begin to seriously attribute these polychrome masterpieces to the ingenuity of Palaeolithic artisans.[40] Furthermore, it would be another century before we would begin to properly recognize how these paintings, as abstract signs of representational behaviors, related to the artisan's own lifestyles and social histories, rather than how these vestiges should integrate within *our* own symbolic histories, cultural traditions, and socio-ethnic paradigms.

This story crystallizes several salient and disheartening conclusions about how we are even able to cognitively, or perceptually, recognize intentional social mark-meaning processes from our own divergent ancestry. However, our principal argument with this analogy is to emphasize how quick we tend to impose our predispositions and interpretations onto perceptually-static artefacts and signs of communication that are the products of a distant intellect, completely and utterly distinctive from our own social histories—sapient minds that took alternative mountain pathways. While there were no other exemplars to contrast these parietal paintings with at the time, experts who together made up the vanguard of these intellectual traditions and disciplinary communities largely castigated the discovery. These reactionary decisions would impede the progress of knowledge into the early hominids of Europe for over a decade. María de Sautuola could recognize the marks as intentional, gestural signs endowed with a signified, representational meaning, given her familiarity with the form of a bison, facilitated by the innate cognitive faculties and cultural histories embedded behind this semiotic insight. But it is questionable whether we, *as an informed audience*, may be capable of recognizing a highly stylized notion (instilled within code-systems of language or semiotical convention, and rendered into material artefacts) that has been cast out into an expanse for intervals of time likely far longer than the short phylogenetic history of our species. There is no precedent for such cognitive and semiotical tasks in intellectual vacuums, only comparative analogies and imaginative theories arising from relatively recent episodes of studying human antiquity.[41]

Despite our ingenuity, we may simply fail to extrapolate any of the signified meaning from disembodied artefacts that have been far removed from their relative "alien" contextual environments. We should recognize that other minds, far removed from *our* familiar social, technological, cultural, and even phylogenetic contexts, may also find it difficult to cognize the indelible fragments of a civilization that they may not be familiar with. We may never know whether our object-archives, and other material culture now comprising the exoatmospheric archaeological record, may indeed be worth a thousand coherent words to minds separated by millennia, or deep expanses of space-time. But this should not be from the lack of serious effort on our part to study, scrutinize, and attempt to reduce the foreseeable problems for prospective observers. Demonstrating such altruism in our message constructs—or responses—may be a commonality we might actually share with distant minds, regardless of our relative divergences in cultural behaviors, attitudes, beliefs, and modes of expressing *what we think we may know* for other unknown minds to one day contemplate from afar.

Notes

1. Analogies should, ideally, be only used as initial ethnographic cross-comparisons, rather than directly applied interpretations arising from other cultural heritage "models." Unfortunately, the latter usage tends to prevail.

2. It is widely acknowledged that the emphasis for human cultures inventing "advanced technologies," such as radar detection and spacecraft launch systems, corresponds with the qualities of Cold War geopolitics, armaments, and defensive requirements. These narratives of conflict, defense, and safeguarding ideologies with political technologies, of course, permeate much of human history and antiquity. The idea of "progress" embedded within these political and technological trajectories is also very much a cultural dogma which correlates with the incremental pursuit of a mythical utopia, or transcendental *manifest destiny* narrative, that underscores the founding principles and theological basis for several Western nations (Bury, 1932).

3. For example, if we are to literally read into the operating factors comprising the now iconic *Drake Equation* for estimating the prevalence of ETI in our galaxy, we can reasonably conclude that our initial measure of an advanced, intelligent civilization lies within its technological capabilities to transmit EM waves that are detectable by *our* counterpart instrumentation; essentially it was a search for relatable elements of humanity.

4. As noted earlier in Traphagan and Smith (this volume), the principle operating assumption for SETI searches were based upon intercepting signals from civilizations that also employ EM communication equipment. While this was based upon our civilization's communication technology limitations, in conjunction with logical prepositions extrapolated from expected contact channels with ETI societies, SETI has since begun searching for a wider range of technosignatures, in addition to the "haystack" of traditional EM channels.

5. For example, why don't we seriously consider how ETI civilizations may choose to downscale themselves away from the envisioned post biological entities we forecast for our own futurity, and how these reduced footprints on their environmental stage may, or may not, manifest in exoplanet biosignatures?

6. Herein, natural languages are defined as any language (including speech and signing) that has evolved naturally through use and repetition over the course of human history, without premeditated alterations.

7. This "bottom-up" model of cognition, or data-driven modeling, assumes that information processing begins with the lowest, fundamental units of a language (phonemes, lexical units etc.), in order to build up a larger picture of the syntactic structure. In contrast to this, a "top-down," theoretical-based approach examines the overall structures of larger lexical assemblages, such as whole sentences, to ascertain how the complete structure, created by a fluent user, lends context to arrangements, and treats each individual unit, though differing dialects in poor corpus samples may introduce noise for both approaches. Modern translation efforts consist of a consistent shift between these two operations for ascertaining meaning (Fodor, 1983).

8. Our point here is that all languages possess different vocabulary for a common object, for example, "Mars" (the planet) which was inscribed in 22 languages on NASA's Curiosity rover *Marsdial*. All these translations culturally signify "Mars" to each of these respective human lingual audiences, though none of these lexical units physically resemble the fourth planet from our Sun. For instance, does Hindi मंगल resemble the planet, or word?

9. A frequently cited analogy for how different human cultures may articulate a common object differently (according to the disputed linguistic relativity hypothesis) can be found in the Yupik and Inuit root-word terminology for describing "snow." At the risk of harmonizing these cultures, languages, and differing dialects into a singular society, we'll simply state that each language possesses many more words for describing "snow," and the properties thereof, than other equatorial-based cultures that infrequently encounter snowy weather.

10. The story of Ishi is often characterized as a noble disciplinary effort to understand the oral histories of a Native American man who was deemed to be the last of his tribal customs, or a "living vestige" from frontier history. However, this colonialist prerogative underscored much of the interactions that scholars had with Ishi as a research subject, and initial difficulties experienced in interpreting his cultural expression through Western customs, while Ishi was also slowly acclimatizing to a new Western social lifestyle. Ishi was also likely not "the last Yahi," as with the mythos of the "extinct" Mayan culture; both regional societies likely still possess living descendants.

11. By using the term technology here, we refer to both the physical technological devices used for augmenting language, but also the mental conventions and habits that facilitate an on-going process of reinvention within a group's knowledge, beliefs, attitudes, and shared customs, alongside means of expressing or symbolizing these.

12. It is uncertain whether the corpus of Indus symbols even represents a traditionally defined language system.

13. The linguist Robert Millar once noted, "When a language is lost, there's also a loss of culture, a loss of the stories that were told in that language." Stories and other narratives provide one form of imperative context.

14. The Rosetta Stone is regularly cited as an example of our capabilities to translate extinct human languages. However, this resource was used in tandem with Jean-François Champollion's pre-existing insights into the Ancient Greek language and prior, corroborating knowledge he obtained from the Philae Obelisk transcriptions.

15. In addition to these evidentiary resources, Saint-Gelais (2014) notes that the common working assumptions behind successful decipherment campaigns stems from our pre-existing knowledge of these scripts as products of rationalized human agents with whom we share many biological and environmental properties. Therefore, we can project some preliminary experiences of familiarity onto some of these enigmatic scripts.

16. The myth over whether natural languages and other symbolic conventions may be capable of accurately disseminating meaning to a distant spatiotemporal agent is likely borne out of the perceived schism between our extensive decipherment experiences of unknown ancient scripts and the often-misplaced confidence we impose upon—human intellect extrapolated over the futurescape, natural language processing systems (a branch of machine learning that statistically analyses and translates corpora of known languages for computers to understand), forecasted AI capabilities, deep learning algorithms, artificial neural network advancements, and a reliance on these evolving technologies to resolve encountered technical issues for us. This predisposed reliance on our ingenuity may also be subject to what David Brin (2019) terms the "low-hanging fruit" technological milestone argument.

17. It is contestable whether a foreign spatiotemporal observer, when confronted with a limited corpus of terrestrial linguistics, may be capable of identifying basic sub structural components such as verbs, pronouns, grammatical rules for crafting sentences, and core temporal contexts that feasibly underscore all natural human languages. Regardless of these analytical capabilities, such approaches still do not equate with translating from enigmatic script to signified meaning, as discussed by one of the authors in the last essay of this volume.

18. It is worth acknowledging that both of this texts' authors also possess divergent opinions about the significance of this manifestation of anthropomorphism, and its relevance for prospective ETI communications. While researchers such as the astrobiologist Charles Cockell (2018) have argued "evolution is just a tremendous and exciting interplay of physical principles encoded in genetic material," and that these common physical forces may theoretically limit the amount of natural variation we could expect for ETI morphological features and cognitive systems, these constraints do not warrant the application of anthropocentrism, nor does it mean that such convergences may develop similar means of expression in limited sensory apparatus, cultural, or social histories. Discussions about alternative linguistic frameworks, in reality, are debates about our own languages and alternative frames of reference, not ETIs.

19. We refer to religious missionaries that were largely responsible for communicating linguistic and anthropological information about indigenous human cultures around the world to Western scholars. It could also be posited that *we are fortunate not to have any interstellar exo-missionaries,* as they were a source of the gradual dismissal of native spirituality, alongside other intricate parts of indigenous cultural customs and beliefs.

20. Astrolinguistics is defined as human language constructs for ETI. Xenolinguistics is defined as ETI language constructs for humans.

21. In addition to these points, Jonas and Jonas (1976) have speculated on how different sense modalities may lead to alternative perceptions of mathematics. The "Olfaxes," for example, may count in chemical concentrations and saturation percentages rather than the definitive, natural numbers we tend to rely upon. Metallurgy and radioastronomy also function on the principles of measuring concentrations and percentages as scientific data.

22. Frank Drake, the creator of the *Arecibo Message*, once acutely noted of Freudenthal's rendition of Lincos that "it assumed that the recipients have brains and logic very similar to ours." It has also been acknowledged that Freudenthal's version of Lincos was created too early to avail of nascent insights into language theory developing during the cognitive revolution or early experimentation within computer programming languages.

23. This is a simple information parsing or punctuation strategy for determining the constitution of individual characters from a perceptible mess of binary EM data. Herein, 8-bit refers to using different 8-character units to distinguish different concept units (e.g., 000000000 = 0, 00000001 = 1, 00000010 = 2, 00000011 = 3 etc.).

24. For example, and in relation to Dunér's initial statement, a copy of this essay, etched in Roman script on a metal plate, does not possess any tangible relationship to the choice of this data carrier. Likewise, the arbitrary choice of a metal plate doesn't provide any interpretive clues to corroborate the exact meaning or intention of this script, and by extension, what our "message" intends to convey as informational context to recipients.

25. These ETI audiences will inherently frame messages under their own familiar cognitive, social, technological, philosophical, and epistemic reference points, alongside their foreknowledge of comparative biomes and biota.

26. Transuranic wastes denote heavy radioactive isotopes, with atomic weights greater than uranium, that result from civilian, military, and industrial fission energy services, defense systems, medical appliances,

and other technologies. Some of the isotopes have half-lives of many millennia, such as Plutonium-239 with 24,110 years, and therefore need to be isolated for intervals of time exceeding 10,000 years for this material to naturally decay into an inert state (Trauth *et al.*, 1993; Lomberg and Hora, 1997; Benford, 1999; Stothard, 2016; RK&M, 2019). While ethically questionable, the uncomfortable experiences of early radiation sickness, as an iconic cue, would perhaps serve as the strongest deterrence to passively safeguard these sites, rather than solely relying on inscribed symbolic media and atomic priesthood customs. This concept of a hazardous "curse" or superstitions as an omen for preventing intrusion on ancestral tombs is comparable to this proposal, though modern scientific excavations have shown little hesitation in ignoring folklore or mythology inscribed across burial sites. Radiation sickness, however, may serve as a "curse incarnate," preventing further excavation should a site be disturbed.

27. For instance, drawing the readers' attention to the pen in front of me, brings forth a mental image of the signified object. The reader actively interprets my substitute sign "pen" (a symbol for a previously experienced object), to understand what I wish to signify to them, despite not directly experiencing my real-world object.

28. Peirce's semiotics theory proposes more than 76 typologies and definitions for signs that may largely be distilled into three canonical typologies, *icons*, *indices*, and *symbols*, which we will discuss here (Peirce and Buchler, 2011).

29. Dunér (2014; 2018) discusses the advantage of indices (as the most basic sign for something else) while coordinating observational behaviors for commonly experienced, interstellar phenomena in METI strategies.

30. A photograph, as an icon, only functions as such if it possesses a familiarity with the material object in the mind of the recipient. It therefore possesses a cultural conventionality for beings that would consider sitting on it. As it is absent from this example, a valid index for this chair may be found within the physical impressions of chair feet left on a carpet surface around a table, indicating a familiar four-pattern geometric void. It is worth noting that such an index can also function as a symbol, as interpreting it is contingent upon the virtue of an associated familiarity with another object (the table). The missing chair impressions also serve as yet another index to denote that someone has been there to remove the chair, likely indicating its utilitarian use elsewhere.

31. Saint-Gelais (2014) has observed that the use of a perceptible similarity is also a form of convention, and that we cannot rely upon such devices to be shared by a distant, spatiotemporal observer.

32. Of course, the conventional familiarity of signs for a chair, or the actual object of material culture, is wholly contingent on the needs of our morphology to use such a device—would we expect the *Star Wars* "Hutt" species to understand the purpose or associated measurements of such a structure, in isolation from bipedal organisms?

33. The *Pioneer Plaques* diagram of the Bohr hydrogen atom has been used within both VGR, The Last Pictures disc, and several other crowd-sourced messages. By contrast, the Cosmic Call transmissions elected to use a Rutherford diagram and low-resolution emissions spectrum to convey, and extrapolate from, this concept. Stephen Wolfram (2018) has also noted that this Bohr diagram could have been replaced with a 21 cm line to provide a better representation of this element's emission wavelength for use within the pulsar map.

34. Another example of this indiscriminate re-use of familiar iconic designs can be seen within Sagan's own declination letter to participate in the WIPP study, in which he advocates the adoption of a "skull and crossbones" emblem due to its historically (and contemporary) negative connotations for Western and nautical audiences.

35. Previous scholarly conceptions of the ancient Egyptian hieroglyphs, and its counterpart Mayan glyphs, considered these signs as symbolic representations of ideas without an associated phonetic language. This "ideograph myth" would, thereafter, inhibit productive decipherment strategies of these symbols for decades.

36. Vakoch explains that these rigorous approaches using EM frequencies are simple, but abstract, modeling for general processes that are akin to behavioral traits of terrestrial species—as opposed to directly representing these complex characteristics which would require an ETI to infer about humans from this symbolism, in addition to the caches of comprehensible redundancy examples. In this respect, these models resemble earlier approaches adopted in Freudenthal's Lincos using meticulous mathematical concepts, examples, and syllogism.

37. This is, of course, premised on the theory that ETI are simply waiting for us to send the first signal. This "send first" policy has been hypothesized by some as a necessary socio-technology benchmark for initially establishing communication, but it also opens the door for a "why should the younger species message first?" counterargument by those who are critical of METI activities.

38. The oldest signs on these cave walls date to the Upper Palaeolithic period of around 34,000 BCE; however, the cave presents hallmark evidence for periodic seasonal occupation over 10,000 to 20,000 years.

39. A digital copy of this manuscript, *Breves apuntes sobre algunos objetos prehistóricos de la provincia de Santander* ("Brief Notes on Some Prehistoric Objects from the Province of Santander"), is available online, courtesy of the *Centro de Estudios Montañeses* (Sautuola, 1880).

40. Sautuola was credited with the discovery of the first Palaeolithic cave paintings in Europe, but he would not live long enough to receive the initial acknowledgment, or indeed Cartailhac's 1902 public retraction, having passed away 14 years prior to the formal written announcement of his family's discovery. María would, however, receive this apology from Cartailhac when he first visited the Altamira cave in 1902.

41. Further to this often-used comparative analogy of parietal art with space-time messages, it is worth highlighting the context of these sites as significant social places for the exchange of ideas, thoughts, objects, waypoints, and customs between tribe members and, perhaps, other visiting groups as a surface for transient socializing practices over seasons, with surrounding paintings arising as a result of these interactions over intergenerational timescales. By comparison, our space-time message deposits seem rather novel in that they are largely rationalized and manufactured during brief timeframes, by small cohorts of like-minded individuals who broadly share in the aesthetic, social, and cultural conventions that frame these intelligible constructs.

The Pale Blue Dot. Featured in *A Simple Response to an Elemental Message* as image #3, along with *Lone Signal* crowd sourced media (photograph from NASA/JPL—Caltech, frames originally processed by Carolyn Porco [PIA00452]).

An ESA technician carefully preparing the Ferrari "Rosso Corsa" paint sample in a specially constructed glass globe—an artefact included aboard Mars Express as the "Red Encounter" project (photograph courtesy ESA/ESTEC and Rien Veefkind [technician]).

The Hubble Ultra-Deep Field (HUDF) captured light emitted by distant galaxies formed about 13 billion years ago, taken over a cumulative period from 3 September 2003 to 16 January 2004. Featured in *The Last Pictures* as image #50 (photograph from the European Space Agency—NASA—STScl/HST).

The "Visions of Mars" silica-glass DVD, containing a rich archive of science fiction material relating to Mars, deposited on the Martian surface by the Phoenix lander. Much of this material was originally compiled for the failed "Mars 96" mission. Re-entered DVD remnants are presently at the bottom of the Pacific Ocean (photograph from NASA/JPL/Lockheed Martin—The Planetary Society).

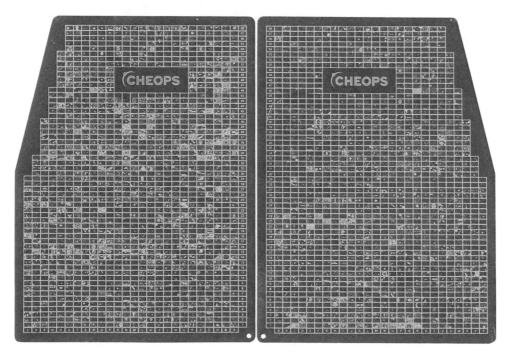

The CHEOPS Plaques—two titanium plates etched with thousands of miniaturized drawings which had been made by children before being fixed to the CHaracterising ExOPlanets Satellite (photograph ©Guido Franz Bucher/Bern University of Applied Sciences and Timm-Emanuel Riesen).

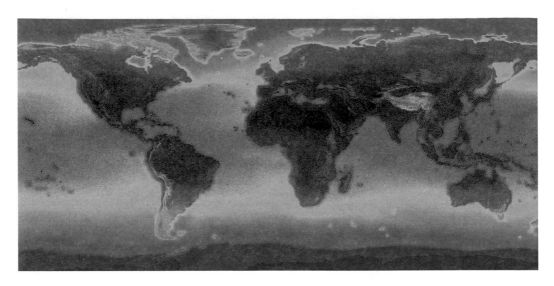

NASA climate simulations which forecast how global temperature and precipitation levels might change up to 2100 under different greenhouse gas emissions scenarios. Featured in *A Simple Response to an Elemental Message* as image #19 (graphic from NASA Centre for Climate Simulation/Vital Signs of the Planet).

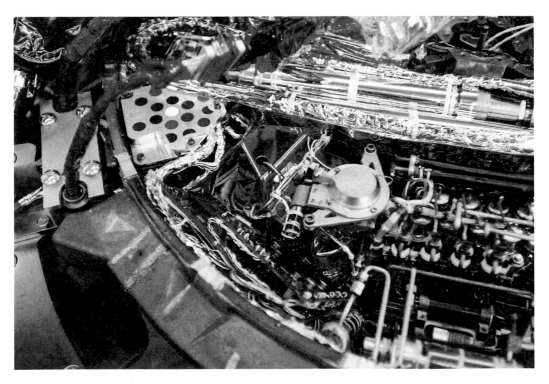

The Beagle 2 color "test card" designed by Damien Hirst at the request of Colin Pillinger as a calibration target plate for the probe's cameras and spectrometers. After entering the Martian atmosphere, contact with the Beagle 2 lander was permanently lost. Later observations by NASA's Mars Reconnaissance Orbiter would reveal that the spacecraft landed safely but could not deploy its communication equipment to receive commands. Since most of the spacecraft unfolded as designed, the calibration plate is likely exposed to the Martian environment (photograph: All Rights Reserved, Beagle 2. Image courtesy Judith Pillinger).

The Sounds of Earth collection, B side of the Voyager Golden Record featuring all musical compositions, earthly greetings, and the sound essay (photograph from NASA/JPL/Caltech and the Library of Congress/"Finding Our Place in the Cosmos" Collection [Item: cosmos000115]).

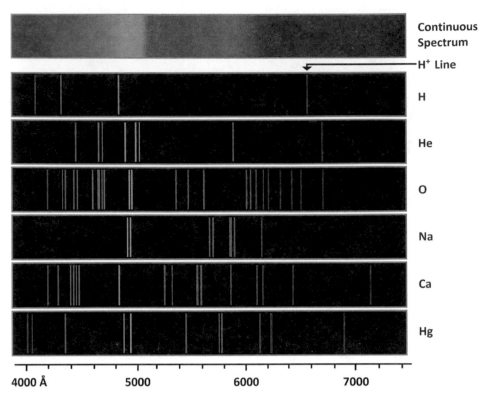

Continuous
Spectrum

H⁺ Line

H

He

O

Na

Ca

Hg

4000 Å   5000   6000   7000

Simple emission spectral signatures for elements Hydrogen, Helium, Oxygen, Sodium, Calcium and Mercury (illustrated by Paul E. Quast).

The polychrome ceiling in the cave of Altamira (©Museo de Altamira, photograph by Pedro Saura).

MER Opportunity rover heat shield discarded at impact site. Opportunity photographed this debris after beginning its mission objectives, but such observations poignantly raise questions about the quantities of material debris (and other forms of contamination) now accumulating across other local astronomical bodies (mosaic from NASA/JPL/Cornell University).

Alan Bean descending the Intrepid descent stage ladder (photograph from NASA/Pete Conrad [AS12–46–6726]).

"Starman" mannequin wearing a SpaceX Spacesuit in the driver's seat of Elon Musk's own Tesla Roadster. According to orbital evolution simulations, modelled by Rein et al. (2018), the vehicle poses long-term collision probabilities with Earth and Venus (≈20% and ≈12% respectively), in addition to close encounters with Mars, Mercury, and the Sun over the next 15 million years (photograph from Space X/Wikimedia Commons, CC0 1.0).

Image of MISSE 3 Tray 1 on the external façade of the ISS, taken on 13 August 2007 after 1 year of space exposure, and shortly before retrieval (photograph from NASA [ISS015E22411]).

Part 3

READING OUTSIDE
THE MESSAGE

Per aspera ad astra

But Does One Really Speak for Them All?

Klara Anna Capova *and* Paul E. Quast

Let us reimagine the *Voyager Golden Record* (VGR) as a museum exhibit (Capova, 2021). The spectrograms are made readily accessible to visitors in the form of a photographic display, along with a transitional soundtrack permeating the gallery, as our spectators view and contemplate this material culture in the original recorded sequence.[1] For the purposes of this exercise, let us also assume the visitors are broadly familiar with our socio-cultural framing conventions and share in our morphology, cognitive capabilities, and other complementary sense modalities (including sensitivity ranges) necessary to perceive and access[2] the subject matters depicted in this media. A sense of familiarity with our mental-material cultural practices might also be advantageous for our visitors during this thought experiment. The visitors have just finished a walkthrough of this museum display, taking time to scrutinize the photographs that document cultural lifestyles and material artefacts that were readily available within American media catalogs from the 1970s. We know from the author's accounts *how* the VGR intended to speak on behalf of Earth (Sagan *et al.*, 1978; see also Scott, 2019). However, from our perspective, and in a rather counterintuitive sense, what subjects, themes or content might the visitors *not glean* from this visual material culture? What aspects of Earth remain opaque, ambiguous, or are simply missing from *our* interpretation of this *cabinet of curiosities*?

Firstly, we do know that the Voyager compendium authors (due to the project's tight deadline, persistent yet tacit risk of censorship by the launching agency committee, and limited data storage capacity) did, in fact, choose to physically omit reference to several complex subjects of human knowledge, alongside electing to present our species under favorable circumstances to the cosmos. People who work within the museum sector are likely familiar with this causality of balancing budgetary constraints, popular public interests (with variable reactions), and academic research, against other imperative resources in order to mount an informative, pedagogic exhibition. Many photographic exhibits and evidentiary resources, in turn, are derived from larger, more extensive collections, which subsequently remain consigned to the storage facility (or cutting-room floor) and, therefore, remain inaccessible for the visitors to interact with—in favor of re-contextualized extracts that may be far removed from the media's original meaning. While these reductive selection criteria are acknowledged within the VGR catalog, such conscious choices within the initial curation of message contents may playfully convey something about our meta-ethics, moral codes, and decision-making processes. But it could be equally revealing of a tendency to be conservative in releasing specific information, or demonstrate a penchant for self-censorship (at least for NASA's preceding administration), leading a visitor to speculate what else may have been intentionally concealed.

We will discuss these lacunae in context with the VGR momentarily. However, it is incumbent of the authors to state that we hold this generational mosaic of Earth in high regard as it continues to serve as an archetype in establishing the broader cultural, democratic, and ethical arguments of the messaging debate and, by extension, the rationale for the great transmission debate (Denning, 2011b).[3] Insights derived from this now popular and frequently acclaimed artefact of material culture may reveal faint glimpses into how terrestrial audiences believe humanity should be known across spatiotemporal distances within *de novo* projects, or as part of sanctioned ETI response messages—coarse, underlying indicators that may subsequently communicate more between (or in this case, outside) the lines than we think.

Therefore, the purpose of this concise study is not to disparage the VGR, but rather to briefly analyze how subtle decisions made during its production may have influenced the subsequent lacunae and reasoning present within the biographies of successive messaging projects. While in principle, the VGR follows in the pre-established footsteps of millennial time capsules (Jarvis, 2003), as opposed to the "postcard" format of its predecessor *Pioneer* and *Arecibo* messages, the contents served as an interdisciplinary discussion for envisaging the challenges associated with representing snapshots of our world—and the conscientious effort that should be exerted in attempting to define our planet rather than solely the Anthroposphere (or hegemonic cultural subsets thereof).

A Walkthrough of the Voyager Compendium

The Voyager compendium, accompanied by a musical selection, originates from various, recognizable, and unfamiliar places and cultural backgrounds, while presenting introductory elements of these heritages and lifestyles as a propaedeutic guide rather than a detailed, encyclopedic account. For example, the limited musical performances on the VGR include a classical "the best of" Western cultural collection, popular rock-and-roll songs, and indigenous compositions from various geographical areas[4] (Ferris, 2017). The audible part of the VGR also includes greetings spoken by multinational citizens of Earth in 55 languages (with one vocalized by whales),[5] and around 21 audio recordings of human activities, biota, machines, and natural phenomena as the *Sounds of Earth* collection (Lemarchand and Lomberg, 2011). As our visitors walk through the exhibit and read the informative annotations superimposed over some of these pictures,[6] they also simultaneously hear this soundtrack and other acoustic recordings playing in the background as a multi-sensory experience of Earth.

The pictorial sequence begins with black and white diagrams and continues through human anatomy, to some colorful depictions that emphasize human activities (e.g., care for other humans, tool production, architecture, nature, animals, and the creation of material artefacts), before expanding into properties of sociocultural technology and concepts of civilizations. The final picture of this sequence was dedicated to Beethoven's *String Quartet No. 13 Cavatina*, according well with the placement of this very composition as the last musical recording. At the end of the exhibition, the visitor may be fascinated by the wonderful diversity of nature and cultures and feel charmed, captivated, or perhaps overwhelmed by the dexterity of this exhibition. Some visitors may think that this purposeful microcosm gives an overview of our world, but this account may not be as comprehensive as we believe it to be. With this in mind, let us now elucidate some of the obvious, absent elements of humanity and Earth from this compendium.

Sifting Through the Cutting Room Floor

Despite the attempt to describe life and its intricate cycles, there is one substantial factor that was missed almost entirely. For clear reasons in wishing to introduce terrestrial biology, a lot of attention was focused on the creation and evolution of life and maturation of the human fetus and birth, yet the VGR does not contain any relevant information pertaining to the concept of mortality, or biological death of human beings overall (Capova, 2008). While this is largely the case throughout the VGR, there are several nuances that abstrusely address this concept of mortality.[7] However, without delineating this fundamental fact of life as a natural part of the maturation cycle, the message cannot be understood as a complete story for the phases of human life, or indeed terrestrial biology. Furthermore, there is much more emphasis placed on the creation of life, gestation, and early developmental phases, but the later stages of aging are not as extensively explored. Rather, it is only suggested in some images, and numerically corroborated in annotations on others.

Terrestrial time capsules embody the stylized, ritualistic behavioral customs associated with the ceremonial entombing of the vessel (or burying of a decedent) in a purpose-built crypt or within soil, for the purpose of serving as a surrogate memory for the author's life (or as posthumous effigies). As discussed earlier in the volume, this symbolic cycle can also be asserted for dispatching fragments of human thought into outer space as much of our messaging activities permanently remove these artefacts from earthly hands, leaving only a residual memory-legacy of these activities in their wake for the minds that they inherently meant something to. Despite this symbolic, material relationship with the concept of death (and nuances of "rebirth" in rediscovery), messaging projects seem to rarely document how terrestrial life tends to shuffle off the mortal coil directly within informative contents (Jarvis, 2003). Outside the VGR, there have been arcane yet appreciable references to this nuance of mortality apparent in language-based submissions to crowd-sourced interstellar radio messages (gestural projects that will likely remain illegible[8]), alongside occasional effigies deposited as votive memorials for deceased astronauts and other individuals. However, the direct notion of death (as discussed, informative contents) remains predominantly absent from the record of comprehendible material culture in outer space—perhaps revealing a comical, psychological denial of this mortal coil as something that only happens to "non-self" others (Dor-Ziderman et al., 2019).

While walking through our VGR exhibit, a viewer may also be confronted with a perceptible absence of the negative characteristics of humankind—an artificially contrived, sanitized perspective that was purposefully engineered to convey, as Sagan explained, "a hopeful rather than a despairing view of humanity and its possible future" (Sagan et al., 1978). The VGR intentionally favors and illustrates positive values,

A memorial plaque for the Space Shuttle Columbia crew, mounted on the back of the Spirit rover's high gain antenna—an example of mnemonic artefacts to commemorate deceased individuals. Spirit's landing site has also been designated the Columbia Memorial Station as a further effigy (photograph from NASA/JPL/Spirit Rover).

attributes, and progressive developments within contemporary human societies as a "smile for the camera" approach (Scott, 2019), while omitting much of the global concerns that citizens of Earth are readily familiar with and experience on a day-to-day basis: war, colonial oppression, prejudice in all forms, biodiversity loss and environmental degradation, famine, disease, and abject poverty. In fact, perhaps the only negative connotations to be found in the VGR may ironically be two sentences in the statement provided by the then-serving president of the United States.[9]

This influence of speaking for Earth in unnaturally positive, golden overtones, or the notion of archiving "the best of ourselves" (Schmitt, 2017), has subsequently been at the forefront of descendant messages and consensus-building projects to date, despite other (theoretical) initiatives (Michaud, 1992) pointedly stating that sincerity should be Earth's best diplomatic policy for ETI communication attempts: "An acknowledgment of our flaws and frailties seems a more honest approach than sending a sanitized, one-sided story" (Vakoch, 2009). The late astronomer and SETI proponent Jean Heidmann (1993) had also voiced a similar preference for including human imperfections in ETI messages, noting that undermining the true nature of human reality is a frivolous attempt to conceal disconcerting aspects of ourselves in pursuit of presenting favorable (and delusional) human virtues. Such decisions inherently possess implications for whom we imagine to be the audience for these messages— agents of fantasy that will only superficially understand humanity *on our terms*, or another rational intelligence that may be equally divided, morally conflicted, and intent on resolving similar social

Operation Crossroads "Baker" underwater nuclear weapons test at Bikini Atoll on 25 July 1946, 10 seconds after detonation. This was one of the first tests conducted after the Hiroshima and Nagasaki bombings in 1945, targeting a fleet of captured and redundant warships. Most survived the blast but became highly contaminated by the radioactive fallout from the water plume. A similar image of this test featured in *The Last Pictures* as image #61 (photograph from the Library of Congress/United States Department of Defense [U.S. Army/Navy]).

and ecological issues as those experienced on Earth today. Furthermore, and in taking the position of the numerous appeals in crowd-sourced messages to seek salvation, or elicit aid, from extraterrestrials to resolve exigent earthly affairs,[10] surely it would be worthwhile to acknowledge the issues that need to be addressed by a benevolent, altruistic alien race?[11]

Outside of the VGR, there have been several modern projects that have addressed the intentional purification of human behaviors, values, and nefarious actions. Perhaps the most discernible exoatmospheric artefact presently conserving visual content of this nature is *The Last Pictures* aboard the Echostar XVI satellite. Paglen's concise collection of 100 significant and "absurd" photographs introduce human historical contexts which chronicle (or alludes to) several moral and social failings, technological travesties, and other dubious legacies arising from modern human proliferation. These subjects include our development of nuclear armaments, wayward genetic engineering, and environmental catastrophes that are all familiar hallmarks of the 20th and 21st centuries, alongside presenting the reality of life for many indigenous and minority populations with which we share this planet.[12] While the human connection is symbolically insinuated through the curation of selective imagery with ecological overtones, the ASREM photographs also document a series of environmental impacts arising from anthropocentric terraforming activities in the terrestrial biosphere, with some of these legacies illuminated in more tangible depictions than others. Without a robust index of curated contents chosen to "speak for Earth," it is difficult to ascertain whether these subjects are even peripherally touched upon within information-dense encyclopedia accounts.

As we continue to cross examine aspects of the VGR exhibit for missing or opaque themes, our visitor may become acutely aware of the subtle, anthropocentric undertones

Grinnell Glacier, Glacier National Park, Montana, 1940. According to the US GNP library archives, the original caption is: "Grinnell Glacier, close-up, side view. Taken during search party for lost person on Garden Wall." Featured in *The Last Pictures* as image #93 (Photograph from U.S. Glacier National Park Archives, with assistance of librarian Anya Helsel).

Grinnell Glacier, Glacier National Park, Montana, 2006. Featured in *The Last Pictures* as image #94 (Photograph from U.S. Geological Survey/Karen Holzer).

associated within the selective use of interspecies relations, depictions of linear evolutionary context, and illustrations of the natural environment as a means of narrating the story of humanity. With some exceptions, much of this visual material was purposefully orientated towards providing vital context for the lifestyles of the specific, sentient species which manufactured the VGR, depicting our curious documentation of other

Standing by a giant Sequoia log near Generals Highway, Three Rivers, Tulare County, c. 1910. A cropped version of this photograph featured in *A Simple Response to an Elemental Message* as image #14 (Image from the Library of Congress/Historic American Buildings Survey/Historic American Engineering Record/Historic American Landscapes Survey [Reproduction Number: HAER CAL,54-THRIV.V,2–17]).

organisms (framing of photographs under Western aesthetic paradigms suggests the focus is on human actions in the shared layouts: photographing, hunting, measuring organisms etc.), human reproduction, biology, sustenance consumption, and technologies, alongside cultural behaviors and interactions with our fellow *Homo sapiens*. Such pictorial representations generally focus on the human agents as conscious entities, intently aware of the spectator's gaze, while conversely framing the other organisms we share some of these frames with as passive automatons, scarcely engaging with the camera lens.[13]

Such depictions are essential for contemplating the apparent stewardship role and apex intelligence of the human protagonists (in addition to presenting some intriguing insights into our prescribed psychologies), but it is likely an independently evolved, extraterrestrial agent may also be incredibly interested in our biome and the broader bio-ecological interactions that have led to the evolution of complex life on Earth. Further to this, the previously mentioned depictions of human biology and lifestyles are often divorced from their broader dependence on interspecies relationships and processes of symbiosis, a discord which illustrates humans as entities that can use, and reside within—but are decoupled from—their broader natural ecologies. This may be revealing of a schism in how we socially assign an elevated value to the modern artificial interactions of our species as something categorically separate from the natural cycles of the biosphere system. But it may also portray humanity as a domineering force within planetary custodianship, or perhaps even as authoritative planetary consumers.

As discussed by one of the authors elsewhere (Quast, 2019), this human-led ecological partiality is also apparent within the caches of other space-time capsules and interstellar radio messages—contents which serve to reinforce humanity's perceived stewardship values, while illustrating other terrestrial biota as a backdrop for our species' evolution and techno-cultural ingenuity. The 1983 *Greetings to Altair* transmission featured multiple low-resolution bitmaps illustrating familiar invertebrates and vertebrates which, while these depictions illustrated select organisms in isolation, predominantly constructed a linear evolutionary narrative for the morphology of *Homo sapiens*. This is quite reminiscent of the "walking primates to modern man" diagrams, depictions which have been recognized as inaccurate, anthropocentric tropes for decades.[14] By contrast, the ASREM library of photographs provided a counter narrative for this grand theory of a "human planet," by depicting dozens of unique organisms and biomes as the focal subject of contemplation, alongside the imperfections of human stewardship activities. However, while access to this visual material is questionable at best, this cache of perceptibly random organisms is far from extensive, nor is it wholly representative of the diverse ecologies, morphologies, and complex webs that comprise even simple biotic networks.

As our viewers finish the VGR exhibit, they may contemplate why this assemblage was diligently compiled, inscribed, encoded, and cast into the interstellar darkness. In context with residual spacecraft properties, they may factually conclude that the VGR served as a unidirectional message to depict humanity and its expansive curiosity, exploration, and desire to learn about the cosmos—how we believe *others should know about us* from afar, conveniently *illustrated by us*. Absent from this autobiography is our innate desire to ask questions about the prospective recipients; their evolutionary histories, lifestyles, biosphere, and cultures; and whether they may read the operating manual of reality in the same way. The majority of our message seems to be only peripherally

Opposite: **Greetings to Altair, bitmaps #5, 6, 7, 8, 9, 10. For last "humans" image, see Traphagan, this volume (bitmap courtesy Shin-ya Narusawa, redrawn from originals created by Professors Hisashi Hirabayashi and Masaki Morimoto. Re-illustrated by Paul E. Quast).**

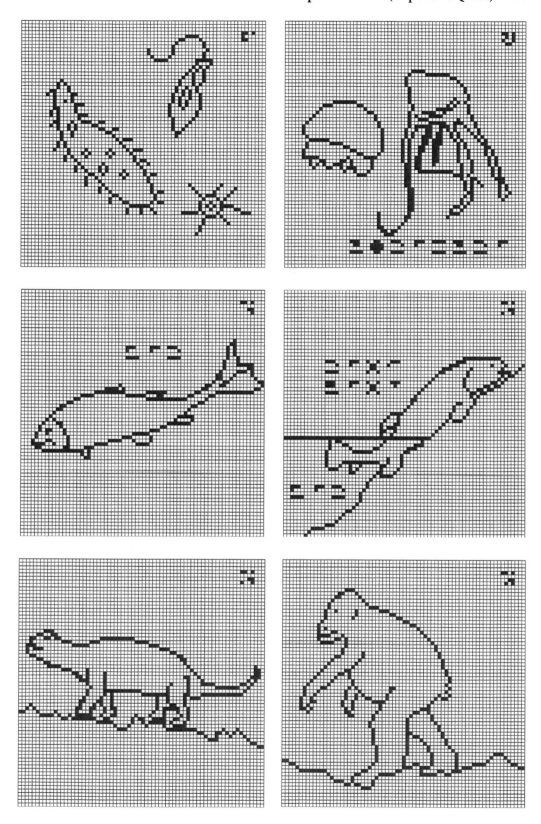

concerned with discussing ourselves, and what *we wish to convey* to the unknown futurity. It is essentially a directive, rather than an inquisitive, information device.

It can be argued that questions form one particular socializing function that, much like Chomsky's comments regarding the roles of particular internal and external languages, is obviously not representative of the entire communicative faculties exhibited by our species. Nevertheless, posing questions to elicit information is an integral aspect of the human experience and our diverse communicative repertoire. Asking questions as inquisitive, socializing devices could perhaps be apparent in ETI cultures—that is, if we continue to contrast exemplar human cultures and equate attributes such as intellectual curiosity and social concepts of reciprocity with ETI's dispositions. But let us momentarily digress to consider that questions, or the processes of information inquiry in general, are not apparent in foreign cultures for a brief thought experiment.

As the VGR was dispatched (as a mnemonic gesture) to physically surf interstellar space for the foreseeable billion years, and there is an infinitesimally small chance that it would even be intercepted, it is perhaps inherently futile to believe that humankind would still be extant on Earth (or elsewhere) to receive hypothetical responses to the record. Ergo, why should we bother posing questions? The lifespan of independently-evolved civilizations, as one of the answers to the Fermi Paradox (Webb, 2002), perhaps do not endure long enough to overlap with one another's expanding spatiotemporal bubbles—especially if *slow* material artefacts are the preferred vehicle for transferring cultural property between stars. By contrast, interstellar radio transmissions have been pioneered to hasten spatiotemporal dialogue with unknown denizens; messages aimed directly at confirmed exoplanets and nearby stellar systems, which ideally aspire to elicit a response within decades, as opposed to millennial timescales.

It might seem obvious to state that sending these "postcards from Earth" implies our intention to commence interstellar dialogue, but our inquisitive desire to receive a response is not explicitly outlined within this transmitted content; it is assumed on the basis of our principle of reciprocity and shared habitual customs, social developmental history, and our theories of mind when interacting with fellow members of *our own species*. It is often misstated that "even children pose questions and reciprocate in conversations," but it is worth pointing out that *children are also residents of these same, intertwined human systems*. John Gertz (2016a), a vocal opponent of the METI enterprise, has argued why we believe extraterrestrial civilizations should feel compelled to send responses as part of a cosmic *quid pro quo* if they have already received a copy of the Terran encyclopedia,[15] while the prominent METI proponent Douglas Vakoch (2012), has whimsically voiced concern that ETI may behave like intelligent cats—"they know we're here, they just don't care."[16] What would the outcome of our messages be in these brief scenarios? Certainly not "contact" as we regularly describe it or portray it in films. Lessons gleaned from simulated exchanges at the Contact Conferences may prove insightful, in the very least for scenario testing and identifying similar human assertions.

While the nature of METI is subject to extensive ethical debates, and disregarding whether an extraterrestrial interlocutor may share in our preferences for a "questions and answers" module of contact, this medium does provide a way of demonstrating modern human inquisitiveness and other embodied behavioral characteristics by directly posing queries (comprehendible or not) to other civilizations—social etiquettes that will arguably stand out from static, dictatorial encyclopedia accounts or other material memorializing activities.

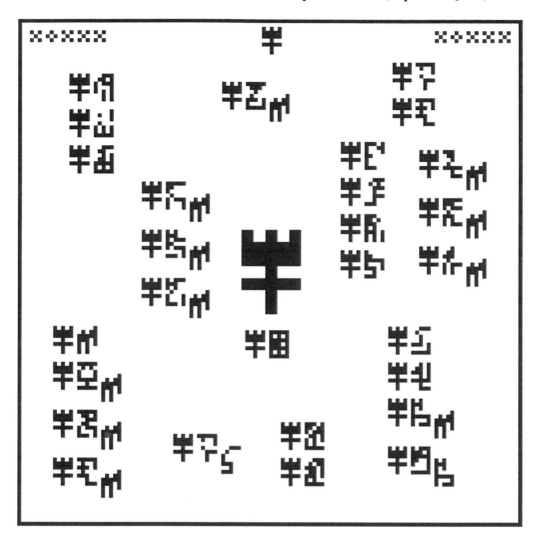

Cosmic Call 1, bitmap #23 (bitmap courtesy Yvan Dutil and Stéphane Dumas. Re-illustrated by Paul E. Quast).

While there is only one known precedent (outside unintelligible, crowd-sourced language-based media) of asking "exoatmospheric questions" seemingly apparent within the Cosmic Call signals, several workshops have proposed similar interrogative devices to practically initiate a dialogic response-exchange using message compositions that are "not complete in themselves." These formats aim to promote mutual exchanges and collaboration between both communicating parties, while capitalizing upon a perceptible common interest (an intrigue in knowing about the other respondent) for further co-extrapolation (Meisinger, 2003; Rosenboom, 2003; Vakoch, 2011a). The strategy of "multi-spectral Lincos" discussed in the preceding essay, may also possess merit as a baseline approach for this application. In theory, this crude form of demonstrating intrigue, altruism, and fostering indirect social collaboration may work. After all, as a matter of human social custom, it is generally acknowledged that overtly talking about oneself in monologues is not a desirable method of retaining mutual interest within a conversation. However, in practice, such strategies are non-existent. Messages from Earth are presently not questions from Earth.

Given the undebatable presence of diverse religious and spiritual beliefs in ancient and modern societies, it is remarkable how such a crucial aspect of human life has been almost entirely omitted—as contents—from the broad range of material cultural deposits emanating from Earth. Mainstream religious convictions and ritualistic beliefs, despite their prominence on a terrestrial setting, have questionably been represented as a coherent subject matter within the contents of intelligible space-time capsules and interstellar transmissions.[17] However, as noted earlier in preceding essays, indirect forms of spiritual expression *often* do influence the production of numerous messaging projects, alongside inspiring candid sentiments for inclusion within crowd-sourced EM transmissions—in addition to frequent astronaut activities, and printed literature deposited on the Moon. While our VGR exhibit visitor will likely be incapable of directly grappling with evidential resources for highly abstract mainstream spiritual expression or concepts of theological belief systems (as this was a subject that was purposefully omitted), there are some visual nuances of faith still apparent in the depictions of known sites of worship, such as the Magdalen College Chapel in Oxford, and the preceding Taj Mahal mausoleum which includes an internal mosque.

Beyond the Writer's Hand: Auditing the Artisans

Clearly the absence of direct evidence for several of these pertinent subjects is not necessarily evidence of their total absence within the exoatmospheric archaeological record. The manner in which a message has been persuasively written can also be as revealing about us as the selection of contents we choose to consciously or mistakenly omit. Taking the subject matter of the previous paragraph as a bridging example, theological and spiritual belief (in a deity, ETI, or cosmic origins) may, in fact, be posited as a galvanizing motivation for human expansion into our surrounding stellar environment, with such "technoscientific prayer" messages (Schmitt, 2017) serving as harbingers for these divine visions in and out of the *Sol* system—why "we choose to go to the stars," as part of something bigger than ourselves.

It is difficult, however, to decouple such convictions from encompassing political and ideological substrates, as they are often (and perhaps purposefully or even erroneously) jumbled within manifest destiny vernacular, used by nations and individuals for advocating space exploration programs, as well as justifying colonization and prospecting efforts (Billings, 2007). In spite of this ideological pastiche, it is generally *asserted* that it is the convictions of science, not religion, which act as a fundamental pillar for recent Western development and, from scientific principles, also originate core sociological beliefs: secularity, rationality, and the idea of advancing progress. These are all cultural attitudes that were imbued within the early Space Age, multinational space programs, nascent SETI and, by extension, Voyager 1 and the VGR. However, are these condensed demarcations in narrative, belief, justification, triumph, and purpose, *truly* an accurate assessment for the driving forces behind activities in outer space?

Firstly, it is crucial to acknowledge here that, despite the frequently cited disparity between religious beliefs and scientific enlightenment, both share a perceptible common bedrock and, to an extent, similar goals stemming from our cultural histories to elevate humankind spiritually, ideologically, intellectually, technologically, or even literally. David Noble (1997), in his thesis on *The Religion of Technology*, has argued that our scientific knowledge and technological prowess, while grounded within practical experience, falsifiable investigations and material knowledge, have never actually diverged far from spiritual belief, religious convictions in salvation, and the prospect of divine transcendence. If we are to try to crudely divide scientific and religious vocations into two contrarian perspectives, we may find it an arduous task to precisely decide where to draw this line as both systems of conviction share commonalities that we may have not previously considered.

At the risk of condensing Noble's extensive research, both systems of articulated belief represent entwined, succeeding stages of human development that share a close, thousand-year-old Western tradition, founded within the social elevation of the practical or "mechanical arts"—later to be known as *technology*.[18] Using this common foundation stone, both systems of conviction seek to varyingly redeem humanities' "lost divinity" as the overseer of nature, recover our sacred, endowed knowledge lost during a fabled fall from divine grace, or achieve a sense of enlightenment through our species' desire to rationally interrogate the platonic properties of our reality (i.e., deciphering the grand operating systems behind nature which have historically been attributed to the divine realm). All these desired outcomes, and their historically patriarchal overtones, underscore much of Western social reasoning and guiding cultural attitudes towards science as an objective, and transcendent vocational practice. Noble also outlines the apparent role of masculine identity that is deeply imbedded within the history of aerospace developments; an inherent infatuation with the creation of phallic rocket technology serves as, perhaps, the most *memorable* of his examples for this virile influence.

If we consider this transition from eschatology (i.e., any system based upon theological doctrines and paradigms) into technology for our rational belief systems, we can see that the progressing concept of "human destiny" towards a penultimate outcome has simply diverged from the spiritual realm into a technical mantel that could be co-sculpted by the scientific mastery of sapient human hands in a "new millennium"—a cognitive shift which we continue to build upon as a mythos of human identity, fate, futurity, and visions of utopia. These guiding principles can be seen to be endowed within global space programs' desire to expand into this literal higher frontier using our own creations of technological artistry, with scientists, engineers, and entrepreneurs metaphorically serving as an authoritative "priesthood" for this new era of humanity. This is not to state that both systems are necessarily commensurable in ideological approach.[19] Rather, we have briefly delved beyond this dichotomy in conviction to highlight the deep-seated congruences of belief and purpose frequently cloaked in exoatmospheric activities, and how this interwoven history contributes to postmodern scientific vocations, and associated practices of material culture.

Following on from this context, and in returning to the contents of the VGR to understand how it was written, it is perhaps not surprising to note that one of the principle drivers behind envisaging the potential future recipients of this golden mixtape stems from the early theoretical musings of SETI and the influential popular fictional depictions of extraterrestrials (which are somewhat founded on the cultural history of intellectual discussions and allegorical reasoning in this imaginative subject). These speculative theories were naturally shaped by the influential musings of popular science fiction literature in the late 19th and early 20th centuries[20]—literature with a strong emphasis on culturally extrapolating from the physical sciences and biology, along with the enchantment provided by advanced technology, but less frequently from the disciplines of the humanities and social sciences. As such, the imagined recipients may be feasibly understood better as a tacit product of this living socio-scientific dialogue—models of extraterrestrials which would still be subject to continuous revision and reinvention, as opposed to remaining as a static scientific nuance of the "other," based upon any known evidence. The cognoscenti of these works as "a self-appointed priesthood of the new era" (Noble, 1997), in turn, hailed from various scientific professions,[21] creating an informative, feedback pattern (McCray, 2012; Capova, 2008; 2013a; Wright and Oman-Reagan, 2018) which would inevitably shape what we pointedly refer to as the "language of science." Scientific theories about ETI, are shaped by the very scientists and writers who imagine ETI.

Early SETI, and this "language of science," are clearly products of the Western scientific tradition, its paradigms, core principles, deep-seated values, and affiliated worldviews, with mathematics serving as a foundation that is often inferred to be a universal, trans-cultural feature of rational thinking (Barker, 1982). Similarly, the very modern notion of interstellar communication practices with our imagined ETI also originates from these Western scientific and technological paradigms, with mathematical fields often described as "a lingua franca for such communication" (Samuels, 2006) that guarantees mutual comprehension across spatiotemporal distances. Consequently, the properties of music that exhibit a strong congruence with these principles of mathematics (such as harmonics, frequency, amplitude, and duration) were perhaps also the reason why an articulate musical selection, introduced as a surrogate example of *cultural technology* (as originally suggested by Barney Oliver), was avidly included in the VGR, in addition to the initial enthusiasm of choosing a nostalgic phonograph record as a suitable data carrier.[22] Indeed, music is still incorporated into modern messaging concepts for this exact purpose, as well as to demonstrate human thinking patterns.

In describing the variety of cultures on Earth, the VGR team favored universalism, while harmonizing generic features of humanity to produce a coherent, linear, and value neutral narrative based within the Western intellectual tradition[23]—albeit while also cautiously proceeding under the likelihood that NASA could partially censor the human form.[24] To substantiate these choices, Western scientific notions of intelligence, curiosity, rationality, and empiricism, alongside other assumed "shared codes" (Samuels, 2006), were also adopted as the foundation stones for the VGR. To further validate these decisions in turn, it seems that only similar, compatible patterns of information interpretation were used and arranged together in order to model a continuous, coherent, and value neutral scientific storyline about the messages' creators (humanity in general). There is much to unpack from this concise description. For starters, this narrational and curatorial process can be better understood if we are to simply infer that the VGR was created primarily to appeal to the immediate *human audiences*, and the surmised commonalities seen between terrestrial cultures, before facilitating access for secondary foreign observers (ETI) who may share in these scientific properties. This assertion accords well with the defined mission outreach and engagement objectives of the VGR, and the lineage of other messages which cite it as inspiration. Despite this prospectively broad, harmonizing appeal to neutrally represent humans and Earth, the philosopher Sandra Harding (1991) has stressed that scientific knowledge is *always* socially situated:

> [S]ocial studies of the sciences forces the recognition that all scientific knowledge is always, in every respect, socially situated. Neither knowers nor the knowledge they produce are or could be impartial, disinterested, value-neutral, Archimedean.

The Western scientific traditions can therefore be perceived as "one kind of culturally specific 'ethnoscience'" (Harding, 1992), in that these customs provide a particularly narrow, and yet authoritative, explanation for our experienced reality, but one that is written for the subset of audiences that are familiar with *our* language of science. In this instance, contemporary Western societies are presumed to be the scientifically literate "knowers" who have a basic command of scientific conventions inherited from their resident socio-cultural environments: "We live in a scientific culture; to be scientifically illiterate is simply to be illiterate" (Harding, 1991).[25] While the VGR was created, as Oliver had suggested, "to appeal to and expand the human spirit" for engagement with modern human audiences, it is notable that we still tend to project such dispositions onto other minds (both local, and foreign) which *we assume* should be capable of abstract, rational and logical thinking:

> We think we know what the first communications [with ETI] will be about: They will be about the one thing the two civilizations are guaranteed to share in common, and that is science [Sagan *et al.*, 1978].

Sagan's optimism in the potential commonalities of *our* science to objectively convey key aspects of humanities' "manual of reality" is enviable, and very much a product of his times and socio-scientific traditions. While his methods are consistent with the established natural scientific *modus operandi*—to provide a systemic understanding of our material reality through persistent empirical observation and experimentation— this approach does not necessarily mean that the VGR offers independent and neutral information to uninitiated cultures. Conversely, together with sociologist Sarah Franklin (1995), we assert that "science is defended so vehemently because it is cultural, not because it is extracultural" as it is, and always has been, situated within a constellation of ethnographic histories, intellectual traditions, and ideological convictions. Thereby, the VGR story, as a vanguard for these customs and traditions, can hardly be described as objective, neutral, or entirely absolved from its encompassing social, cultural, philosophical, and political bonds. As the philosopher Joseph Pitt once remarked (on science):

> If science is to do the job [of describing empirical reality], then it must be *our* science—that organized activity created by men which examines the limits and depths of our understanding of causal relations between objects. So to imagine another science just won't do ... there is no necessity for us to think this way—we obviously aren't born thinking this way [Pitt, 1982].

Disregarding the patriarchal overtones, such epistemic arguments may be understood better through the historical and symbolic perception of our science as a gradual, accumulative process of cultural knowledge acquisition, stretching back to the defined European Age of Enlightenment (and much older intellectual traditions such as that of Graeco-Roman antiquity). But how these Western paradigms and intellectual customs acutely shape our distinctive worldviews, observations, and musing about the imagined "other," should also be treated as a broader extension to the cultural relativism argument.[26]

Our science, and its epistemic foundations including recognized limits, are a suitable means for modeling our understanding of reality, and it does this quite well. But we should understand such vocational convictions and traditions as a product of, and in context with, the creator's cultural histories, rather than as a mental landscape that is innately representative of the qualities of other rational minds (even that of many of our fellow human societies). This idea was pointedly alluded to by Harding (1991); "science produces information, but it also produces meaning." Yet even this signified meaning within the microculture of the "scientifically literate," does not remain static for "knowers." The philosopher Peter Barker (1982) raised this exact contention when observing how polarizing scientific understanding may become once long-established "common" knowledge and theories (or personal conceptions thereof) are challenged over shorter, transformative timeframes:

> People on one side of a revolution may be quite incapable of understanding people on the other side—they may even fail to recognise that what goes on after the revolution is still science.

This model of viewing our surrounds under Western scientific traditions, and categorizing our shared reality into segregated disciplines, can not only be found within the VGR pictures and diagrams, but also in the use of several definition captions (mathematics and chemical formulas as conventional designations) which underscore our preferences for describing properties of our experienced world. There is nothing wrong with this; the VGR was purposefully created using the best available scientific information from these select fields in a perceptibly objective manner at the time of creation. But, as mentioned above, our socially accepted knowledge and systems of expression (much like the "meaning" we associate with linguistics and semiotics), do change over intergenerational timescales. After all, our scientific theories looked quite different less than a century ago. An example being the determinism of classical physics in respect to the emergence of quantum theory, as seen through the menagerie of succeeding atomic diagrams used across multiple messages (in some instances, older models are continually used in preference over contemporary valence-shell diagrams).

Another strikingly obvious example of such a period piece of knowledge, which is indicative of the recent paradigmatic shifts in astronomy, can be seen with Frank Drake's representation of the Sun with nine planets on the VGR and elsewhere. In 2006, the International Astronomical Union drafted a new definition of what constitutes a planet and consequently downgraded Pluto as it does not meet these properties. The scientific paradigms simply changed, and therefore the Solar System proportions were reduced to eight planets.[27] Consequently, the models of our Solar System placed on the Voyager and Pioneer probes (alongside depictions transmitted as part of the *Arecibo Message, Greetings to Altair*, and *Cosmic Call* bitmaps) are no longer considered scientifically valid by our present-day definitions.[28] Does this change invalidate the information on the VGR? Or does it exemplify our changing attitudes to scientific knowledge as an acknowledged cultural practice which is reinvented over successive generations of research? Do we tend to change our definitions and metrics arbitrarily? At the time of this writing, Pluto is still estranged from the Sol System.

In general, the degree of epistemic changes is not only as a result of the gradual, cumulative enhancement or acquisition of *newer* knowledge piled upon older theories. As the philosopher Thomas Kuhn (1962) argues, it also occurs through scientific revolutions, whereby older theories are discarded to accommodate newer rational interpretations. It can, therefore, be asserted that our enhancement of amassed scientific knowledge may serve as a very human (or supra-cultural) "central project"; the pursuit of phased understanding and progression across a corpus of disciplines, focuses the energies of a population on attaining another, higher epistemic step in this socio-scientific

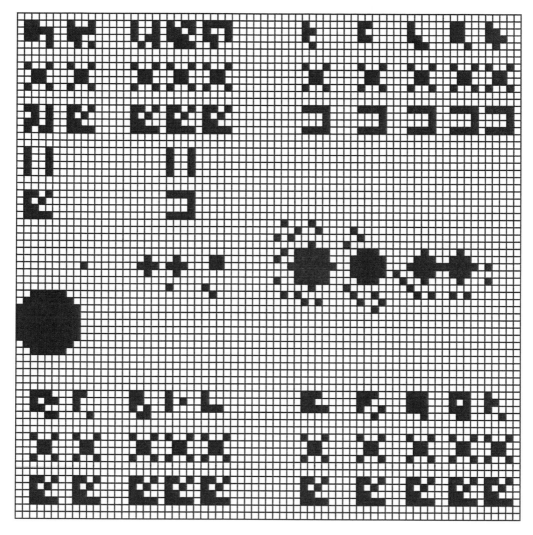

Greetings to Altair, bitmap #2 (bitmap courtesy Shin-ya Narusawa, redrawn from originals created by Professors Hisashi Hirabayashi and Masaki Morimoto. Re-illustrated by Paul E. Quast).

ladder. In the meantime, we continue to send our evolving theories into space, hoping for praise, sharing, annotation, or amendments by other similar-trajectory civilizations—should anyone wish to peer-review our unique observations.

It is obvious that more than displaying objective, valid epistemic information, the VGR compendium communicates a generational, scientific portrait in segregated disciplinary categories, rather than a more holistic approach towards science as advocated by the theoretical physicist David Bohm (1980; 1985). The story of earthly life on the VGR is therefore written as a scientific fairy tale (Martin, 1991), anchored to a unique time, space, and worldview. Given this provenance, the VGR can be described as both a spiritual vessel as much as a technological space-time capsule, an artefact arising from specific cultural-intellectual practices, more so than an objective depiction of phenomenology represented by human hands (if one could even exist). This is, perhaps, also true of succeeding lineage messaging artefacts.

In spite of this "ethnoscientific" approach, the VGR's scientific narrative about human beings is considered an archetypical message. Not only does it visually reference broader homogenous populations, dominant cultures, and hegemonic lifestyles using generalized supra-cultural examples, but it also opts to provide some concise insights into a number of differing minority and indigenous populations scattered across our planet's surface. Much like the Pale Blue Dot photograph and Sagan's eloquent words on this image (Wright and Oman-Reagan, 2018), the VGR attempted to symbolically speak on behalf of some communities that you've *never heard of*, evoking "the marginalised, oppressed, erased and forgotten" cultures and other terrestrial organisms that typically do not directly benefit from our Space Age technologies (Gorman, 2009c).

Terrestrial material deposits, such as time capsules by comparison, are frequently evocative of the attributes and conventions of the encompassing societies and nearby ethnic communities, alongside the dispositions of the individual person(s) that created them. Seldom do they allude to other modes of thought, tradition, or belief outside their familiar community boundaries (apart from a handful of millennial time capsule contents). This focus can certainly be plainly seen within the broad range of "higher frontier" objects that arose prior to the VGR as an initial consequence of the national rivalry of the Space Race, in addition to the ilk of lineage depositing activities still practiced today.[29] However, it is questionable whether we should continue to project such techno-colonialist extensions of nation-states outwards into the "commons of all humankind," as insinuated representations for "all humankind," or hegemonic factions thereof. Perhaps, it is an easier task to contemplate *whose Earth* is, in fact, represented in this material, with several dominant cultures, ideologies and other perspectives already preserved at the expense of our planetary diversity (Western, scientific, and heterosexual prerogatives are, by far, the established "norms" across messaging projects). Alas, such communicational activities in EM signals and physical artefacts still need to be recognized *by us* for what they are, prior to someone choosing to answer these cosmic calls— prospective conversations between individual or collectives, not that of their encompassing societies, communities, organizations, governments, or civilizations (see Traphagan, this volume).

Often, when looking at the blankets of material culture dispatched into outer space, the aggregate of pre–VGR content serves to perpetuate traditions, tropes, and nuances from the creators established background or mainstream cultural identities and ardent beliefs. The ideological stance of VGR's representative ethos for including elements of world heritage outside established Western tradition (although still under the umbrella of its scientific-cultural ontology), opened a door into how "speaking" could become a more encompassing and, perhaps, even democratic, practice—a clear departure from earlier Space Race artefacts, which is likely one of the overarching reasons for the time-tested popular appeal of the VGR. However, this threshold has rarely been eclipsed by subsequent message constructs that evoke the VGR as inspiration. One of the closest, analogous projects to this "multi-ethnic" ethos (and lineage) of the VGR was the *Portrait of Humanity*[30] diamond wafer—a robust disk which was due to be installed within a specially designed holder aboard the Cassini-Huygens mission (on ESA's Huygens lander for Titan). This "postcard from Earth" would be subsequently cancelled due to internal project disagreements and, consequently, the holder remained empty. Perhaps, this status metaphorically signifies humanity, as the mission scientist Tobias Owen quipped:

[T]he absence of a marker on Cassini provides a tiny, tiny hint of the dark side of the human psyche: That we don't get along with each other nearly as well as we would like (our cosmic neighbors) to believe. We are trying (as usual!) to tell ourselves how wonderful we are, hiding our blemishes as best we can.[31]

Voyager: A Touring Exhibit of the Anthropocene?

Of course, different people at different points in history would have made disparate choices in VGR contents, expressions, and beliefs for representing Earth and human identities from afar in revised variations of this project (Helmreich, 2014; Garcia, 2016; Sharp, 2016). Such is the oscillating nature of human diversity of opinion in democratic representation and our frequent discordance on many earthly topics. In presenting this appraisal of the VGR and the authoring committees' foundational work,[32] our discussions about the nature of content selection, and lacunae, gives us a point of reference in approaching messaging to extraterrestrials (or posterity) as a sociological-scientific practice *for studying ourselves*. Such a perspective allows us to observe how we may counterbalance "accuracy" versus "the best of" accounts of Earth, and quality of information over quantity arguments, while reflecting upon meaning, sincerity, and our deep-seated convictions in establishing hypothetical discourse.

It is also an opportunity to understand why the creators of the VGR chose to depart from the nativist cultural perspectives common within earlier Space Age deposits and instead adopted the surrogate customs and paradigms of Western scientific models, which would subsequently inform the generational decisions of many lineage projects. Rather than portraying the variety of lifestyles, dispositions and cultures on Earth, these intentional messages presented a rhapsodizing story of contemporary Western societies, its scientific foundations, and how the represented minority populations fit under the rubric of what we try to narrowly define as a collective human "manual of experienced reality." Whether the reader agrees with these decisions or not, the Voyager committee's choice was to send what a perceptively objective description of human life meant to them at the moment of this project's inception—*and they did just that*.

Maybe one day in the far distant future, the VGR and other harbinger messengers that have been sailing across the universe for thousands of centuries—Passive-METI artefacts of material culture that will reside far away from the inevitable changes on planet Earth—will find an audience to tell their stories, thoughts, and ideas about *our* understanding of Earth in a similar manner to our museum exhibit. What might these *others* glean, or recognize to be notably absent from this chance encounter with an envoy of ancient Earth? Perhaps, seeing our planet in the eyes of such a distant *other* may paradoxically enable us to recognize how we regularly warp and distil this profound terrestrial story in our favor—and the inaccuracies in interpretation, or confusion, that subsequently arise as a result of these purposeful choices, oversights, sanitizations, omissions, and narratives underscoring scientific languages.

Regardless of this interpretation, a partial fragment of our story as a species directed toward the stars sends a strong message to contemporary earthlings. Understanding and describing a complexity and diversity of life on Earth is a challenging and cumbersome task, with accounts prone to error, social assumptions, extreme distillation, and unavoidable trivializations. But such tasks also support opportunities for us to cooperatively learn and engage with one another, while exploring the differences between our own varied cultures, psychologies, value judgements, and underlying beliefs systems that altogether mix, reinvent and contribute to the unique cosmic heritage for our evolving Earth.

NOTES

1. While assembling a phonograph record player using the included stylus and abstract instruction manual is not a trivial task, the protective "Interstellar Envelope" cover featuring these instructional notations should ideally be presented as an inscribed artefact to function as this exhibition's "preface" text or preamble prior to displays.

2. We use this term to acknowledge informative contents that have been purposefully rendered into signs to hypothetically provide accessibility of the signified meaning for an unknown recipient, though this definition does not attempt to distinguish whether these commensurability stratagems will be successful or not.

3. The *Pioneer Plaques* provided a limited level of public intrigue and engagement in representing our world over spatiotemporal distances. However, the VGR would expand this discussion with its variegated corpus of cultural and scientific media, scored onto a mass-consumption medium from the immediate era(s).

4. Taken as a whole, the compilation of selected music is considered to meet a high standard in classical Western tradition, not only in the quality of the musical repertoire (*The Magic Flute* by Mozart, *Symphony No. 5* by Beethoven, *Brandenburg Concerto* by Bach), but also within the reputation of included interpreters and conductors (Glenn Gould, Edda Moser, Sviatoslav Richter, and Igor Stravinsky). Native musical selections, with some inaccurate accreditation (Gorman, 2013; Ferris, 2017), featured a melodious and insightful set of compositions, with the audience offered Aborigine songs, Azerbaijan bagpipes, Georgian chorus, percussions from Senegal, a Pygmy girls' initiation song from Zaire, *Night Chant* by Navajo Indians, and music from Java, Mexico, New Guinea, Japan, Bulgaria, China, and India.

5. Given the varying contents of these "greetings" messages, there is not enough lingual data to support an analysis of languages. However, these expressions do demonstrate a varied range of human vocalization capabilities.

6. The annotated features on these images denote a number of informative properties such as the messages' place of origin, dates, sizes, composition, and quantity measures, which cover a plethora of social, biological, chemical, and environmental topics needed to understand the basic principles of earthly eco-bio-cultural processes.

7. This may be perceived through a simple cross-comparison of the physical forms of pachyderms (depicted in pictures 67 and 98), in context with the hominid skeletal structures presented in earlier human-biology visuals.

8. As a cynical example, from reviewing the ASREM script-based submission archive (a project that was developed around environmental ethics), there have been 84 unique mentions of term "die," 74 for "destroy," 18 for "extinct," 17 for "kill," 16 for "killing," 15 for "dead," 12 for "death" and 13 for "remains" (term is un-assorted).

9. As cited from President Jimmy Carter's speech on image 118: "We are attempting to survive our time so we may live into yours. We hope someday, having solved the problems we face, to join a community of intergalactic civilizations." This limited content, given its dependence on natural languages, will likely remain illegible.

10. This desire to "reach out" to extraterrestrials in order to solicit aid or advice was also a motivation for the recent *Sónar Calling GJ 273b*—a viewpoint that was at least voiced by the Sónar directors (McCarthy, 2019).

11. This concept of presenting the imperfections of humanity for ETI to resolve was theatrically explored using the *Golden Archive Drive* device in the 2018 science fiction film *The Beyond*. Did we deserve ETI's gift at the end?

12. It is worth noting that these moral and ethical failings of humanity, and their accurate interpretation as such, is expressively reliant upon the recipient's foreknowledge of terrestrial historical contexts and how diverse cultural histories, to varying degrees, intersect with and depart from this popular (and condensed) "narrative of Earth." The VGR, and many other propaedeutic messages, chose to represent tangible characteristics of a value neutral human civilization, as opposed to cultivating complex global and local historical narratives as a means of introducing humans to the cosmos (such experienced stories of social history would likely remain illegible).

13. It can be argued that at least some of these frames correspond to the guiding principles of wildlife photography; to observe and document, not interfere or be actively seen (or generally detected in any way).

14. This overly simplified diagram is a product of Western scientific pedagogy and has been historically

employed as a linear social-evolutionary means of distancing *Homo sapiens* from other members of the hominid family tree. Additionally, such imaginative "appropriated pasts," provide revelations into our common prejudices, and reveal deep-seated social attitudes for how we sometimes categorize fellow humans as inferior specimens of our own species (McNiven and Lynette, 2005). It also misrepresents phylogenetic branches of the tree of life; one organism does not simply cease to exist when the successive organism evolves. Similar presentations are likely also apparent within encyclopedia accounts of Earth, as the overriding objectives of such projects is to export as much information as possible, as part of a quantity over quality approach. A similar hominid skeletal diagram, penned by the natural history artist Benjamin Waterhouse Hawkins, was also submitted by members of the public to the *Lone Signal* transmission project, alongside other human evolutionary diagrams.

15. The premise of Gertz's argument is that information diversity may be tantamount to cosmic-capital trading.

16. It is curious to note that cats seem to be a recurring motif and analogy within SETI literature and messaging projects. Some examples of which include: Marven Minsky's proposal (1985) to dispatch a living cat to extraterrestrials as a common referential object, the "cat piano" depiction within *The Last Pictures*, the Lone Signal crowd-contributed "pet pictures," and genealogically similar felines in the ASREM images, alongside the new *MEOWTI project*. This may be a coincidence, or a curiosity, indicative of message author preferences, but it could also be seen as comparable with the proposed Schrödinger's cat experiment, metaphorically representing the ambivalence of whether ETI are extant, or not, in context with our observations—Schrödinger's cosmic cat?

17. It is worth highlighting that, in order to communicate such allegorical and nuanced materials in a legible manner, the recipient of these messages will need to be supplied with extensive social, historical, theological, and philosophical contexts for complex human behavioral phenomena from a plethora of diverse intelligible samples.

18. Noble expands on this "technics" definition of mechanical, practical, or "useful" arts to include painting, decoration, gardening, metallurgy, and later feats of technical prowess such as the initial construction of mechanical devices, alongside the advancements of fields of science such as biology, chemistry, physics, and engineering—which the Harvard Professor Jacob Bigelow would later term "technology." It can, therefore, be argued that the VGR, and Voyager spacecrafts represent this tradition of crafting highly complex and detailed mechanical arts, to be revered as an ambassador of Earth as a "best of us" approach in the higher frontier.

19. It can be asserted that one field is occupied with understanding "the creator," and the other "the creation."

20. These works of fiction, in turn, are arguably based upon the pioneering foundational works of earlier extraterrestrial communication theorists who hailed from Western scientific traditions (Raulin-Cerceau, 2010).

21. In this case, scientists (as products of their surrounding academic and cultural traditions) write pioneering science fiction stories from their respective expertise about ETI and contact scenarios which, in turn, feed back into inspiring SETI theories and guide development of scientific searches over successive generations.

22. The perceptible interrelationship between the fields of science and music (alongside the neurobiological interconnections within individuals who work between both professions) is extensively documented elsewhere. These properties also accord well with the characteristics of EM frequencies for interstellar communication.

23. This neutral narrative of representing humanity in general biological terms can be seen within perhaps the first serious multi-ethnic depiction of the human form to be dispatched into outer space, the *Pioneer Plaque* couple drawn by Linda Sagan to intentionally represent non-specific human racial characteristics. Despite this, the depicted binary-gender couple received criticism on these very grounds and is often described as a clear manifestation of "heteronormavity" (Warner and Collective, 1993). In addition to this, it is worth acknowledging that the same non-specificity principles were applied within Paul Van Hoeydonck's earlier *Fallen Astronaut* sculpture (as requested by the astronaut David Scott), deposited as a memorial effigy during Apollo 15.

24. The censored image in question was a George M. Hester photograph of a naked man and pregnant woman published in *The Classic Nude* in 1973. NASA's position was due to criticism previously received from the drawn couple depicted on the *Pioneer Plaques*. Despite this conservatism, there are several depictions of semi-nude figures and diagrams of sexual organs still apparent on the VGR. NASA's concerns can therefore be asserted to be based on the inclusion of *seemingly* amorous imagery, arguably from a patriarchal perspective.

25. Harding's comments specifically reference feminist theory and the historical exclusion of women's contributions to, and gender disparity within, scientific culture. However, her arguments also possess tangible significance for the historic exclusion of contributions from other minority and ethnic populations to these fields of "collective" human knowledge.

26. Cultural relativism is the idea that an individual or group's beliefs, values, convictions, and practices

should be understood based on that groups' own culture and associative customs, rather than be judged against the criteria of another's. The theory is frequently applied (as a thought experiment) in discussions of universal ethics.

27. The IAU members at the 2006 General Assembly agreed that a "planet" is defined as a celestial body that (a) is in orbit around the Sun (b) has sufficient mass for self-gravity to overcome rigid body forces so that it assumes a hydrostatic equilibrium (nearly round) shape, and (c) has cleared the neighborhood around its orbit. These criteria are defined in *Resolution 5A: Definition of "planet,"* IAU 2006 General Assembly, Prague.

28. Despite this novel change in planetary definition criteria, the recent *Sónar Calling GJ 273b* tutorial file, from the May 2018 transmission series, still featured Pluto as an accepted member of the Sol system.

29. An obvious, residual legacy of this pseudo-colonialist expansionism may be perceived in the inclusion of a U.S. flag as a wrapping medium for the phonograph, both of which are placed inside the Interstellar Envelopes.

30. This project featured a stereograph of a multi-ethnic community of people posed on a beach scene. Similarly, this project's digital progeny *One Earth Message* followed in line with this ethos of presenting diversity in human culture, thought and ethnic heritage, alongside wishing to accurately represent other biota outside of their relationships with the Anthroposphere (according to the project's FAQ). The initiative was due to uplink this *Golden Record 2.0* as a fifth interstellar message aboard the New Horizons probe. But, at the time of writing, it has seemingly become inactive—perhaps, ironically reflecting the same frequent social conflicts, and changing values, ethos, and intentions of such goodwill projects, created as questionably accurate metonymy for Earth.

31. This sentiment has been extracted from a longer Tobias Owen statement about this project (Benford, 1999).

32. It is acknowledged that our appraisal predominately contends with the visual media of the VGR and, as such, the chosen audio may impart alternative or conflicting perspectives to our pictorial lacunae discussion.

Finding Meaning
in Terra Nullius and Beyond

Uncovering Clues in the
Exoatmospheric Archaeological Record

Paul E. Quast

> While swimming in a river—a tributary of the Biryusa River in eastern Siberia—a local boy hurt his foot on some sort of piece of iron. When he retrieved it from the water, rather than throw it into deeper water, he brought it home and showed it to his father. The boy's father, curious as to what the dented metal sphere contained, opened it up and discovered this medal inside…. The boy's father brought his find to the police. The local police delivered the remains of the pendant to the regional department of the KGB, which in turn forwarded it to Moscow … the appropriate KGB directorate found no threat to state security in these objects, and after notifying Keldysh as president of the Academy of Sciences, this unique find was delivered to him (Chertok, 2006)

In what has been categorized as a very "strange but true" incident from Space Age history, this account by Boris Chertok, in his memoirs about the Soviet space program, refers to the failure of the *Tyazhely Sputnik* (heavy satellite) to depart Low Earth Orbit—a mission that was intended to be the first human object to reach the planet Venus, while also transporting cultural and political iconography to this world. Instead, the stranded satellite broke apart while re-entering the atmosphere in February 1961, scattering debris across a large swathe of Siberia in what would become one of the USSR's scarcely documented, failed missions.

The Siberian boy in this story discovered a submerged artefact, but something convinced him that this "piece of iron" was unusual—*the right kind of strange*, warranting investigation. Expanding on this curiosity, his father then managed to pry open the sphere to reveal a very-specific context to the now clearly artificial object—"1961 Union of Soviet Socialist Republics" imprinted on the inside, along with a map of the Earth and Venus orbits around the Sun. Recognizing the context of this message from living memory, and with official stately connotations implied by the now identified contents, the object was then hastily passed to the authorities, eventually reaching the end-link of the Soviet governmental directorates, in this case, Keldysh. In the summer of 1963, Keldysh then handed it to Sergei Korolev, the Soviet rocket engineer and supervisor for all USSR probe launches. Korolev then gifted the asset to Chertok, whereby he remarked, "I was

Tyazhely Sputnik pendant. The disc was apparently inserted into the northern hemisphere, perhaps as a counterbalance weight for buoyancy in case the artefact landed on water in the inferred tropical forests on Venus. Planet locations accurate to date of launch window for the spacecraft (illustration courtesy of Paul E. Quast, composite based on original layout for the successor Venera 1 pendant [also from 1961], design created by unknown Soviet illustrator[s]).

awarded the medal that had been certified for the flight to Venus by the protocol that Korolev and I signed in January 1961." Sarcasm and irony for the pendants fate aside, my departure point from this story is that this object was only deemed significant when a cascade of contextual happenstances had proven it to be anthropogenic—a burden of proof all artefacts must fulfill when scrutinized under modern archaeological assessment criteria. If the piece of iron hadn't been initially investigated as a curiosity—an anomaly within its found environmental context—it would likely have remained at the bottom of that Siberian river.

The field of archaeology can be broadly defined as the scientific study of how human populations interacted with their social and natural environments across regions of Earth's surface over definable slices of time, using material artefacts and other found residual traces. Beginning at the traditional "trowel's edge," or in the mind's eye of this tool's wielder (Malafouris, 2018), the profession and its diverse research methodologies are generally occupied with understanding relationships between human behavioral patterns and the vast phenomenology of material culture interactions (Hicks and Beaudry, 2010), arguably "in all times and all places" (Reid *et al.*, 1975). Rarely (if ever) are these remnants studied in isolation outside their broader found contexts. Provisional conclusions about an artefact, for example, are crafted through ethnographic comparisons or contrasts with present human societies,[1] surviving biographical materials from past civilizations, historical planetary records, and evidentiary materials from surrounding site conditions (amongst other contextualizing tools and direct ex-situ analysis of the object), to support qualitative interpretations for further examination. The overall emphasis of this field is to avoid bolstering the familiar grand syntheses of human culture typically available in popular history books, instead opting to preserve the intellectual wealth of heritage, lifestyle experiences, ritual expression, and cultural diversity available within our species' multigenerational repertoire for further scientific scrutiny.

On Earth, there are many environmental clues that serve as a great storehouse of information for archaeologists to deduct background context about an excavated relic. The encompassing substrates of soil surrounding an artefact contain clues about human activity and inhabitation, through the analysis of this medium's texture, color, depth, isotopic composition, and proximity to other tentative artefacts. In addition to these indicators within "the dirt," there are several local evidentiary resources available such as dendrochronology and other dating methods, historical climate data, trace fossils (such as footprints, refuse deposits, microbial colonies, and excrement), and nearby affiliated sites that allow us to tentatively peak into our ancestor's early settlements and regional lifestyles. The *absence of contextual evidence* is also a vital indicator, as seen in the above pendant story.

In contrast to this terrestrial praxis, space archaeology (as commonly defined) investigates the range of exoatmospheric material culture arising from the recent Space Age,[2] a relatively modern archaeological record of technological objects and debitage which is distributed in varying concentrations across geocentric orbit and on, or around, other astronomical bodies (O'Leary, 2009a). An archaeological assessment of our radiosphere, and types of *inadvertent* messaging patterns arising from purposeful EM signaling activities, has been separately discussed at length elsewhere (Quast, 2021). Derived from the trace analysis of societal waste by-products (Rathje, 1999) and the four strategies of behavioral archaeology (Schiffer, 1976), this subfield focuses on investigating the symbolic interactions and relationships between recent space material culture and people, in order to study how modern human civilizations have manifested from the successive historical, technological, geopolitical, philosophical, and scientific phases of our species' pioneering leaps off-world (Staski, 2009).[3] The goal is *not* to corroborate the rhetorical tropes of mainstream sociological narratives which have traditionally defined the Space Race, Cold War, and contemporary world histories. Rather, space archaeology endeavors to understand how various societies (or subsets) (McCray, 2012) and their belief systems have conceptualized their surrounding spatial environment, using the assemblage of contemporary material objects (or more accurately an *armchair survey* of in-house reports, often incomplete public documentation, and replicas of these distant artefacts).

Items of social and technological infrastructure, as the archaeologist Michael Schiffer (1987) has posited, naturally transition from a systemic context (i.e., the active, operational phase of space hardware) into this realm of archaeology. Therefore, these inactive remnants, or their debris fields, may be studied under modern archaeological methods to understand the contemporary regional lifestyles and social stratification of launch-capable populations—*and*, by omission, the often conflicting, residual legacies of social inequality experienced by indigenous and minority populations who did not directly benefit from these aerospace technologies (Gorman, 2009c).[4] Aside from the power dynamics and clear sociological implications, these studies also stem from a growing, academic intrigue within the decline of modernist architectural sites, alongside how varying populaces socially assign symbolic value(s) to objects and, upon fulfilling their utilitarian purpose, how we choose to interrelate with these remnants of this material culture (i.e., as waste deposits, significant social documents of antiquated human heritage, or as an amalgamation of both types of legacies) (see color insert).

Outer space presents us with a radically alternative archaeological landscape to that we are readily familiar with: the two-dimensional Euclidean geometry of discrete site coordinates are substituted for complex orbital equations with hyper-velocities in four-dimensional matrices; geologically-active soil strata is exchanged for (contestably) inactive regolith; cubic volumes of atmospheric pressure are replaced with vacuums, or extreme variants of Earth's thin blue line; and interference produced by terrestrial erosion processes are swapped for a constant bombardment of micrometeorites and high-energy particles from cosmic sources. To push the limits of our epistemic envelope further, all these extra-archaeological phenomena are sculpted under the empire of gravity from our local star, in addition to tugs from other latent sources (Gorman, 2009a; 2014). The Earth's own grip can be fairly tentative. In contradiction to this inhospitable ether, celestial environments seemingly preserve artefacts well over intervals of deep time and, therefore, this archaeological record of material objects in outer space may far outlast the remnants of launch facilities and documentation available on Earth (Gibson, 2001; Gorman, 2009c). How might space archaeology as a retrospective, "hands-on" disciplinary practice that begins at the "trowel's edge" approach such regions of outer space, sculpted by alternative natural formative processes, to speculatively study future sites and objects of human heritage? This position is now fast becoming a practical research stance for interpreting early space exploration heritage, residual infrastructure, and retrospective legacy investigations.

In such an extreme environment where all our reference points may be relative, it can be difficult to ascertain whether the traditional, informed tools of archaeological fieldwork may uncover a meaningful context for the various artificial objects we cast into our stellar environment; considering such hands-on, inductive reasoning may be the only means available for futurity to study these material remnants, in lieu of surviving terrestrial records. How will these observations, in turn, influence the re-interpretation of space-time capsules or other material culture deposits that are intentionally created as a direct product of techniques employed by communication theorists, anthropologists, archivists, conservationists, and museologists? After all, these purposeful messages are unique in that they do not solely arise from the cultural formative processes of deposition (i.e., discarding of waste materials), loss, abandonment, and the votive burial of artefacts which are traditionally the purview of archaeological investigation. There is a clear intentionality to the artefact's design, choice of contents, and relational contexts, which are largely the result of artificial selectivity by the numerous individuals and micro-cultures that choose to craft such assets. A large part of this intrigue in messages is also centered on the prospect of reclamation, or reuse, and how this recovery process may, therefore, shift these artefacts from an archaeological context back again into a systemic context for new audiences (i.e., as mnemonic devices, commemorative relics, votive applications, archival projects, or other communicative functions etc.).

Following in the Footprints

While practical experiences of conducting fieldwork beyond Earth are sorely lacking and subject to speculation, we do, however, possess one (albeit brief) instance of an in-situ, space-based "archaeological" investigation on Earth's natural satellite for insights and discussion (O'Leary, 2009a; Capelotti, 1996; 2009; Staski and Gerke, 2009; Gorman, 2019). As part of the objectives for the Apollo 12 lunar mission, the astronauts

**Pete Conrad studying the pre-existing Surveyor 3 near Statio Cognitum prior to disman-
tling components. The astronauts originally brought a shutter timer, to enable them both to
feature in some photographs together, but this unused item may likely now be found nearby
in the ascent stage "toss zone" deposit (photograph from NASA/Alan Bean [AS12–48–7133]).**

Charles "Pete" Conrad and Alan Bean were tasked with landing their "Intrepid" lunar
module near the pre-existing Surveyor 3 (henceforth S-3)[5] landing site to undertake a
preliminary assessment of the residual spacecraft, before bagging hardware samples for
further ex-situ analysis back on Earth.[6] The astronauts produced an ample catalog of
photographs, documenting the lander's hardware, position, and footpads in the dirt,
before removing the remote sampling arm, television camera and other paraphernalia
lia for material-exposure testing on Earth. The astronauts would also install experiment
ment modules nearby and leave behind several of their own communicative artefacts for
votive applications on the designated *Statio Cognitum* site.

Sticking with this same site for convenience, let us imagine that a future archaeologist in 8000 CE[7] has been granted an opportunity to revisit Statio Cognitum, and
re-re visit S-3 (or preferably to conduct a three-dimensional archaeological survey to
maintain site integrity). What interpretive context might our space archaeologist glean
from this peripheral material culture in-situ, and how can the aggregate of this scientific

heritage aid, or inhibit, a rational understanding of the Intrepid message deposits? For the purposes of this hands-on thought experiment in "the regolith," I will try to refrain from conducting an exhaustive appraisal, but this archaeological site is ideal as a brief case study of early fieldwork for six identifiable reasons:

1. The Moon, as it is clearly incapable of hosting its own indigenous population, has accumulated material culture deposits as a by-product of interactions with several variegated multicultural nations over successive generations. Statio Cognitum is but one small, and early, example in a larger futurescape of multinational hardware on the surface.

2. Statio Cognitum is regularly cited as an important heritage site of the 20th century, though usually overshadowed by the Apollo 11 and Apollo 17 sites (NASA, 2011) and, as with similar zenith sites from the U.S. space program, served as a milestone in advancing human spaceflight alongside political and military aspirations. This particular site has also recently been the subject of several independent discussions by parties interested in planning return missions.

3. As with all Apollo missions, we have a catalog of visual documentation to inform pragmatic speculations about this site from a first-person perspective, alongside subsequent appraisal work performed by the U.S. Geological Survey using this documentation and remote observations. Moreover, S-3 had also separately taken 6,315 photographs of the pre–Intrepid *intact site* for comparison.

4. As noted above, the mission and landing site would demonstrate the only instance in which there has been documented human interaction or "multiphase occupation" on a previously extant lunar landing site—a unique occurrence in space exploration history. This interaction changed the original historic properties of the S-3 landing site directly, while also absorbing the perceived boundaries into Statio Cognitum.

5. The astronauts, as active participants, modified the integrity and material culture of this site by discarding dozens of their own artefacts as part of their planned operations, while leaving thousands of trace fossils across this interlinked site.

6. All five of these points may augment the interpretive context of the inscribed message plaque left as a commemorative deposit by the latter Apollo 12 mission.

Without speculating about the present material condition of these artefacts, there is an abundance of evidentiary resources available across this site despite the astronauts' brief, 31.5-hour choreographed occupation. Perhaps the most visible and iconic features of Statio Cognitum that can be seen via remote observations are the 2,300 meters of boot prints which collectively weave a trace fossil site boundary, lunar tracks which are frequently stated to be evocative of the ancestral hominid footprints in the Tanzanian Laetoli region (Heidmann, 1992; Capelotti, 2004; O'Leary, 2009b; Koren, 2018; Hanlon, 2019). It is questionable whether each of the two separate Extra-Vehicular Activities (EVA) pathways would be distinguishable in these mixed thoroughfares (in remote observations, I found the trajectories difficult to resolve). However, under ideal preservation conditions, each of the footprints *could* conserve intricate clues that may allow us to reconstruct basic morphological characteristics for each astronaut, in much less the same way an experienced hunter may acquire iconic indicators from animal footprints.

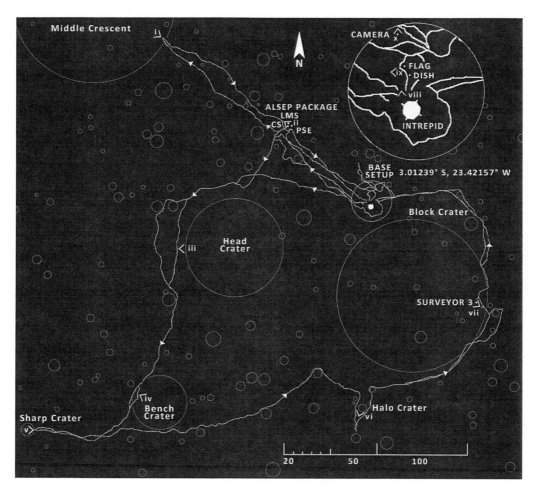

Statio Cognitum site layout map, denoting observable trajectories for walkways, alongside material deposits, studied lunar features, and some photograph locations: AS12–46–6841 (i), AS12–47–6928 (ii), AS12–49–7214 (iii), AS12–49–7225 (iv), AS12–49–7263 (v), AS12–49–7303 (vi), AS12–48–7133 (vii), AS12–46–6726 (viii), AS12–47–6953 (ix), AS12–47–8899 (x) (revised diagram based on Apollo 12 traverse map prepared by the U.S. Geological Survey and published by the Defense Mapping Agency, re-illustrated by Paul E. Quast).

Despite the relatively clunky space suits and reduced lunar gravity, it may be technically feasible to ascertain approximate height for each separate individual via footprint movement patterns, upturned regolith layers, leap-strides (Alan Bean was slightly taller), imprint size differences, and average weight distribution in depression depths. There are many in-situ photographs taken of these impressions as, in addition to data collected from Apollo 11 footprints, these trace fossils were documented as a scientific experiment (NASA, 1969), serving as a test for measuring the chiaroscuro, material reflectivity and density of lunar regolith in the basaltic plains of the Mare regions—alongside fulfilling ideological purposes.

In addition to these EVA tracks, there are several larger artefacts and discarded objects scattered across the entire site: the Intrepid descent stage, S-3 lander (~133 m away from Intrepid) along with its retrorocket (unknown location), and the various experiment packages (~128 m away from Intrepid). If our future archaeologist forgets to

Low-altitude remote observations of Statio Cognitum (photograph from NASA/GSFC/Arizona State University/LROC spacecraft [M175428601R]).

bring her camera on this survey, she will still have various photographic paraphernalia available at Statio Cognitum to choose from, including several forgotten (exposed) film rolls.[8] As traditional with Apollo landing sites, a highly bleached, store-bought flag may possibly remain erect nearby, alongside discarded attire, bags of human excrement and emesis, tools, plastic bags, empty food ration packaging, arm rests, sanitary tissues, and at least 60 other known, documented artefacts (Capelotti, 2010; NASA History Program Office, 2012). Furthermore, fragments of the aluminum and mylar foil from the descent stage may be scattered throughout the site due to the proximity of this surface-lining material to the roaring ignition source of the ascent stage.

We can only surmise the approximate location of dozens of these artefacts within the "blast zone," and what Lewis Binford (1978; 1982) termed a "toss zone,"[9] as the site obviously itself did not materialize with future archaeological analysis in mind, clearly demonstrating an *a posteriori* legacy for Statio Cognitum. However, this cursory settlement contains a unique snapshot of the design and engineering considerations that went into the creation of each particular tool, object, and seemingly unimportant relic, reflecting the most advanced fabrication technologies available in the 20th century, and perhaps the technical limitations of early spaceflight capabilities (e.g., the travel-distance restrictions associated with early life support equipment, calorie intake for duration of expected stay, the calculated need for astronauts to shed mass from the ascent stage prior to attempting lift off, etc.).

Understanding how the archaeological record of Statio Cognitum initially materialized across the fourth dimension can also provide intuitive clues into why the astronauts constructed the material substrate of this landscape as active agents, rather than as passive automatons mechanically operating on an itinerary across an environmental stage. In the case of all "Apollo culture" (Capelotti, 2009) landing sites, there is a clear, chronological sequence of events preserved within the directionality, orientation, and deployment of equipment along this trajectory of trace fossils as the astronauts created and interacted with islands of natural formations and discarded materials.[10] The "toss-zone" is a simple example of an unintentional, pre-departure depositing behavior materializing from the indiscriminate discarding of refuse; however, the meandering pathways of footprints reveal an intentional mental-material relationship associated with some dispersed artefacts over the duration of the astronaut's occupation.

Depending upon the state of footprint preservation, it seems plausible for our archaeologist to assert that the ALSEP[11] experiments originated as a material element of the Intrepid lander, due to the differences in footprint depression depth arising during the struggle to deploy this weighty equipment when compared with "average" astronaut tracks. This may seem like a trivial conclusion; however, the same changes within deployment depression depths—or lack thereof[12]—possess implications for initially recognizing the nearby S-3 lander as an independent, pre–Intrepid artefact; a perspective which is further corroborated by footprints overlying the older blast marks and scattered debris emitted by this hardware's own propulsion system. Our archaeologist should be able to reasonably conclude that Statio Cognitum was intentionally established at this seemingly arbitrary location to interact with this pre-existing hardware, as opposed to any other predetermined site-selection criteria for the experimental packages or other unique lunar landscape criteria. This determination could have strong implications for interpreting the remainder of the site, its contents and, by association, the emphasis early space pioneers placed on explorational activities. The *selenography* of this material landscape developed with this interaction in mind, but establishing why S-3 founded this site (in lieu of all removed scientific instrumentation) may be harder to ascertain. To further corroborate the prior establishment of the S-3 site, the post–Apollo 12 report (Mitchell and Ellis, 1972) delineates several "abrasion events" caused by the erosive sandblasting of lunar dust uplifted during the landings of S-3 and Intrepid—a patina that our archaeologist may likely find on all surface-dwelling artefacts after the Intrepid ascent module launch.

As all EVA thoroughfares depart and return to the descent stage module, a future archaeologist could also rightly determine that this central node contained a stable life-support system when compared with smaller, discarded units, and also served as a launchpad due to the scattering of regolith and debris outwards as a result of the launch "plume effect." Boot prints nearer this blast zone will likely be heavily eroded, layered by debris, or even obliterated in comparison to more distant impressions preserved in the lunar dust elsewhere.[13] However, more elaborate studies on lunar environmental processes have suggested how solar-activated, electrostatic properties of the lunar dust may degrade the physical features of these trace fossils over extended time intervals (Heiken *et al.*, 1991), thus destroying our finer regolith impression clues. As the Apollo 12 astronauts were not expected to apply archaeological fieldwork methodologies to the S-3 site, we cannot definitively conclude the potential adverse implications of these non-terrestrial natural formative processes on footprints from their report on recovered materials, leaving this selenological mystery to be resolved by future, longer-duration lunar missions and surface experiments.

We are not only left with these artefacts and traces of astronaut movements, but also plenty of other biological matter. For example, excreted products (for those who wish to keenly study such "proto-coprolite" remnants) may provide useful insights into the dietary preferences and digestive systems of each astronaut, but also serve as a guide for assessing the holobiont condition of each person at the time by assessing the remnants of microbial populations in this matter, as well as inferring about the overall state of organism health by examining the quantities of unabsorbed nutrients (in addition to also ascertaining how human physiology may have responded to early spaceflight pressures).[14] These isolated "deposits" may also serve as long-duration experiments, enabling researchers to study the deep time impacts of radiation exposure on microorganism viability, as with similar multi-century extremophiles tests operating on Earth (Cockell *et al.*, 2019). Further to these resources, material artefacts such as branded object insignia (i.e., Hasselblad camera paraphernalia), instruction manuals and other samples of literary resources (if apparent), could—to a limited capacity—permit a crude philological analysis (i.e., understanding the range and assemblages of individual Roman script characters, rather than interpreting strings of baseless English words or sentences), alongside a rudimentary introduction to our pictorial framing conventions and visual preferences. This *may be* gleaned through the forgotten film reel imaging surfaces, and the discarded Hasselblad's internal Reseau plate (as astronaut helmets were so clunky, no optical viewfinders were placed on the Zeiss-Biogon camera model) (see color insert).

If our future archaeologist takes a closer look between the ladder rungs of the Intrepid descent stage, she may notice a relatively inconspicuous, lightly dusted rectangular plaque, featuring inscriptions of alpha-numerical characters on a brushed metal background. Unlike other Apollo Lunar Plaques, there are no stylized globe projections

The Apollo 12 plaque attached to the ladder on the descent stage landing gear strut of the Apollo 12 Lunar Module (photograph from NASA/Wikimedia Commons [S69–53326]).

or an epigraph describing the purpose of this object; we are limited to the message comprised of 25 unique characters (and hand-written variations of some symbols—if recognized as such) surrounded by a double border.[15] Clearly the symbolism remains equivocal; does the plaque commemorate an apotheosized burial in an envisioned heaven? Do the stylized signatures serve as a cartouche for identifying three important rulers of Earth? If these columns identify three individual beings, why are there only two distinctive trails of footprints? In context with other sites, why did "President Richard Nixon" not sign this version like two other Apollo Lunar Plaques? Was the astronaut "Nixon" on other missions, but not this one? In conjunction with these amusing enigmas, some admittedly far-flung, our archaeologist may be truly perplexed if she also finds the space art object *Moon Museum* allegedly affixed to a leg of the descent stage (under the thermal blanket), though this claim has been widely disputed. This small ceramic wafer "museum" contains six drawings by well-known Abstract Expressionist artists, including a sketch of a tiny penis by Andy Warhol.

At the risk of drawing any firm conclusions about accumulative site processes and enigmatic evidence that may contextualize message deposits, let us instead briefly discuss some tangible elements that may perceptibly inform our archaeologist's opinions. Interpreting a votive message—and by extension the entire archaeological site—requires careful observation of signs from this surrounding microcosm in order to craft a tentative meaning, and subsequent narrative, for further investigations. The astronaut footprints, while limited to a binary set, provide a conceptual bridge for our archaeologist to decipher *some* of the utilitarian mystery of this site (i.e., the scientific investigation and technological manufacturing capabilities of the launching populations). The relative patterns of these trace fossils across the surface of the Moon, and interlinking between artefact islands, *may* provide an inkling into the purposes of this visit (to set up ALSEP experiments and acquire older S-3 hardware), thus allowing our archaeologist to construct a mental impression of the technological ingenuity, and logistical energy requirements, necessary to mount a human excursion to this nearby astronomical body for scientific fieldwork. In the very least, our archaeologist can, therefore, rule out early space tourism as a *prime motivation* for this site's initial establishment, though this depends on actions *we undertake* from the 21st Century onwards.

While these observable properties and residual human presence *may* allow our archaeologist to surmise a crude interpretative context for deciphering one layer of the plaque as a votive epigraph for a scientific mission, it is doubtful whether this implied interpretation, the numerous custom-engineered artefacts, and trace fossil chronological record, will provide insights into the second-order interpretation of this site as an extension to the intense technological rivalry and propaganda conflict between the United States and the Soviet Union. This is, of course, the defining ethos which underscores the purpose of Statio Cognitum and, by extension, the Apollo missions and equivalent Luna (Луна) uncrewed program. Our future archaeologist may also conclude that, due to the technological and logistical resources necessary to mount this early human spaceflight expedition, the aggregate of purposeful, communicative media available solely on Statio Cognitum is representative of the broader international community, as opposed to a deposit of Americana artefacts.[16] In this respect, despite the mission serving as a sequel to Apollo 11's defined "We came in peace for all mankind,"[17] the physical legacy of this particular site possesses less tangible significance for the international community due to the lack of discernible multinational epigraph sentiments,

Top left: Apollo 11 plaque (graphic from NASA/Wikimedia Commons [S69–39334]). *Right:* Apollo 13 plaque. Note: this plaque re-entered with the Lunar Module, plunging into or near the Tonga Trench in the Pacific Ocean (graphic from NASA/Wikimedia Commons [S70–34685]).

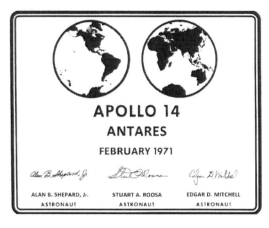

Center left: Apollo 14 plaque (re-illustrated by Paul E. Quast, based on NASA/JSC photograph [S71–16637]). *Right:* Apollo 15 plaque (graphic from NASA [S71–39357]).

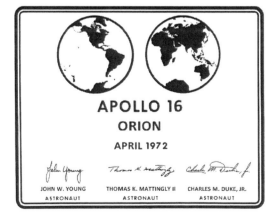

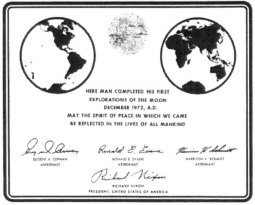

Bottom left: Apollo 16 plaque (re-illustrated by Paul E. Quast, based on NASA/JSC photograph [S89–36956]). *Right:* Apollo 17 plaque (graphic from NASA/GRIN service [72-H-1541]).

alongside the use of specific nation-state emblems that do not extend their cultural prestige to multinational audiences. Most scholars across disciplinary divides tend to disparately share in this perspective on the rhetoric associated with the Space Age; however, in some circles, this may still be seen a controversial conclusion. While our archaeologist may lack this contextual foresight in her investigations, she may unintentionally deduce one accurate motivation for the creation of this Space Age site and plaque: to imprint a specific ideological meaning on the Moon—perhaps, for ceremonial or "shamanist" purposes. Indeed, as suggested by one of our reviewers, it would perhaps also be a worthwhile exercise to dwell on how such a culture of misinterpretations may arise with the loss of this information, after the passage of *enough time* (for instance, the notion of "marking one's territory" was raised in reference to astronaut feces).

This assessment partially undercuts a promising conclusion on how the surrounding archaeological site of Statio Cognitum can optimistically provide material clues that may aid in wholly interpreting the layers of meaning behind the Intrepid Lunar Plaque. Nevertheless, such speculation about this archaeological record is beneficial for improving inferences, further technical discussions, and inciting debate on the perceivable interpretation of these sites through independent, legacy investigations. It is

Photograph of Charles Duke's family left by the astronaut on the lunar surface. The reverse side of this picture features all the protagonists' signatures, preserved thumb impressions, and the brief message: "This is the family of Astronaut Duke from Planet Earth, who landed on the Moon on the twentieth of April 1972" (photograph from NASA/Charles Duke [AS16–117–18841]).

unlikely the appropriate context is available at Statio Cognitum to ideologically interpret the Lunar Plaque, in lieu of any pre-existing knowledge and known context. Yet, such sites are never studied outside of their regional relationships. Assuming the affiliation between specific hardware could be established, Statio Cognitum possesses close, material and temporal relationships with several dispersed mission objects, such as the provisionally identified Intrepid ascent stage (which seemingly impacted over 72 km away), alongside the tentatively identified third-stage Saturn IV rocket body (otherwise known as J002E3) that remains to this day in a helio-to-geocentric orbital cycle. At present, J002E3 returns to haunt the vicinity of Earth orbit every 40 years or so.

Concurrent to this, there are also another five "Apollo culture" (Capelotti, 2009) landing sites[18] which possess hundreds of additional human spaceflight artefacts, alongside six comparative specimens of intact Surveyor hardware, and other commemorative deposits such as the five additional Apollo descent stage plaques, *Apollo Goodwill Messages*, the *Fallen Astronaut*, Jim Erwin's eclectic caches and other astronauts' contraband to support an interpretation of our Intrepid Lunar Plaque. It may be possible to also cross-examine this subset of American 1969–1972 heritage with similarly-aged, yet comparatively poorly studied, Soviet sites (Finney, 1992) that possess their own distinctive design traits and obscure pennant artefacts, in addition to deposits left by several other nations (Stooke, 2007), and the forensic legacies of intact crash sites (Spennemann, 2009).[19] However, these missions are only one small sampling of the wider density of intergenerational, and ever-advancing, material culture from the Space Age residing on these nearby astronomical bodies and wandering the expanses of our local stellar system.

One of Life's Little Problems

Perhaps one of the most significant material relics that we *have already deposited* into outer space, which could foreseeably contextualize our curated message libraries, is terrestrial life itself. I do not refer to the theoretical debates on defining life criteria (usually through the widespread usage of analogous reasoning from terrestrial life when re-mapping earthly experiences. For a detailed discussion on the philosophies underpinning this aspect of astrobiology, see Persson, 2013), or the familiar vertebrates who unfortunately endured our early spaceflight tests, but rather the seemingly inconspicuous microbial zoo or "culturable matter" that, in minute traces, stowaway aboard aerospace hardware. The DNA of these single-celled organisms presents a dense library documenting billions of years of irreducible computation within phylogenetic evolution, redundant mutations, genetic mistakes, and biological adaptations to specific ecosystems, alongside serving as a hereditary blueprint for all ancestral microorganisms. However, these microbe's resilience to extreme environmental conditions can also be ethically problematic.

This challenge raises the specter of the S-3 camera, recovered during the Apollo 12 mission, which was found to be internally contaminated with the bacterium *Streptococcus mitis* on samples of polyurethane foam insulation stuffed between circuit boards (Mitchell and Ellis, 1972). Instances of contamination in this hardware were isolated from one exterior and five interior locations, including samples of *Aspergillus pulvinus,* and another *Aureobasidium* fungal species. Whether this study verifies that these microbes were able to survive on the lunar surface for over two and a half years is accurate or not, Earth microbes have proven to be a tough bioburden to evict from space hardware, no matter what sterilization procedure is employed to pre-treat spacecraft (Glavin *et al.*, 2004). In this context, the reader can imagine the bioload presently inhabiting the deep innards of the recently launched Tesla Roadster (Szondy, 2018) (see color insert),[20] or populating the recent Beresheet impact site alongside the shipped colony of deceased Tardigrades—after all, these organisms are symbiotes containing microcultures (Sample, 2019). At present, we have only one dataset available from re-visiting prior space hardware to contemplate the astro-zooarchaeological impact of our growing bio-footprints off-world, but *this contamination is expected to also reside elsewhere.*

In principle, I agree with the established, international consensus on ethical practice governing the purposeful introduction of invasive life into non-terrestrial landscapes, specifically in regard to the balancing of our modern technological capabilities to frantically do so, against the moral dilemma of whether we should be carelessly conducting panspermia experiments—the "just because we can, should we do it?" argument. Certainly, the history of invasive "neobiota" across Earth's habitats should give us cause for concern for the ensuing impacts of such hasty or inadvertent actions. Given this author's research focus on examining deep-time legacies and prospective impacts, the proliferation of microbial contamination on other celestial bodies is highly relevant for future scientific investigations, not to mention good planetary management policies. There have been several legitimate scientific enterprises involving life experiments in Earth orbit,[21] and on other astronomical bodies,[22] alongside accidental inclusions (S-3), and cargo ejected to make a human spaceflight return feasible (as a precaution, the bags of Apollo emesis and excrement were apparently isolated within the equipment bays of the descent stage). All these objectives are reasonable, given the fact that they are either isolated experiments, or unfortunate accidents, and it is likely that harsh ultraviolet rays have already sterilized many of these samples into dry dust. Certainly, planetary protection policies should work *both ways*, and force us to mitigate these instances as best we can, while also documenting where anthropogenic contamination events may have taken place. However, it is troubling to see an emerging practice of purposefully forward-contaminating other environments outside these circumstances, with questionable justifications, or ethical oversight by those who contribute to mission engineering or objectives.

Such notable actions present an unscientific, slippery slope precedent that can only serve to embolden those who wish to further "spill" life on to other worlds as a personalized "second genesis" (not too unlike our *gestures to eternity* category), technological demonstration for using DNA as a cheap data carrier for encoded human heritage resources (Zubrin, 2017), or perhaps to "repopulate Earth, rebuild humanity, and kickstart humanity's outward migration as we spread life in the Universe."[23] The Biblical story of Noah's Ark, a frequently cited metaphor for the inspiration behind such lifeboat projects, posits that it is humanities' divine role as a global custodian to preserve and shepherd other biota into the future. But, it is unsettling to see how rational minds—some of which are cognoscenti in their respective disciplines—approach this exigent challenge of our Anthropocene age by proposing to scatter DNA samples into likely irretrievable locations, as opposed to investing within actual biodiversity conservation programs. After all, in the last 50 years, we have only re-visited one such site for retrieval applications, a journey with a heavy price tag. It would be truly discouraging for future conservation efforts to only know of endangered or recently extinct species as cell cultures residing on a lunar petri dish. This all too brief assessment, of course, does not take into account the established biocentrism arguments, value judgements, ethical choices, and spiritual tropes of human attitudes we export alongside such decisions when deciding what life *we deem useful* for such preservation projects—contexts which our future archaeologist will already possess through her experienced living world (or remnants thereof).

There is an extensive body of literature pertaining to the ethics of planetary protection protocols (Cockell and Horneck, 2004; Cockell, 2005; Kramer, 2014; Schwartz, 2016; Kminek *et al.* 2017) which must be counterbalanced against any future, exogenesis decisions to needlessly damage prospective "planetary parks" on other worlds. While *astro-environmentalism* is a controversial topic within space communities for separate ethical and philosophical reasons, such rational, *a priori* discussions should at least be an ethical hallmark of a sapient, conscientious species while we attempt to conceptually reconcile this approach with other competing agendas. It is unclear how apparent such ethical discussions and prospective contamination risk assessments have been within the designing phases of biotic messaging projects and data storage technologies for space to date. Having needed to complete a mountain of paperwork to previously work with simple genetically modified organisms (in a laboratory setting), I would expect similar precautions to already exist when *intentionally* transporting microorganisms off-world. Per contra, and in conflicting with my opinions on the matter, it is equally understandable how beneficial these approaches of using DNA archival solutions would be for conveying both information and crucial meta-semiotic context across spatiotemporal distances, providing a complex, biological archive as a common referent object for the sender, and prospective recipient, to mutually hinge semiotical assumptions upon, ensuring "the medium is the message" as discussed earlier in this volume. Regardless of our variegated beliefs and technological capabilities to achieve these applications, such messages "on behalf of humanity" should ideally be openly discussed by humanity, or at least mission ethical boards that actively contribute to these standards, prior to unilateral launches of these intentional and irretrievable bio-footprints.

Understanding the Recent Past, Protecting the Distant Future

Whether the futurity of space archaeology depends upon incomplete antiquated records,[24] Apollo 14's lunar golf balls and other dusty artefacts, indirect perceptions of relic satellite infrastructure, ancient microbes, or fragments of these material legacies found on the bottom of off-world rivers, *alongside the continuity of our theories and scientifically-verifiable methods*, the conclusions that will be drawn by these investigators will be dependent on observational data, colored by the panoply of presumptions they may bring into such analyses. By many accounts, these brief case studies, and sources of preliminary data, provide some evidentiary value for celestial archaeological sites in contextualizing resident message artefacts as purposeful, symbolic effigies. However, such examples are likely more imaginative and optimistic than what will be encountered by the first space archaeologists to visit these slowly degrading sites.

Similar to languages, the commemorative or votive comprehension of these messages will likely become abstruse as posterity nations, cultures, traditions, and heritage customs change in unimaginable ways over intervals of geological time. One thing we can be sure about, however, is that such messages and any ancillary environmental contexts lent by the surrounding material remnants, will signify *different things* to *different people* at *different points in time*—connotations that are subject to the observer's *own historic and contemporary experiences.* We may find that understanding *our* exoatmospheric sites like Statio Cognitum, or mobile remnants of material culture such as Voyager 1, entails the finding of relatable elements of *ourselves, our* mental relationships with transformative matter, *our* value judgments, and *our* beliefs reflected within these highly emotive and intricate artefacts, contexts which may not readily translate across space and time to other minds. Unlike our future archaeologist in these brief scenarios, we are fortunate to be deeply embedded within the same socio-cultural milieu that initially established these recent sites, and can rely upon our shared anthropological, notational, and cultural histories, alongside a familiar semiotic context with supportive documentation of these missions, to comfortably impose our labels on almost any asset we have launched during the Space Age. Simply being human even confers many contextual advantages but, should our future archaeologist be extraterrestrial in origin, they will likely require much broader socio-cultural, bio-sensory, environmental, and technological contexts to inform their own provisional, studious determinations about these messaging artefacts or archaeological sites (*a posteriori* locations or otherwise).

In the absence of best-practice agreements and multinational recognition of these sites or other mobile artefacts of archaeological significance, we are left on tentative footing as to whether these material deposits will remain intact from immediate human intrusion over the foreseeable futurity (Barclay and Brooks, 2002). Our archaeologist may get her permit to survey Statio Cognitum, or the more iconic Tranquility Base, in the 81st century only to discover excessive activity from twenty-first-century space tourism has all but erased the distinctive heritage properties of these sites—*one wrong step for (hu)mankind.* Perhaps, Kilroy-inspired graffiti may be found on the Intrepid descent stage, personalized gestures to eternity may be deposited across the site, payload objects may be extracted[25] for lucrative distribution on the collectors' market, commercial rover tracks[26] may have encroached upon much of the remaining trace fossils, or the entire site may be levelled—presenting an intellectual jigsaw for salvage anthropology efforts, or even forensic architectural assessments. Perhaps, doing "as much as is necessary and as little as possible," as per the tenet of the Burra Charter (2013), may aid or inhibit this deterioration. However, we cannot definitively rule out the impacts that future tourism will sporadically have while such measures are debated. The documented destruction of Palmyra sites, the Buddhas of Bamyan, and other ancient monuments are salient examples of our modern capabilities to intentionally erase others' social, cultural, and historic heritage for the sake of an ideological difference, in addition to the callousness of past colonialist transgressions for similar aspirations. The bioinformatician Thomas S. Mehoke (2009) has even quipped that our defunct space probes may, one day, be used as target practice by Klingons, as per the movie *Star Trek V: The Final Frontier,* or perhaps as further surface-to-space weapons test targets by posterity nations who wish to flex their geopolitical muscles.

Given the Moon's remarkable, close history with human symbolic and material behaviors since antiquity and prehistory, we should also be mindful of how Western exploration and disposal of waste on its surface may also impact upon the customs, stories, and beliefs of other cultures we communally share this natural satellite with. No one nation or people own the Moon, despite frequently misguided claims. We would hardly condone such littering and destruction of UNESCO World Heritage sites or Blue Shield protected historic structures by international tourism, so we should hardly expect it to be communally acceptable for off-world sites of heritage intrigue—or planetary parks. There is also a case to be made for *inadvertent* damage to existing heritage deposits, as exemplified by the accidental erasure of some previously unidentified parietal art in the Grotte du Cheval cave.[27] Accordingly, a sanctioned human presence may even be a cause for concern in maintaining site integrity, as we have plainly seen with the nearby Intrepid plume effects and astronaut excursion to S-3. The ensuing missions to the Moon will serve as a bellwether in this regard for the rest of the inner planets.

In the absence of agreements over the recognition of these celestial heritages, the "commons" of these sites and state-owned hardware may simply end up mirroring similar heritage locations of natural or historic beauty on Earth. Our vital storehouses of knowledge necessary to comprehend these pioneering leaps off world and, by extension, the messaging artefacts of the Anthropocene on other worlds may, like most of the twenty-first-century culture, be consigned to the forgotten corridors of time or serve as nondescript waste deposits that require exhaustive deductions for even basic archaeological determinations. Advocacy groups for preserving these sites—on other astronomical bodies, in resident orbit(s), or as repatriated museum displays[28]—vary far and wide, often supporting conflicting visions, opinions and forethought on how these historic event remnants may be feasibly conserved for posterity (Fewer, 2002; Spennemann, 2004; Rogers, 2004; Stooke, 2007; Brooks and Barclay, 2009; O'Leary, 2009c; Gorman, 2017b; Westwood *et al.* 2017).

Whether heritage conservation costs and associated legacy resources are also required for maintaining—or preventing access to—these prospective off-world heritage sites, alongside who will be responsible for providing these financial, technical and logistical services, is yet another, particularly challenging dimension to this custodianship (Barclay and Brooks, 2002),[29] especially when counterbalanced against overstretched funding for other deserving causes, like simple eco-maintenance activities on spaceship Earth. Lessons from maintaining the active ISS, Hubble Space Telescope, and incarnations of older space stations like Salyut, Skylab, Tiangong, and MIR, may provide a logistical inkling into the economic costs associated with preserving mobile museums in orbit *and beyond*. After all, significant elements of culture will likely incur significant conservation costs. This is a truism in contemporary museum sectors, which already need to allocate substantial resources towards the planned future conservation of assets, prior to even acquiring an artefact for their collections. However, the best-laid plans can only stretch so far into the future and become further exasperated when the distant asset constantly moves at hyper-velocities (relative to the conservator's base of operation). In light of this, a fundamental paradigm shift within our frame of reference for exoatmospheric archaeological sites and conservatorship is required before any decisions can be feasibly made, otherwise trust degrades in the face of ensuing acts of failure, alongside with reputations of contributing parties. For the moment, such responsible discussions are actually outpacing the technological and logistical demands, but we have an ever-shrinking window to enact such prospective decisions or conservation soft-laws.

On the Precipice of a Future Archaeology

From its origins in William Rathje's colorfully titled *Projet du Garbage*—who also initially coined the term "exoarchaeology" as "the study of artefacts in outer space" (Rathje, 1999; Rathje and Murphy, 1992)—space archaeology has become a remarkably generative field of international study. The field has migrated from the relatively obscure fringes of Space Age history to take center stage in contemplating how our successive

Astronauts Steven L. Smith and John M. Grunsfeld during an EVA to replace several gyroscopes contained in the Rate Sensor Units inside the Hubble Space Telescope as part of the servicing mission STS-103 flown by the Discovery shuttle on 19–27 December 1999. While ST-103 was specifically a repair mission, it is predicted that objects like Hubble will also incur significant costs for safe returns to Earth (as assets for museum displays), though there are some smaller precedents with the return of Westar-6, Palapa-2 (both re-launched), LDEF, and EURECA. As Hubble incurred 5 of these crewed servicing missions, the object is also considered another documented example of "multiphase occupation" in space (photograph from NASA/JSC).

steps off-world may have left an indelible impression on modern human behavioral patterns. This impact has subsequently manifested outside of the initial launch-capable populations to encompass the broader customs and interactions of various cultural lifestyles that we canopy define as a global human civilization. The field is a relatively nascent discipline of study; however, it has already gravitated a number of experts from the historical, industrial, legislative, scientific, and technological communities involved within contemporary space exploration—parties who continue to compile charts, collect spacecraft manuals, and establish legislative precedents, or other theoretical frameworks, necessary to formally recognize these objects outright as *a posteriori* heritage deposits in *terra nullius*, worthy of investigation and, perhaps, conservation measures for their geopolitical, technological, cultural, aesthetic, scientific, and spiritual significance. I will leave it to the reader to decide which category the bagged Apollo biological samples fall under.

In the ensuing decade, there will be an impetus for substantial growth within this multilateral discipline as these various parties continue to study and identify sites of socio-historic significance,[30] apply new working theories for further scientific examinations, create sufficient documentation for these regions as part of heritage resource management strategies, draft a menu of conservation guidelines to ensure the preservation of these legacies,[31] and advocate for the formal recognition and ethical management of this common human material heritage—hopefully as a multinational commission if it is to be realized (Spennemann, 2004).[32] Whether such activities will enact long-term change is acknowledged to be highly dependent on our civilization's other coinciding priorities and exigent agendas, most of which contend with contemporary terrestrial and custodial affairs that require crucial resolutions for the benefit of future generations who may eventually undertake such imaginative space fieldwork studies. Despite this, such a natural, protective extension for humanities' archaeological record in outer space is a tantalizing prospect that could, in theory, only be beneficial for our future archaeologist to study the long-term societal changes and legacies resulting from Space Age technologies on the story of *homo sapiens* and, by extension, the messages we leave behind to consciously preserve fragments of human thought for anyone who may encounter them.

NOTES

1. This is a systematic cross-comparison with these societies, as opposed to directly applying isolated analogies from these cultures to explain other idiosyncratic behaviors, customs, and values, of past human civilizations.

2. Due to the nature of this volume and the author's own research preferences, I have chosen to focus specifically on off-world resources; however, there is a rich body of literature that contends with the terrestrial remnants of Space Age heritage infrastructure on Earth, i.e., historic launch pads, communication stations, space debris etc.

3. This includes how societies react to telecommunication systems and remote observations of Earth's surface.

4. Such power differentials between societies with access to technology who write a nation's history, and those who cannot do either, should not be overlooked as this issue also extensively manifests within the record of message contents; the concept of representing "us" in messages becomes much narrower in this regard.

5. A pathfinder spacecraft that landed on April 20, 1967—two years prior to *Intrepid* (November 24, 1969).

6. There were other scientific objectives for this mission such as setting up the first long-term lunar experiments, which remained functional until they were deliberately shut down on September 30, 1977, due

to lack of governmental funding. Several planetary scientists regard this as a spoilt resource and an apt metaphor for the political ethos governing the Space Race and national space programs in general.

7. As noted by Jarvis (2003), this is a standardized chronological benchmark for millennial time capsules that was initially established by Thorwell Jacob's *Crypt of Civilization*, based upon an initial "zeta" starting point of human civilization in 4241 BCE (the earliest recorded calendar date in Egyptian bureaucratic-priestly lore). The *Crypt of Civilization* (initially conceived in 1936), in turn, projected this figure of 6,177 years into futurity, giving a targeted capsule-opening "omega" date of 8113 CE. Given the extended timeframe, and for the sake of this exercise, let us assume our archaeologist is also not familiar with the cultural, symbolic, or linguistic paradigms of the populations that launched these artefacts, and that she is approaching this site in a similar manner to an archaeologist surveying an early human settlement.

8. While this photographic film was accidentally forgotten and has likely degraded due to high-energy particle exposure over the last 50 years on the lunar surface, it is intriguing to contemplate how such an *a posteriori* time capsule deposit would be interpreted by our future archaeologist.

9. Binford's original term applied to disposed material waste from Nunamiut hunting camps, not the lunar sites.

10. Perhaps, it could be argued that these footprints can tell a personal story for each of the astronauts, providing archaeologists with an intentionality to their behaviors and activities on the Moon. (They may have been steadfast in taking direct pathways for completing mission objects, as per their training, or impressions could depict instances where the astronaut meandered off-trail into relatively vacant regions, in pursuit of curiosity over mission.) Capelotti (2010) raises a rather amusing example of this subconscious manifestation of human behavior, apparent within various Apollo expedition photographs; astronauts seem to walk alongside recent lunar rover tracks, imitating the manner in which a pedestrian uses a separated footpath beside a roadway.

11. ALSEP is an abbreviation for the "Apollo Lunar Surface Experiments Package" units placed on the Moon's surface by Apollo missions 12–17. The remnants of the Apollo 13 package are still at the bottom of the Pacific.

12. After conversions, the (intact) S-3 spacecraft, with a landing mass of 296 kg, would have weighed 49 kg in the lunar gravity. If this lander was manually deployed as equipment by astronauts during the Apollo 12 mission, the astronaut footprint depressions would have been far deeper, in a similar manner to the ALSEP tracks.

13. While the descent stage may have provided some shielding for several of the Apollo mission footprints, these impressions are likely to be still heavily degraded as a result of the ascent stage launch. The "first-footprints" from Apollo 11 may also be heavily degraded when compared to the iconic photographs of these features.

14. These biological products may also enable our archaeologist to (if necessary) perform Carbon-14 dating to confirm the age of the site, in addition to other comparative resources such as measuring the rate of decay for products within the radioisotope thermoelectric generator that powered the ALSEP experiments. The quantity of excrement in these vessels may also denote how long the astronauts were resident on the lunar surface.

15. By comparison, even the *Project Gnome* epigraph on Earth (marking an underground scientific nuclear detonation) contains a richer corpus of symbolism and redundancy material on an oxidizing copper plaque.

16. Technically, our archaeologist's judgement can be construed as right and wrong here. While the deposits on Statio Cognitum are artefacts of Americana, it is worth noting that the Apollo program is cited by NASA to have consisted of over 400,000 people, and 20,000 industrial firms and universities—many of which subcontracted to international collaborators, alongside seeking other international partnerships for global tracking systems, etc.

17. As cited from the engraving on the Apollo 11 Lunar Plaque. The phrasing was intended to be inclusive of all humankind. However, a number of critiques have noted the patriarchal semantics tend to alienate a large portion of the human population along gender lines, in addition to also commenting on how the rhetorical frontier tropes underpinning this message have historically misrepresented others' cultural, political, and ideological beliefs. In a rather poignant criticism, Carl Sagan had also questioned the veracity of this message, in context with 1969 geopolitical world history events simultaneously occurring while the Apollo 11 mission was underway.

18. Sites are located at distances of 1,425 km (Apollo 11); 181 km (Apollo 14); 1,187 km (Apollo 15); 1,186 km (Apollo 16); and 1,758 km (Apollo 17) from Statio Cognitum. Each site also possesses close associations with their lunar ascent modules and Saturn IV rocket bodies that crashed elsewhere on the Moon. The Apollo 13 Saturn IV rocket body is the only material heritage artefact from this aborted lunar mission to be deposited.

19. It is regularly cited that human missions to the Moon have deposited over 187,400 kg of artificial materials on this natural satellite's surface as landing equipment, crashed probes, experiment packages, or other debitage.

20. This car is Elon Musk's personal Tesla vehicle, which was frequency driven down the Los Angeles motorway. According to a collision probability assessment (Rein et al., 2018), there are several contamination possibilities.

21. The International Space Station hosted the EXPOSE-E, EXPOSE-R, and EXPOSE-R2 experiments, which allow exposure of chemical, material, and biological samples to outer space while recording data during this display.

22. The Chinese *Chang'e 4* lander reportedly germinated cotton (with visual confirmation), rapeseed, potato, and *Arabidopsis thaliana* seeds within an enclosed container on the Moon's surface (Wong, 2019). Unfortunately, the fruit fly eggs failed to hatch during this brief window in the lunar month.

23. Quoted from the *LifeShip* website—a space project for saving DNA samples, accessed on February 28, 2020.

24. As an example of the limits of incomplete records, NASA engineers (when looking through past spacecraft blueprints in order to redesign scoops to remotely acquire soil samples for future missions) noticed that the S-3 diagrams were missing. In a rather ironic twist, the engineers had to allegedly base their designs off the actual S-3 remote sampling arm, which had been retrieved from the lunar surface during the Apollo 12 mission.

25. Lockheed Martin has already expressed interest in acquiring additional artefacts from the Apollo 12 site, likely as samples for further research into aerospace materials and technologies tests.

26. The *Google Lunar X Prize* (Clery, 2016) offered a "heritage bonus" to lunar pioneers that could visit and photograph these Apollo landing sites from within the defined "safe" boundary of these delicate areas.

27. The walls of the Grande Grotte at the Arcy-sur-Cure cave complex were partially cleared of modern graffiti, but this work inadvertently removed pigment from a few unidentified parietal art deposits under these marks.

28. Technically, recovering space hardware for display in the origin nation's museums is a feasible way forward in compliance with international space treaties. However, recovery is an arduous and costly proposal, especially considering the retirement of the U.S. Space Shuttle program (an average cost per shuttle launch was $775 million).

29. From personal experience, the economic costs associated with maintaining appropriate conservation measures for artefacts is frequently eclipsed by the initial desire and advocacy of attaining custodianship of these relics, sites, and monuments. The true cost of a heritage assets, however, lies in the ensuing upkeep over time.

30. The *Lunar Legacy Project* (O'Leary, 2009b) has extensively documented Tranquility Base, *ISS Archaeology* is a group studying the "microsociety" of the International Space Station, while the Beyond the Earth Foundation has produced an interactive *Heritage in Space* platform as an active catalog for peer-research and engagement.

31. Preliminary conservation guidelines have suggested to formally draft an external boundary around protected regions to ensure present day landings/launches do not unintentionally interfere with these sites (O'Leary, 2009b; Westwood *et al.* 2017), while space archaeologists have also proposed to use synthetic radar aperture data from remote satellite observations, much like those pioneered for studying the Nazca Lines (Comer *et al.*, 2017), to monitor for natural or artificial degradation.

32. The non-profit organization *For All Moonkind* is devoted towards formally establishing the legal recognition of these heritage sites, and currently possess observer status on the United Nations' *Committee on the Peaceful Uses of Outer Space* (COPUOS). This group has also established and circulated a best-practice guideline document to protect early Space Age lunar archaeological sites—a document which has accrued a number of signatories from commercial and governmental leadership teams, intending to launch future lunar hardware.

Who Will Speak for Earth?

*Stakeholders, Scholars, and Stewardship
in the Future Exoatmospheric Record*

PAUL E. QUAST

Over 40 years ago, Carl Sagan (1980) popularized the question "Who Speaks for Earth?" as a ruminative foundation for encouraging future discussions on this multifaceted challenge, in retrospect of his decade-long expertise in personally contributing to the exoatmospheric archaeological record.[1] Sagan was not the first to pose this complexity but, in many respects, he and his group of collaborators embodied the ambassadorial role envisioned within the fantastical novel *I Speak for Earth* (Brunner, 1961), inviting us to consider other worlds, other minds, and other timeframes during a precarious period in our geopolitical history,[2] while also establishing some of the basic messaging principles that would be employed within lineage space-time capsules and some interstellar transmissions (at least until modern METI communication theories proliferated). Of such gestures to eternity, Sagan keenly remarked:

> For those who have something they consider worthwhile, communication to the future is an almost irresistible temptation … it expresses great hope for the future; it time-binds the human community; it gives us a perspective on the significance of our own actions at this moment in the long historical journey of our species [Sagan *et al.*, 1978]

Answering Sagan's quandary seems intuitively easy at first, but it has proven to be deceptive in practice. The future of deliberations over myriads of space-messaging practices raises an abundance of important ethical and sociological questions which continue to draw upon the social, democratic, philosophical, theological, cognitive, and epistemic threads that presently define our diverse conceptions of the "human condition," and our inherent responsibilities outside this anthroposphere. Crafting accounts for a plethora of different applications can be liberating but also contentious, existential as well as jubilant, personal as much as egalitarian, command veneration or spark chagrin debates, and serve as a modern critique for our civilization, celebration of life, or paradigm shift within long-established worldviews. But few answers to this complexity have been satisfactory, and most approaches have embodied centricity in some form or another. If looked at as a statement, this quandary, and our decision(s) on how to address it, are inextricably bound to our deep-seated motivations for why we think we should speak, who we believe these messages are representing, and what content we ought to communicate, alongside recognizing how the value judgements we commit today may theoretically influence the human archaeological record. Examples of these influences can range from self-documentation of our social and cultural interactions during the unfolding Anthropocene era, preservation of essential information for future steward-ship applications, and contamination of outer space environments alongside the habitats of other astronomical bodies, to prospective asymmetrical communication with an ETI (an inherited cosmic phone call which will need to be answered by unborn generations).

Drafting a collective message could be a lethargic, difficult, or laborious process, with any reasonable way forward necessitating us to embrace diversity rather than uni-lineal evolutionary narratives and exercise intersubjectivity within our thoughts, along-side actively recognizing that we cannot resolve all encountered issues with a single strategy. Such abrasive ideas also tend to transition us from a curatorial context to an ethical one; do we necessarily need to speak as one, for all, or even at all for that mat-ter? Moreover, reflecting on such "time-binding" decisions may also grant us pause to critically assess the "state of affairs" on our planet *now*, and challenge us to be bet-ter by demonstrating our capabilities to sustain conversations upon which the future of Earth now depends. Such considerations may seem overwhelm-ing, however, failure to grasp these big-ger lessons, and questions associated with representational practices, may unintentionally lead to messages that serve as monuments to the hegemonic egos of bygone generations or sterilized depictions of our complex, heteroge-neous world—unfortunate accounts "on behalf of humanity" that subdivide us,

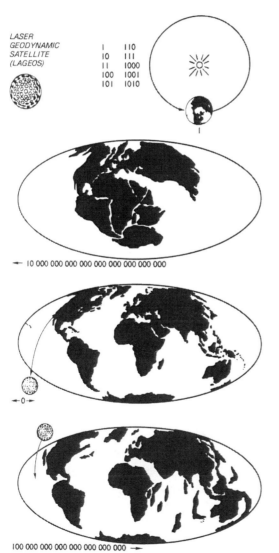

The LAGEOS Plaque, designed by Carl Sagan at Cornell University in 1976. The plaque features a stylized LAGEOS-1 satel-lite, along with a representation of Earth's orbit with a conventional arrowhead demonstrating direction of motion. The upper area of the plaque establishes binary arithmetic as a counting scheme, which is then used to approximate continental drift of Pangea (from the Permian period, about 268 million years ago) to our present tec-tonic configuration (middle projection), alongside a future prediction of Earth's configuration 8.4 million years from now (when the LAGEOS orbit is predicted to decay). The location of Vandenberg Air Force Base (the LAGEOS launch site) is also shown in present and future maps. Also featured in the Voyager Record as image #39 (photograph from NASA/Goddard Space Flight Center).

rather than unify. Yet, we continue to meet this innate urgency to have our voices heard or "yoohoo-ing" (Brin, 2014) across distances, "to represent ourselves and connect with the unseen" (Denning, 2011b), our minds (or subsets thereof) contemplated from afar by recipients with qualities we may not distinctively recognize as human. Indeed, from reviewing these accounts, this author barely recognizes the planet, or species, that we often describe with such enthusiasm.

At present, "Who Speaks for Earth?" using *de novo* messages is synonymous with the freedom of speech as, in the words of the late psychologist and SETI advocate Albert Harrison (2014):

> [T]here is no real way to control efforts to speak for humankind. As more and more people gain access to powerful computing and broadcasting equipment, there are few if any practical ways to keep free-lancers off the air [radiosphere]. Any government, scientist, theologian, entrepreneur, or hobbyist who has access to a powerful radio transmitter, laser, or spacecraft could send an irretrievable message into the void.

At present, these irretrievable messages manifest across an extensive array of material categories, for an equally vast number of reasons and motives. As a sample menu of some: EM emissaries to distant civilizations, envoys for tortilla chips, opera invites to Qo'noS, gambling-addicted chatbots like *Ella* (Oberhaus, 2019), the cacophony of secretive messages using amateur equipment, plaques, DVDs, books, Lego, human hair, DNA, questionable biota samples, novelty memorabilia, and electric sports cars as a symbol of "wasting excessive wealth" (Gorman, 2018). These are but some examples of diverse ambassadors emanating from present-day Earth, as this accumulative history of exoatmospheric messaging has largely been a result of small, disconnected cohorts of individuals who wish to implement their thoughts and designs upon the "final frontier." METI has, so far, bore the brunt of this criticism for obvious reasons we shall see later. However, as with every technological endeavor that far outpaces social adjustments, ethical foresight, and legislation not yet written (Herkert, 2011; Marchant, 2011), this vagueness in guidance, oversight, and critical scrutiny can also turn into a nightmare for satellite services, ground-based observation programs, planetary exploration missions, and the usage of these regions as a prescribed *common province of all humankind*.

Given the ambiguity within the budgetary requirements, planning, and fabrication phases of messaging ventures (not to mention, the frequent failure of such crowd-sourced, goodwill projects), it is acknowledged that the following discussion is highly speculative—owing to this definitive lack of reliable evidence, but also the inherent unpredictability in assessing future voices within exoatmospheric material culture deposits. My intention here is not to provide a manual for creating acceptable messages (at this point in time, I believe that no such consensus could possibly exist, and merely wish to highlight this persistent bone of contention), best-practice guidelines, or render any reliable interpretation on historical projects that "speak for Earth." Rather, I will discuss some of the complexities and common failings associated with current incarnations of the messaging practice, alongside examining some social attitudes that contribute to this enigmatic and intangible material culture. For the purposes of this exercise, and to buttress my discussion, I'll outline some contemporary research avenues and recent project announcements, while framing these initiatives under general practical and ethical quandaries that are frequently associated with constructing intelligible messages.

What Should We Say?

Heidmann (1993) once suggested that the *Encyclopædia Britannica*, transmitted into space with little concern for decipherability, would suffice as a communicative medium, capable of indiscriminately relaying comprehensive quantities of knowledge about our diverse terrestrial civilizations to anyone who may be listening. Since then, many similar proposals that share this perspective have suggested the U.S. Library of Congress, World Wide Web servers (Shostak, 2011), or Wikipedia[3] (Wikimedia Deutschland, 2016; Freeman, 2019; Spivack, 2019) could also suffice as a modern "Encyclopedia Terra," capable of facilitating access for any recipient "sufficiently intelligent" to analyze syntactic patterns of redundancy information to uncover archetypical universal grammars, or perform translation campaigns for "meanings."

To briefly unpack Heidmann's argument, this author finds the concept of an Encyclopedia Terra tantalizing, given the fact that the transmission of such a hefty repository of techno-cultural knowledge in an undifferentiated stream of binary bits would inherently and haphazardly relay copious amounts of metadata, information-theoretic, and epistemic scaffolding for how we organize our entwined minds, bodies, and relational worldviews. The recipient could, thereafter, plausibly apply algorithmic signal-processing procedures to determine features such as corpus lexical structures and frequency of grammatical repetition and deduce the complexity of information, alongside assessing whether there may be any affinity to their own language faculties (Elliott, 2011; Elliott and Baxter, 2011; Doyle *et al.*, 2011). Such variegated *gestalt* can reveal subtle, idiosyncratic glimpses into the intellect of the sender(s)—even if the sanctioned information remains opaque, to which Marvin Minsky remarked, "the process by which you parse a language and understand a grammar is much more important than the grammar itself" (Sagan, 1973). In this respect, these encyclopedia are representative of the socio-symbolic dimensions and cultural interactions of the specific code-system creators. But such brash *quantity over quality* approaches may also be fallible given we often, sometimes uncritically, utilize technological vernacular when comparatively describing the conceptual architecture of human brain functions directly with modular computational operations—often with Cartesian demarcations openly implied (Quast, 2021).

Let us consider this discussion under a worst-case yet, given the time intervals for recovery, pragmatic scenario, whereby future terrestrial languages, or pictorial conventions have radically diverged or become extinct in favor of newer constructs that bear no resemblance to contemporary framing habits (and there is a lack of discernible records about our civilizations as per oft described "catastrophe" scenarios). We'll assume that human posterity—descendants of the species that crafted these contents and who share some familiar contexts—are the intended recipients of an encoded EM signal, however vague or impossible this scenario may be. Let us also accept the argument that applied analytical tools will plausibly be capable of accurately interpreting our encoded encyclopedia as alpha-numerical scripts and bi-dimensional bitmaps, before formatting this information into visual displays, similar to the layout of a physical space-time capsule. It is worth again acknowledging that this identifiable hierarchy of information structuring does not equate to a full semantic analysis or translation of the meaning infused within these abstract, symbolic scripts (see Quast and Capova, this volume), nor does it unravel the connotations imbued within our abstruse photographs (see Gillespie, this volume).

Reaffirming this challenge does not bring us any closer to a probable solution, but, in the very least, it enables us to interlink this issue with the foundational cognitive and cultural dissociation we have historically experienced with temporally situated parietal artworks and extinct language scripts. Accordingly, human posterity or an extraterrestrial recipient may understand how the body of information is *structured or represented*, but this useful mental notion does not equate to an understanding of what this information *signifies*, a readability obstacle which we have been unable to fully resolve within ancient human texts or graphemes (Pope, 1999). We cannot rely upon the imagined recipient possessing such interpretative ingenuity, even if we include elaborate cribs, primers, and pidgins to promote accessibility for defined "intelligent" recipients. It is impossible to bypass these plainly semiotic problems using such analytical methods that flagrantly violate the interpretive chasm between a detached material sign, signified meaning, and the unknown signifier (i.e., the observer). Interpretation, after all, is a triadic relationship, seeped in many living contexts, and we should refrain from inferring about the capabilities of the recipients based upon empirical data (we have none). In the words of the astrophysicist Yvan Dutil (Perkins, 2012), "You can't just start with a complex message…. You have to do a lot of teaching before you can talk." Similarly, the philosopher William M. Schuyler (1982) also pointedly remarked of this enduring semiotic challenge:

> In language, there is a trade-off: the more explicit information given, the longer the message. The usual strategy for avoiding excessively long messages is to let a great deal remain implicit. But if too much is left unsaid the message will not be understood. So you can trade brevity for clarity and vice versa.

Putting aside the various professional dispositions and diverse schools of thought actively contributing to the theoretics of this academic discussion, it seems logically *wrong* to conclude that such a proposed library of knowledge would be wholly representative of humanity in a single brushstroke. Given the glaring fact that natural languages and pictorial conventions are hardly even considered universals amongst present-day human populations (Vakoch, 1998b; Denning, 2014), these metonymy may serve as presumptive tropes of (a version of) Earth that may not be accurate or exist, as outlined earlier in this volume. These encyclopedic accounts are delineated in dominant alpha-ideographic-numerical-visual systems and are primarily developed as pedagogic tools for domestic consumption by present-day human hegemonies that possess the same cognitive and social toolkits necessary to assimilate this content—or at least understand other living cultures well enough for societies to function. Our analytical approaches, in turn, are also shaped by this innate foreknowledge. An encyclopedia of (what we at least deem to be) rational material may possess no intuitive meaning, and little foundational basis, for anyone outside those already educated within the conventionality of these contents, nor initiate the desired outcome we hope for (Traphagan, 2017; 2019).[4]

These repositories also pose the rather unique argument of how this limited sensorimotor material embodies the broad range of complementary sensory perceptions available to humankind and other terrestrial organisms; how do we articulate gusto and olfactory sensations across millennia? Do we *scent* or *flavor* message contents for "consumption"? When constructing such intelligible artifices for imagined sensory channels, we must be mindful that our narrow decisions are inherently representing (for a recipient's channels) a representation (our own phenomenological comprehension) of *our* representation (our neurobiological and sensory perceptions) of phenomenology—a transitional chain of signifying meaning, embedded in our cognitive conventions and a panoply of other cultural, historical, and ecological worldviews. Michael Arbib (2013) discusses elsewhere some alternative sensory communication expressions as a departure point from the narrow confines of human sensory dispositions. Similarly, the zoologist Arik Kershenbaum (2021) contextualizes the range of other sensory ranges and channels available on Earth for SETI, while the *MEOWTI* initiative creatively imagines ETI communications from the viewpoint of our feline friends. However, these conversations have, so far, neatly ordered sensory experiences into perceptional categories (sight, smell, taste etc.), without providing any meaningful engagement with cross-relational sensory experiences (such as synesthesia, as noted by one of our reviewers).

Furthermore, it is widely recognized that the curation and classification systems for contents to include within such volumes is subjective, never value-neutral, and also shaped by the cultural histories, methodological traditions, and philosophies of disciplinary knowledge-acquisition models, with each operating on divergent principles of concepts, truth, and empirical data. As the rhetorician Ekaterina Haskins (2007) notes, such paradigms are further modified by the value judgements committed by archive editors, information system curators, or indiscriminate algorithms, each of which possess their own idiosyncratic expertise, interests, desired outcomes, and competing agendas.[5] An obvious example of this can be seen in how our conceptions of human history are subjected to extreme cases of distillation (Denning, 2006; 2011c; Stiegler, 2009), shortening narratives that have shaped world events into concise, popular, linear storylines to the point where they may not be distinctively accurate, possess erroneous discontinuities, or contestably reflect the identities of the cultures they may evoke.

> [E]ven when the broad contours are agreed upon, we often cut history up into pieces to better understand it, or conversely, paste together disparate events into a single story (Zerubavel, 2003).

The difficulties encountered within interpreting this media is further crystallized by the ever-increasing spatiotemporal disunity between the sender and recipient cultures which, given the millennial timescales likely associated with retrieval of messages, will only heighten the lack of a common contextual basis and introduce disruptive noise experienced in decoding efforts.

It is hard to imagine whether an egalitarian corollary of information that may be genuinely representative of everyone, and comprehensible for anyone—including *those you have never heard* of—can in fact exist. As Schuyler previously surmised, the dilemma of such a pre-emptive selection process inherently lies within our reductive attempts to represent the whole for unknowable observers, using the smallest number of constituent units and how to, in trying to evoke all contents that may matter in explanations to an envisioned recipient, avoid cultivating such a Terra Encyclopedia as an identical twin of Earth's biological and technological living memory—comparative in data size, density, and complexity. The tradeoffs for each argument are abound. Nevertheless, from my reading of such notions, all message contents to date broadly segregate into either these privately curated contents (or third-party curated libraries designed for alternative applications), sporadic crowd-sourced media, or an amalgamation of both approaches (as often employed within public interstellar EM messages).

In attempting to answer this query of "what to say" to profoundly unknowable denizens, several interdisciplinary organizations and individuals have instigated crowd-sourcing experiments to establish further consensus on how to address this impasse for further analysis. In assessments conducted on the SETI Institute project *Earth Speaks* (Lower *et al.*, 2011; Vakoch *et al.*, 2013), several common themes emerged within crowd-sourced submissions which may serve as indicators (not empirical measures) for any future messages from Earth, with "We are humans of the planet Earth" being identified as a common, shared topic (but interpretations of this topic drastically vary). However, the authors cautioned that the participants emanated from a small demographic sampling of national populations and social dispositions inherent within the "commons," the database largely being representative of young adult males from regions of the British Commonwealth.

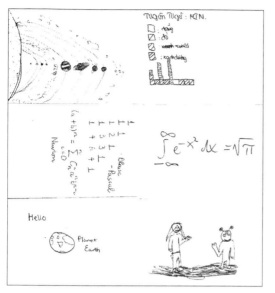

Message ideas and general concepts developed during the Royal Society Summer Science Exhibition 2019 (images courtesy of "Messages from Afar" [fromafar.world] and UK SETI Research Network).

Preceding this initiative, the late Canadian academic Allen Tough (2000) insti-gated *Invitation to ETI* which, in part, invited the global public to issue greetings via the project's website to solicit a response from extraterrestrial probes that may be resid-ing within the confines of our Solar System.[6] As part of the Royal Society Summer Science Exhibition in 2019, the UK SETI Research Network hosted an exhibit which included a consensus-building project for message construction.[7] While these exam-ples are solely consensus-building platforms, there is already a diverse assemblage of crowd-sourced, script-based submissions from completed transmission projects avail-able to study: the Australian Space Week project *Hello from Earth* (25,878 messages), National Geographic's *Wow! Reply* (20,000 Twitter messages), the Lone Signal public library (8,201 messages), and the 3,775 ecological messages from ASREM (Quast, 2017) being but a few examples. The popularity of the VGR as a frequently acclaimed repre-sentational medium could also, in theory, be subjected to similar examinations as part of a much broader effort to elucidate how contemporary earthlings (likely emanating from these same, limited demographic pools) may wish to hypothetically respond to ETI messages, information that may be highly useful for studies by the IAA SETI Per-manent Committee.

Given the lack of observational evidence for extraterrestrials to date, for better or worse, there are technically no right or wrong answers to this conundrum of what we should send, or whether to send messages at all—only opinion, speculation, and selective analogies that corroborate what we ourselves would like to receive from such spatiotemporal encounters. It is, perhaps, more beneficial in the long run to openly acknowledge and accept our collective ignorance in this issue, rather than dwell on presumptions and fantastical extrapolations from ourselves. Perhaps, an infrequent, omnidirectional EM "blip" outwards every quarter century using special frequencies may be enough to "speak for everyone," while also bypassing the intelligibility con-cerns often associated with biographical "Encyclopedia Terra" accounts, to simply attract ETI curiosity towards Earth (and in the process, shifting the remaining com-municative burdens onto such foreign cultures to investigate the significance behind our synthetic cosmic phenomenon)? Moreover, we should be mindful of how often intentions meld together; public opinion on contacting ETI now ranges from person-alizing gestures to eternity for others to know about *our personal lives* to formal acts of demarcating human existence.

Much like the Onkalo deep geological waste depository marker proposals (or lack thereof), it could be equally as important to *not* rely on such messaging devices and contents to transfer information about us across deep space and cosmic time. It may be preferential for such civilizations to perform *their own* searches, on *their own terms*, to garner *their own* rational understandings of humans and Earth from our residual mate-rial legacies and technosignatures—as opposed to favorable autobiographies on how "great we are." It would be debatable whether such an approach by ETI could cultivate a more accurate archaeological portrait of Earth, but certainly they would rely on these observations and metrics when contemplating any contact—a sobering thought for modern Earthlings. Both perspectives represent differing stances on the ethical scales of METI activities and the perennial lack of an established consensus on the often-praised benefits, or self-inflicted harms, arising from this enterprise. Whether the reader agrees with these opinions or not, both proposals warrant further scrutiny in context with other competing agendas that address prospective deliberations with ETI.

How Should We Preserve Information?

Answers for this question are, perhaps, the most matured out of any of the posed challenges. Without sticking my neck too far out into a field that I have only been peripherally involved within, conserving digital or analog information possesses many social, democratic, and interpretive challenges, but the technological fabrication of physical data carriers capable of writing, storing, and decoding this content for end-user reading after intervals of geologic time is by no means a trivial task either (Manz, 2010; Elwenspoek, 2011). Such functions are now of practical rather than fictional interest. Needless to say, so too are our choices of materials and locations for such significant, physical archival missions (Guzman *et al.*, 2016).

The second law of thermodynamics dictates that a closed system always gradually increases towards a state of maximum entropy (or equilibrium), whereby energy (and by extension, highly ordered material structures) gradually degrade towards a maximum disorganized state along a vector of time. Simply stated, there are more *disorganized configurations* for the matter comprising our message to tend towards than organized structural ones, and so this matter will likely transit towards the former *disorganized states* over time, given how unequivocal the latter *configuration* needs to be in order to function in its messaging capacity. Given we do not have any cosmic librarians or archivists actively maintaining distant message caches, it can be assumed that our data carriers will face this process of degradation over time, causing an un-recoupable loss of organized information, both in terms of the physical architecture of the object and the signified contents stored in, or on, the medium. The vacuum of space presents us with an ideal environmental arena, free of geological erosion processes to store a data carrier but at the price of exposing these artefacts to the bombardment of high-energy subatomic particles from the solar wind, or other cosmic radiation sources; extreme temperature fluctuations; impacts from micrometeorites, and anthropogenic debris fields; and the gradual abrasive ebb and flow of cosmic dust that physically compromise storage devices.[8] We use the term "vacuum of space," but this is a misnomer as space is clearly not empty, with damage likely incurred as a gradual, accumulative process, as opposed to solely occurring from transient incidents. Given this, it is therefore necessary to create highly stable data carriers that may offset this entropic decay over longer intervals of time.

There are several institutions and other consortia presently advancing research and practical experimentation within long-term data carriers and communications technologies for use within future messaging projects. Perhaps, the first organization to seriously engage with the modern enterprise of corporeal information preservation technologies for application within deep space was the Long Now Foundation. For the purposes of practically testing new materials for mobile archival applications in space, the Long Now Foundation (alongside their partners Applied Minds) placed a small laser-etched piece of black oxide coated titanium (Rose, 2008) on board the International Space Station's MISSE (*Materials International Space Station Experiment*), hardware that is reminiscent of NASA's *Long Duration Exposure Facility* (see color insert). This is the same medium coating the surface of the *Rosetta Disk*,[9] and the recovered specimen will likely lead to new innovative research avenues in these preservation materials. As an extension to this archival project, a wearable extract of this language database has also been micro-printed by the company NanoRosetta/StamperTech as part of a "lots of copies keep stuff safe" strategy.[10] Another organization called Arnano, based within Grenoble, has also recently unveiled a corrosion-resistant 20 cm sapphire disk as a medium capable of storing micro-etched information for approximately 1 million years—durable material artefacts that have already found purpose within the memory-retention activities of Andra (the French national radioactive waste and management agency) for their transuranic waste depository sites. This technology is also the medium of choice for the similar *MoonArk* and "*Sanctuary*" lunar archiving projects.

The *Human Document Project* is a consortium of loosely affiliated researchers, academics, and enthusiasts who gather for biennial conferences to scientifically discuss how key aspects of contemporary culture can theoretically be preserved for 1

The Rosetta Disk v1.0—a three-inch diameter nickel disk containing approximately 13,000 pages of documentation in over 1,500 human languages, c. 2008. The original, similar-looking prototype version, launched with ESA's Rosetta spacecraft to the comet 67P on 2 March 2004, contained about 6,500 microscopically etched pages of language translations—presenting a comparative lingual guide for over 1,000 languages (photograph ©Long Now Foundation).

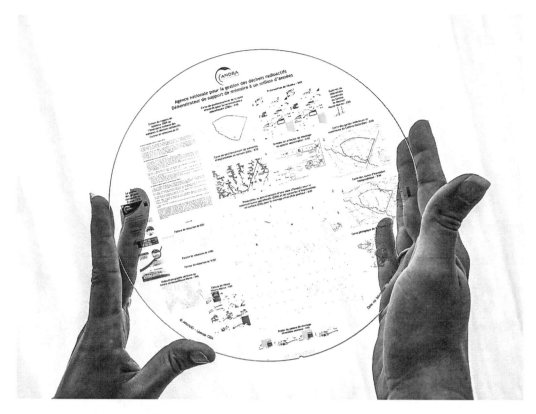

An Arnano sapphire disk containing etched information about the waste depository practices of the French National Radioactive Waste Management Agency (ANDRA) (photograph ©ANDRA/Jean-Noël Dumont, with assistance of Alain Rey/Arnano).

million years. While this group has yet to establish an archival project or contents fit for this experimental purpose, since 2010 the organization has focused on creating a multi-disciplinary scientific community dedicated to addressing the social, technological, and philosophical challenges facing the long-term archiving and storage of information. In recent conferences, several speakers have discussed the uses of nickel nano-etching, ceramic micro-film and tablets, synthetic amber, artificial sapphire and "5-D quartz optical storage devices" or "Superman crystals" (Kazansky *et al.*, 2016) as data carriers capable of conserving information for billion-year intervals.

In addition, this consortium has also mused about the suitability of utilizing non-replicating, synthetic DNA or artificial genes in desiccated microbes or other vessels. However, there are several technical limitations associated with the retrieval of sequences of information from known DNA coding schemes, alongside determining the deep time durability of this data carrier when persistently exposed to cosmic radiation sources. A recent paper by Davis (2020) and his collaborators presents some forward momentum in addressing these complexities by encoding archives in the halophile, *Halobacterium salinarum,* which is subsequently placed in stasis within crystalline mineral salts. However, the use of DNA, alongside several of the other methods listed above, still requires practical experimentation in blind tests for information retrieval and recognition of signified meaning after decoding. (It is easy if you *already know what you are looking for,* and *contexts for the information*). In addition, the intended end-reader will need to initially interpret some of these data carriers as purposeful message deposits and, of course, feasibly re-engineer observational technologies for reading.

Similarly, the Arch Mission Foundation is also advancing development of data carriers that can preserve contents for the foreseeable billion years. This organization is currently experimenting with several of the aforementioned technologies while investing within the improvement of these fabricated media for the benefit of future archival projects. In collaboration with the University of Southampton, this group recently launched a 5-D optical storage disk containing an encoded copy of the Asimov Foundation Trilogy in the glove compartment of Elon Musk's infamous Tesla Roadster. As there is a write-speed restriction presently associated with this technology,[11] subsequent "Archs" have come to adopt the established micro-etching technologies of NanoRosetta, in which this group has subsequently invested to create a higher-resolution variant of this archival medium they call "nanofiche." This format is capable of densely storing thousands of analog pages that may be read under an optical

The "Arch" quartz storage device containing an encoded copy of Isaac Asimov's Foundation Trilogy, which was placed inside the glove compartment of the Tesla Roadster (photograph courtesy Nova Spivack/Matt Hoerl/Arch Mission Foundation).

laser, with layering capabilities for storing other compressed, digitally encoded contents (Spivack, 2019).

In choosing to employ particular data carriers in message designs, it can be surmised that any of these decisions inherently reveal the message creator's underlying assumptions about whom the imagined recipient(s) may, in fact, be. For example, the use of sophisticated, highly dense data carriers that require specialist equipment and knowledge to discover and decode contents (Zhang, 2013) reveals our tendency to extrapolate current human civilizations along a unilinear, techno-evolutionary trajectory usually associated with notions of perpetual progress and other cultural dogmas that underscore an envisaged utopian future. Storage technologies do, and often have, evolve in the unpredictable hands of those that wield it, but devices have also reverted to prior (and simpler) formats, diverged along completely alternative trajectories, employed newer forms of data compression techniques, or even reached discontinuities over short intervals of time. For instance, Microsoft's recent Zune media devices have now been out of production for *longer* than the full lifespan of this product line. Moreover, technology changes as a result of a panoply of human motives, not just because of our concepts of efficiency, accessibility, and similar logistical qualities.

Traditional inscribed data carriers—often employed by the diversity of pre/antiquity human civilizations—have been proven to successfully transfer intent (and occasionally signified meaning) to foreign cultures who may not be familiar with the precursor's heritage. Instances of this have been documented across this volume, notably the Assyrian familial depositing activities which are some of the oldest documented examples. But employing these information carriers, as *a priori* cultural exchange systems, can allude to a prospective, fragmented futurity with little surviving records from the pioneering civilizations. This fall was certainly the case for the Assyrian empire,

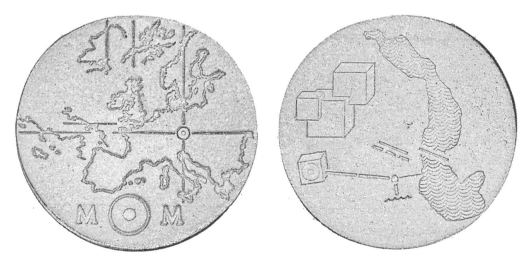

The A and B sides of the Memory of Mankind ceramic token. The left-hand side (A) provides coordinates for Lake Hallstatt using geological features of the European continent, while the right-hand side (B) establishes a unit of measurement and distance from the lake to the archive inside Mount Plassen, along with a representation of salt crystals. The project is also developing a ceremonial gathering custom for token holders—a long tradition intended to assist in memory retention for the archive and its location (photographs/token courtesy of artist Martin Kunze).

but we continue to find their fragmented historical records, with one material inscription being rediscovered under the ruins of the recently demolished Tomb of Jonah. The *Memory of Mankind* token is an excellent example of such a low-tech, structurally simple, and readable artefact of material culture that should, in theory, enable a range of human posterity with diverse capabilities to immediately recognize a purposeful intent in the ceramic disk (and perhaps the semiotic mapping system to locate the nearby information vault secreted beneath Mount Plassen). The token creators have also sought to forge a long-term cultural tradition for their token bearers as a deep time memory-retention scheme.

My takeaway point here is that there is nothing wrong with envisioning a particular futurity, which may plausibly stretch from stone to silicon ages. But we should be mindful that simpler inscribed or etched media, that have been traditionally employed for a number of applications by a broad range of ancient civilizations, have historically proven to be successful at conveying intent, or even signified contents, to an unknown futurity, in comparison to more sophisticated, culturally-dependent formats from antiquity (e.g., Incan Khipu strings) or even recent technological-compatibility issues that require further clues and extensive troubleshooting.[12]

How Can We Get There?

In the second decade of the 21st century, there have been several proposals made by international organizations, citizen science communities, and other parties to launch thousands of new telecommunications satellites into geocentric orbit as part of expanding commercial interests within these regions—an era which the mainstream media has inevitably dubbed as "the new space race." Commercial constellations such as Amazon's *Project Kuiper* and SpaceX's *Starlink*, as well as the equivalent Boeing, Telesat, and OneWeb networks, are already in various phases of production to corner the increasing market shares for satellite services. These are paving the way for a much more crowded, de-nationalized geocentric orbit, albeit with a probable increase in possibilities for global audiences to potentially launch lightweight novelty items, nano-archives, or cars as spacecraft counterbalancing mechanisms or public engagement activities. Despite the potential message opportunities for their customers, such accelerated expansion by commercial sectors into outer space is already creating challenges for the continuity of ratified space treaties, in addition to posing severe risks for stewardship activities and, sometimes, also eroding trust in future cooperative tasks in outer space affairs. Recent promotional stunts, such as the hollow optics and rhetoric of the "Billionaires Space Race," certainly exemplify how discouragingly unbalanced our local orbit it likely to become without proper international scrutiny, guidance, and legislation to manage this environment. The sobering activities of the company *Sent into Space* also highlight the profoundly strange materials that people are already sending into our outer atmosphere on a temporary basis.

All these denoted (orbital) launch opportunities are presently subject to the articles of the OST (United Nations, 1967); signatory nations to this accord are responsible for supervising activities (conducted by their governmental and non-governmental entities) so that they conform to these legal provisions (VI), while also ensuring jurisdiction (VIII), and financial liability (VII) for these objects (and messages) remains with the launching state party who must oversee such initiatives.[13] We have seen how the commons of our terrestrial environment can quickly become congested with discarded material, and outer space—during this Space Junk Age—already presents us with another cultural futurescape, populated by escalating fields of debris (such as satellite fragments, separation bolts, flakes of paint, aluminum particles, fuel slag, snap-frozen masses of liquids, etc.) and larger geopolitical objects consisting of passivated satellites, launch vehicle stages, dummy-payloads, etc. (ESOC, 2020). Future satellite infrastructure and constellations will need to operate within the boundaries of interagency debris mitigation guidelines in order to avoid entombing our planet in fields of space junk—a high level of international coordination within orbital operations which may inadvertently cultivate our own "Clarke Exobelt" as an observable technosignature (Socas-Navarro, 2018). Moreover, there is also a growing reliance on enacting Article VII of the OST in the last several decades to contend with liability issues arising from space debris raining down across our continents. Recent encounters with former Soviet, American, and Chinese debris come to mind.

In addition to the geography of Earth orbit, several commercial entities have set their sights on launching spacecraft to land on the Moon, likely as a result of Google's recent Lunar Xprize that initially drew over 30 competing teams from over 14 countries.[14] Most of these Xprize parties actively integrated message payloads as engagement, outreach, or funding ventures. This competition likely catalyzed the recent revival of planned national human spaceflight programs,[15] but it has also promoted Kilroy-ists to place tributes, memorabilia, and other customized novelty plaques on the Moon, Mars, and beyond through initiatives such as the *AstroGrams* project, initiated by the Apollo 16 astronaut Charles Duke. Several other private companies are continuing to advance plans for lunar exploration, including the *dearMoon* project facilitated by SpaceX, in addition to Elon Musk's well-known intentions to pre-emptively establish a Martian colony—likely before technologies are rigorously tested. There have also been international discussions about several other hard locations that may be used to establish isolated information depositories within our Solar System (Guzman *et al.*, 2016). In addition to the previously discussed legislative considerations and environmental ethics, such future missions and messages are also subject to interagency planetary protection protocols (Kminek *et al.* 2017; NASA, 2019), guidelines that intend to responsibly mitigate astronomical contamination from spacecraft bioloads, as discussed earlier in this volume.

The *One Earth Message* project introduces us to an alternative, albeit electronically dependent, vehicle for projecting extensions of ourselves beyond our terrestrial stage in the form of "uplinking" our mosaic of Earth to the radiation-hardened MIPS-3000 microprocessor encased within New Horizons integrated electronics module.[16] In relation to this approach, there have been several successful attempts to re-activate "zombie" geocentric satellites (for scientific purposes) however, assuming the internal memory could be reactivated after elongated periods of idleness (and remain uncompromised), future schemes may propose to uplink electronic messages into decommissioned space hardware that temporarily remains functional within geocentric orbital ranges or abandoned in the highly stable Lagrange points.[17] Further parties have also suggested locating communications satellites a few light years away from Earth as remote transmission platforms to conduct METI activities (Harrison, 2014) or establish distant sentinels capable of downlinking contents to disconnected human societies—though it is unclear how such hardware will remain viable during transit and deployment over spatiotemporal distances. Insights gleaned from operational problems with comparatively local space hardware, like the Hubble service missions, should be incorporated into such plans.

In addition to these physical message deposits, several organizations are in the midst of advancing and optimizing transmission technologies for use in the next generation of communication (and transit) applications. As part of the prominent *Breakthrough Initiative* funding for scientific and technological exploration in SETI, the *Starshot* project is undertaking development of a "light beamer." This system is comprised of a phased array of high-power, near-infrared laser beams used to accelerate fleets of miniature probes with light sails (called *StarChips*) to 15–20 percent of the speed of light in order to flyby neighboring exoplanets such as Proxima Centauri B within 20–30 years. The reader will be no doubt aware of this initiative as it was announced with much scientific acclaim in a prominent press event though, to a lesser extent, the SETI community have speculated how the *a posteriori* leakage radiation from these photo-gravitational assists to dispatch fleets of StarChips may be detectable by ETI (Guillochon and Loeb, 2015; Benford and Benford, 2016).

I highlight the potential use of this beaming technology to conduct METI activities using such high-powered lasers as the organization *Interstellar Beacon* have voiced their intention to do so in order to deliver their four-program message to prospective interlocutors.[18] As this StarChip project is advancing in tandem with the Breakthrough Message competition,[19] it can be assumed that these centimeter-sized probes may inevitably host a physical mosaic of Earth encoded within digital, micro-etched, or DNA data carriers for interstellar Passive-METI applications. Proponents of METI have already highlighted the potential of these interstellar craft for their communication enterprise (Osborne, 2016) as such projects will likely serve as the furthest, physical (object) extensions of humanities' material culture—far surpassing the foreseeable distance to be covered by Voyager 1 over the forthcoming millennia.

The microwave bandwidth will also likely continue to represent our planet at any meaningful distance through the historical weak leakage radiation and contemporary radio profile emanating from terrestrial infrastructure, in addition to the powerful, tight-beam messages we craft for a range of METI and outreach activities. While presently only accepted as best practice guidelines, the *SETI Post-Detection Protocol* (IAA, 2010) serves as an outline of formal actions which should be undertaken by governmental and other non-governmental entities following the detection of confirmed signals from extraterrestrial civilizations. Article VIII clearly outlines the need for international authorities to responsibly discuss the prospects of whether we should respond to a received signal, preferably prior to anyone transmitting such an EM message on behalf of a fledgling Earth. However, the language of this protocol does not explicitly preclude *de novo* activities—to much present-day contention. In spite of this protocol, it is anticipated that any such ETI transmission may attract dozens of unsanctioned responses from a plurality of professional and non-expert parties with access to these technologies, an issue about which the former United States diplomat Michael Michaud (2003) once remarked:

> Having humankind speak with many voices may be representative of diversity, but it also may be bad policy. Imagine yourself in the place of an ETI that receives a barrage of messages from the Earth. How could you conduct a rational dialogue with such mixed signals? Who would

you believe, those humans who seek an exchange of scientific information, those who desire to convert you to the true faith, or those who announce their intent to exterminate you?[20]

In the continued absence of any clear, formal, or legal agreements,[21] the SETI community remains polarized on the moral and democratic (but largely political, ethical, and social) implications of conducting *de novo* METI activities. Some proponents of METI (Zaitsev, 2008; Haqq-Misra *et al.*, 2013; Vakoch, 2010b; 2016) argue in favor of the benefits for our cosmic, cultural, and political awareness granted by immediate experiments with neighboring stellar systems, while opponents (Billingham and Benford, 2011; Brin, 2014; 2019; Gertz, 2016a; 2016b) highlight the menu of diverse impracticalities, discord within implicit claims of ETI benevolence without evidence, prospective unknown risks, and wisdom of incorporating precautionary principles into this enigmatic enterprise. Given the general impetus surrounding concepts of contact, this author sees wisdom in the old Russian proverb "trust, but verify," before creating responses at the very least. There is also the claim that METI is pseudoscientific in nature—owing to the range of assertions committed about an observing ETI's reception, interpretations, and responses. However, we should also be mindful that SETI, as a passive endeavor, also possesses its own unfortunate history of such accusations and refutations of this ilk.[22] Michaud (1993) has further remarked, "In effect, sending deliberate communications to another intelligent species would mean conducting relations with that species," and several other opponents have also signed a statement opposing METI on a number of these grounds (SETI@Home, 2016)[23] in order to encourage vigorous international debate on the subject prior to such diplomatic activities. In spite of these logical protests, several projects have already circumvented these deliberations and entered into intractable, exo-social experiments, conducted "on behalf of humanity." METI remains as a controversial enterprise owing to, as the anthropologist Kathryn Denning (2011b) excellently summarizes:

> different models of the scientist's role as citizen and/or leader; disparate ideas about society's readiness to cope with frontier science; variable political substrates, particularly ideas concerning individual freedom and state control; competing ideologies of globalization; and the perceived relative risks and benefits of contact … derive partly from different thinking styles, including tolerance for risk, and partly from inferences based upon … contact on Earth.

The range of these perennial issues, and spectrum of dispositions that contribute to arguments, in turn, are extensive, spanning several schools of thought and many decades of speculative insights—all contributing to the modern dynamics outlined in the above academic discussions. Furthermore, several researchers (Denning, 2011b; Musso, 2012; Korbitz, 2014; Smith, 2019; McCarthy, 2019) have critically examined the underlying motivations, ethics, and driving psychologies of the contributing author stances as personal intuitions rather than scientifically verifiable positions in this debate, in addition to developing metrics such as the San Marino scale (Almár and Shuch, 2007) for assessing risks associated with deliberate transmissions from Earth. I have also argued for the proper implementation of this scale for general context, but until someone takes up this challenge, it is likely we will only ever possess a small piece of the puzzle for understanding our expanding human profile across other stellar systems. Indeed, the texts of this book provide much food for thought on the diverse range of practices that collectively contribute to these *signs from Earth*, alongside revealing the conflicting and illogical legacies for messaging signals already "out there." But the entirety of these deliberations remains both conjectural and academic due to the absence of another intelligent civilization to hinge our broad assertions upon. This all too brief characterization for the dynamics of this debate has not been summarized to state that there is anything implicitly wrong with METI or, in these cases, the fragmentary rendition of this practice our "advanced civilization" presently operates. Rather, I wish to raise critical awareness of the diverse opinions and conversations that need to continue to guide the decisions of such hasty activities–actions that will be inherited by the great, silent majority of unborn human generations (Krznaric, 2021) who cannot readily "have a say." Similar insights could also have bearing on several Passive-METI initiatives that stand a chance of recovery as ancient envoys of Earth.

Speaking into the Expanse

Given the focus of this study, I am often asked what these autobiographical tokens from present-day Earth inherently mean to me, our homeworld, and the dynamic populations that will continue to live out their lives over the unfolding futurescape of our extra-terrestrial communique. There is no easy, or indeed singular, answer to such a cryptic yet captivating quandary, owing to the profoundly unknowable futurity for these artefacts of material culture and the broad, perplexing ambiguity instilled within these caches of representational media. Leaving prosthetic traces of ourselves to cascade down throughout the ages, after all, seems like a natural process of succession, based upon our established principles of bequeathing and inheritance, alongside our desires to simply be remembered beyond the lifespan of the last individual who may recollect *our* life experiences and our own small contributions to the evolving story of humanity.

Given the increased frequency of multinational launch capabilities, broadening economic access to advanced communications technologies, widespread interest within the human inhabitation of space, and the colonization of nearby astronomical bodies, it is likely that we are beginning to see the next phase of the Space-Heritage Age (Hays and Lutes, 2007) which will increasingly preserve the abstract philosophical, cultural, and psychological thoughts of modern human societies, or individuals, within the archaeological record of the defined "higher frontier." This material heritage, including messages and encompassing aerospace infrastructure and related debris, will essentially function as an ambassador for this human enterprise within a new environmental setting. It would be remiss of us to fail to recognize this emergent ecological context and, by affiliation, our unfortunate history of engaging with newer habitats. In light of this higher-order interpretation, messages (as bellwethers for our material agency) can no longer be perceived as an inconsequential, cursory matter, or afterthought for human proliferation within its surrounding cosmic environment. These relics instead arguably serve as an experimental testing platform for envisioning the unique technical, conceptual, and philosophical challenges associated with facilitating deep time planning and trans-millennial planetary custodianship, while also inspiring zeniths within artistic, cultural, archival, and technological material culture (not to mention social critiques for our era).

Despite this, I share Jon Lomberg's (2004) concerns regarding the possibility of losing sight of ourselves, and the spirit of these activities, by crafting a legacy of intentional pastiche artefacts with no serious consideration towards their reinterpretation by posterity or ETI—high-tech effigies which serve as dishonest reflections of their author's intents or well wishes. While an archaeological record should ideally contain anecdotal evidence from a variety of sources, it is questionable whether such imperative context may be accurately gleaned from the broad range of artefacts presently emanating from Earth. Furthermore, there is an increasing trend of eclipsing the message intent and prescribed benefits of informational contents in favor of sensationalizing the messenger agents(s)—self-defeating idolization which simply adds noise to the intended ethos of these legacies while, in my personal opinion, complicating the formal documentation and accurate archiving of these contemporary projects. Yet, such accounts are ubiquitously evident throughout messages from antiquity; the voices of long-dead rulers and gentry figures, who were capable of commissioning "insane" messages, are a crucial contextual resource for informing archaeological inquiries. But what does this self-serving material, and the legacies it may create (and reinforce), tell us about the accumulate story of humanity? Whom do we imagine to be the recipients for our variegated stories?

We should not lose sight of *why* we choose to expend such efforts, materials, and time within creating intelligible prostheses of our minds—even if they may never be discovered and understood for what they are. The biographies of such archival and metaphysical projections can epitomize the best intentions and ideals of the human spirit and what it means to be an actual sapient species, or export the worst, presumptive tropes that have typically not been inclusive of minority populations we share this planet with. Perhaps, this may be comparable in accuracy and imagination with Ruth Benedict's study *The Chrysanthemum and the Sword* (Traphagan, 2014). It is sobering to envisage how some reputed "time-binding" public relations feats, thinly veiled money-generation schemes, overt visions of personal grandeur, or sensationalist "stunt" messages may be seriously contemplated by our distant descendants, who will likely already possess significant knowledge about their forebears. Keep in mind, these recipients will already possess crucial contexts from their inherited ecologies and contemporary chapters of human history or, in the case of ETI, acquire similar resources if they continue to monitor perceptible changes within Earth's weak technosignatures for context. The futurity of space, "as a province for all [hu]mankind" (United Nations, 1967), may end up mirroring the worst, techo-colonialist qualities of human history, with these ill-fated constructs also serving as a bellwether for the nefarious, representational behaviors that we may unwittingly intend to export into this frontier as unfolding, indelible Terran legacies.

If our own investigations into our species' antiquity have taught us however, however, it is that first-person subjective accounts will form only one evidentiary tread for distant minds to study. Our windows into human history would indeed be quite contorted if archaeologists simply relied on such intentional, superficial profiles of dynasties, kingdoms, and religious orders. In this sense, and per contra to my noted concerns, the menagerie of pastiche artefacts may also serve as an accurate reflection of the broad socio-cultural disagreements and material fetishes that are hallmarks of our contemporary ages as opposed to the sanitized intelligible autobiographies we dispatch across our cosmic neighborhood. In the interest of brevity, I will leave the reader to contemplate whether this potpourri of eclectic messaging practices detailed in the Appendix could provide evidentiary value and symmetry for future archaeological investigations into our early Anthropocene era.

In drifting away from the spirit of the original question, perhaps the most profound

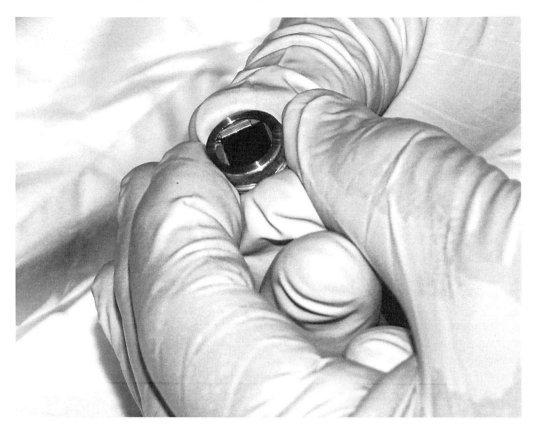

Close-up of a microchip with 1.2 million names on board the Mars Curiosity Rover (photograph from NASA/JPL/KSC [PIA04851]).

deep-time message we are fashioning at the beginning of the third millennium is the emergent state of the Earth itself as humans begin to physically and mentally depart our fragile world. This unique, biotic island has long nurtured the emergence of hominids in mutual symbiosis with countless generations of trillions of other substantial organisms, and will no doubt continue to play an umbilical role within prospective colonization attempts on other planetary bodies. This "next step" in planetary insurance has traditionally been defined "a matter of utmost practicality," as the rhetorical tropes of space exploration, much like our surrogate messages, begin to conceptually place *the notion of home* behind the prospective, transcendental "frontier" that lies ahead as a human "destiny." We are indeed already transforming our embodied minds in advancing material designs for the future of humanity off world, perhaps at the expense of our planet's imminent future—choices which may already foreshadow how our species, or "the best of us," may conduct itself if we find a "Planet B."

The pragmatic decisions humanity makes now about our home world and how we choose to interrelate with other terrestrial life, our in-situ Solar System, and the interstellar space beyond this will influence how our descendants follow in our custodial footsteps if attempting to colonize other worlds. The word "colonize" is already ubiquitous in the literature surrounding our quest to move on to other planets (there are indeed entire debates surrounding these loaded semantics terms at present), while conceptually *de-terraforming* the supportive embrace of Earth's biome in the same fatal

Close up of the names, images, and signatures etched on a microchip, installed on the OSIRIS-Rex mission to 101955 Bennu (photograph from NASA/GSFC—The Planetary Society).

swoop. The virtues instilled within goodwill messages to the future may become unpalatable when the observer is presented with abhorrent, conflicting circumstances, and inherited Anthropocene legacies—unfolding oversights that posterity will definitively know how to read and properly interpret. After all, well-wishing messages dumped alongside our refuse will likely not inspire much admiration by an inheriting posterity.

Moving from Sagan's version of "Who Speaks for Earth?," to the ecological rendition of this quandary originally posited by Barbara Ward (1973), our changing planet as a message[24] may inadvertently communicate the crucial background context for our present civilizations far beyond the subjectively-curated, ancillary autobiographies we dispatch into outer space to ensure tailored segments of our species are contemplated—and remembered—from afar. Scientific Earth-observation satellite data, alongside our studies of *Earthshine*, may perhaps be the most revealing indicators for this accumulative indelible legacy to the stars while also providing us with a mirror to witness how anthropocentrism can materially manifest across deep space and cosmic timescales. Could we recognize *these signs for Earth*—or is it already an exoplanet? (Gorman, this volume). Therefore, the pale blue legacy of our fragile biosphere cannot simply be disregarded or considered a vacuous backdrop devoid of meaningful context for the entwined processes of human behaviors, mind, matter, message, meaning, and future. As the economist and philosopher Kenneth Boulding once excellently opined:

> Man is finally going to have to face the fact that he is a biological system living in an ecological system, and that his survival power is going to depend on his developing symbiotic relationships of a closed-cycle character with all the other elements and populations of the world of ecological systems [Boulding, 1965].

Perhaps, in a way, Boulding's prediction has already come to fruition. Perhaps, our preconceptions and cavalier attitudes towards space as the "frontier" solely for transcendental human destiny, and the blurring of ideology, religion, political exigency, and technological superiority (along with our concepts of Earth as merely "one small step" from the cradle to better homes), has thus far inhibited our growth and survival off world—counterbalancing our moral agency against our expansionist desires for wrong or ill-considered reasons (Smith *et al.*, 2019). Whether our immediate futurity resemble aspects of our contemporary civilizations or versions of our current ethical foresights, moral compasses, or pretentiousness for only homo's *sapience* should rightfully be cause for concern at this moment in time. In particular, we should exercise caution when indiscriminately exporting such aging tropes of "space escapism" as a common inheritance for imagined "advanced" heterogeneous populations of similar mind, or space as a vacant region for the proliferation of succeeding post biological entities (which will receive both virtues and imperfections from a broad smear of human behavioral domains). In this respect, the philosophies and rational substrate contained within the symbolism of messages provide but one clairvoyant window into how we may come to envisage our much broader steps off-world, and enable us to question how the virtues and lessons learned from such time-binding exercises may be harnessed to further resolve these poignant terrestrial affairs—firstly on Earth, *our only real home*.

Regardless of this expansionist argument, or how we wish to construct the contemporary mythology of *Homo spacians*, as we depart the *Critical Decade* (Steffen and Hughes, 2013), we are now entering a new exigent phase of our planet's history. Earth has become an artefact of human material culture in much less the same manner as the futurescape of these biographical messages—the "marvelous" scientific and technological harbingers that frequently far outpace our ethical foresight. Both "messages" will serve as our generation's indelible legacies to unknown recipients who may only know of the rich cultural, heritage and social histories of our civilizations superficially, from afar. It's plausible to assert that this "eco-message" may also function as a very poignant barrier for the very ETI communication our technological harbingers strive to establish. Clarke's fable of "advanced technology," required to initiate ETI discourse may take second stage to the planetary ethics we now openly transmit as imperative background context in our passive technosignatures. Simple monitoring (on ETI's part) of these freely emanating signals could allow observers to glean valuable insights about their prospective interlocutors—for example, how humans fail to balance the ecologies of their homeworld, and choices made in continuing these harmful trajectories. This, of course, is also subject to whether our technosignatures or messages are indeed found at all—it *only* took 13,000 years to re-discover the parietal artworks at Altamira, another decade to recognize an ancestral significance, and a further 80 years to cease projecting *our cultural ideals* and *our social customs* onto the distinctive minds that rendered such wondrous depictions from life.

Many of these brief essays outline highly theoretical subject matters that are already several disciplines wide, and many centuries deep, in thought, yet our musings should be treated as appetizers and an invitation for the reader to participate in future discourse. At the risk of cherry-picking facts or drawing any firm conclusion to underline these enigmatic, multifaceted essays about cryptic questions, without verifiable answers, now in the exoatmospheric archaeological record, we believe it is now imperative to pass this project and the APOH catalog to our readers. We hope this resource may provide an informed departure point for our readers' imaginations to investigate how our ideological portraits of Earth may be interpreted (or not) by another rationale being residing within the unfolding futurescape of our local Solar System. Or perhaps, consider how this expanding and experimental profile *from* humanity may be someday studied by an independent, extraterrestrial civilization residing somewhere "out there" in the great cosmic expanse that only receives the faintest, biographical glimmers of a comparatively adolescent Pale Blue Dot.

NOTES

1. Sagan, over the course of his message-crafting career, would come to largely author the *Pioneer Plaques* with Frank Drake, Linda Sagan, and faculty from Cornell University, influence the contents of Drake's *Arecibo Message*, and draft the *LAGEOS* epigraph, while also spearheading committee development of the iconic VGR in the "language of science." In addition, the VGR also had numerous collaborators from universities, research disciplines, and members of the public realm. Further to this, Sagan would co-write the VGR publication *Murmurs of Earth* (1978), alongside other later science fiction literature based around ETI communication.

2. Several authors have previously noted that the exigency of the Cold War, threat of thermonuclear war, and dissolution of Western ideals were strong driving forces behind our exploration of the frontier and our techno-scientific attempts to preserve posthumous memories of humanity beyond this perceived "era of apocalypse." Sagan, in discussing the VGR, opined: "No one sends such a message on such a journey without a positive passion for the future." It can be argued that the rhetoric of the 21st century and human recognition of the Anthropocene have produced similar circumstances which have catalyzed the broad range of contemporary messaging activities—initiatives that seek to preserve mnemonic fragments of ourselves *from ourselves.*

3. This author is apprehensive about using the Wikipedia repository as a basis for conveying cultural

properties across intervals of deep time and space, given that the repository is peer-developed by users who disparately but broadly share convergences within cultural, epistemological, and social conventions within an integrated global community—accounts *for* recipients who embody these same principles, and share in an environmental context. We may entertain the thought that these systems of representations are broadly shared, but they require situated contexts and mental bridges for interpretation, functions that are unconsciously mobilized without effort by present, interconnected societies, and supplemented by common material culture. There is no guarantee that our future descendants will share fragments of these pre-established conventions gained over the course of our socio-cultural history. Therefore, encyclopedia archives may impede, or exclude, comprehension for these recipients, even if they share morphological and biological histories.

4. The main antagonist *V'Ger* from the 1979 movie *Star Trek: The Motion Picture* also comes to mind here.

5. As a rather poignant example of these competing and conflicting agendas, this author has continually argued against the mis-categorization of the *A Message from Earth* project on Wikipedia. The project initially possessed its own encyclopedia page before being poorly re-categorized under the targeted stellar system. This author has since reinstated the project page from the original, arguing that the *Arecibo Message* and VGR could also be re-categorized under "M13" and "interstellar space" pages respectively. Sadly, information loss is also already a reality for documenting such interstellar transmissions and similar projects (Quast, 2022). For instance, several cited resources on the *A Message from Earth* Wikipedia page now link to defunct articles, an issue that is now also manifesting across the entire APOH catalogue, a resource that is becoming the only primary evidence source.

6. The premise of *Invitation to ETI* detailed how such probes may be covertly monitoring our civilization through the World Wide Web and were, in theory, only awaiting our awareness and invitation to appear.

7. Together with the *From Afar* website (www.fromafar.world), the group acquired over 13,000 SETI-METI survey responses (at the exhibit and online), while also receiving over 1,000 physical message designs and ideas during the week, documenting a range of thematic approaches for bitmap constructs.

8. Similar dissipation and interference issues arise for our EM signals as they transit through the vast clouds of ionized gas which spread the signals bandwidth (i.e., Doppler shift), in addition to also interacting with absorbing cosmic dust and other exotic interstellar phenomena, as these pulses traverse towards the targeted exoplanets and stellar systems.

9. This *Rosetta* [prototype] disk consisted of a micro-fabricated archive of about 6,500 text pages which were etched using a process developed by Los Alamos Laboratories and Norsam Technologies, circa 1997. The newer *Rosetta Disk (v1.0)*, featuring about 13,000 pages, requires use of an optical microscope with X650 resolution.

10. Reading the 1,000 pages of this newer *Rosetta 300 Language project* wearable disk would require an optical microscope with about x100 resolution—contents would be legible using a Leeuwenhoek microscope from 1666.

11. The crystal can store 360 terabytes of information, but the write speed was limited to 13 kilobytes per second (Anderson, 2016). Assuming continuous writing, it would take 320 days to reach the crystal storage capacity.

12. For example, flight engineers at NASA had to undertake "software-archaeology" to reactivate Voyager 1's thrusters after 37 years of dormancy. This saga required engineers to review decades-old data and examine the software that was coded in an outdated assembler language system (for proof-reading coded commands).

13. According to international space law, space infrastructure developed by commercial operators—and, by extension, messages included aboard these spacecraft—will remain the property of the supervising state party that is a signatory to the OST (United Nations, 1967); governing bodies and future successor nations will remain liable for damages incurred to other space hardware by their active or decommissioned satellites.

14. The main *Lunar Xprize* was not claimed; however, SpaceIL did receive an award for being the first team to officially hard-impact the Lunar surface with the Beresheet lander in 2019.

15. The most prominent proposals being the futuristic international *Moon Village* envisioned by Jan Woerner, along with the international *ARTEMIS* program, and *Lunar Orbital Platform/Gateway* spearheaded by NASA.

16. There are two Mongoose-V radiation-hardened microchips stored within two different integrated electronics modules on *New Horizons*. The author is unsure whether *One Earth Message* proposed to store the crowd-sourced message in one or both units as a redundancy strategy.

17. In addition to legislative issues arising from the OST and other principles of international space law, there are also several technological compatibility issues associated with reactivating dormant space hardware.

18. While this author is aware of this proposal, I am admittedly unfamiliar whether Interstellar Beacon is concurrently funding development of a separate beaming technology or adopting use of one of the aforementioned systems to begin optically transmitting their four-program message. Similar low-powered systems have been tested by both ESA and NASA over recent years; a test using a transmitted bitmap of

the *Mona Lisa* to the Lunar Reconnaissance Orbiter currently holds the distance record for such experiments.

19. According to the press material about *Breakthrough Message* (November 29, 2019), the initiative "aims to encourage debate about how and what to communicate with possible intelligent beings beyond Earth. It takes the form of an international competition to create messages that could be read by an advanced civilization. The message must be in digital format and should be representative of humanity and planet Earth."

20. By contrast, the late physicist Freeman Dyson saw such a discord of responses as accurately representing the social dynamics of humankind, insofar as it draws into sharp focus our regular *imbroglio*, or miscommunication, with one another and lack of agreements on many fundamental earthly issues.

21. The *Second Protocol*, developed to address *de novo* messages to ETI, has largely failed to establish agreements or guidelines as a result of ongoing dissonance and lack of engagement between METI opponents and proponents.

22. A principal argument of this pseudoscience case is the falsifiability of a scientific hypothesis; SETI searches our surrounding cosmos for verifiable signals, a search which could theoretically extend *ad infinitum*, while METI requires a response from an unverified inhabited exoplanet/stellar system—a response that may or may not happen, but resolving this issue would also necessitate us to devote instrumentation for the continual monitoring of these increasing number of candidates, again theoretically extended to *ad infinitum*.

23. The statement signatories all hail from technological industries, academia, facility managers, and scientists who predominantly possess affiliations with the UC Berkeley SETI program.

24. This "world message" is elucidated within the geological strata of our planetary crust including the broad usage of concrete and plastic materials, atmosphere isotopic composition, fluctuating and extreme weather patterns, elevated background radiation levels, water table impurities, lineage GMO organisms/gene cycles, artificial belts of satellite debris, static bubble of EM radiation propagating outwards from our planet, and other localized, indelible legacies which are currently being investigated as part of the *After the Horizon* research program.

Afterword

Challenges

CORNELIUS HOLTORF

Sending messages beyond the Earth and into deep space is an occupation that many people find fascinating and inspiring. Space messages evoke powerful themes: the variety of human legacies on Earth, the uniqueness of humanity as such, and the remote corners and distant futures of the universe. This volume provides a useful introduction to the history and practice of space messages sent from Earth, exploring a range of key issues and their social and cultural contexts.

Given the mind-boggling distances in time and space of the universe, space messages transcend what we can know, and perhaps even imagine, from our specific vantage point on Earth. In practice, these messages which are outwardly directed at other intelligent beings are effectively about communication with human beings like us, on Earth. As the authors discuss throughout this volume, there is much to be learned from designing and sending space messages about specific characteristics of our own civilization. Space messages give answers on who is speaking for Earth, what they choose to speak about, and how they speak. Space messages also manifest how their authors conceive of the relations between present and future human societies. They often imply a contemporary interest in presenting our world favorably to future audiences, whether that means favorable at face value or favorable in terms of meeting our responsibility to convey (certain) important facts. Invariably, it appears as if those sending information into deep space hope to be appreciated as future generations' honorable ancestors. At the same time, all this reveals a sincere expectation that anticipated future audiences will want to use our messages to learn more about us at all.

In addressing imagined futures, deep-space messages often draw on our heritage. Heritage transforms space into a cultural field, and it has been doing so since the very beginning of space exploration (May, 2020). From the American flag which Neil Armstrong proudly planted on the Moon, to the astronauts' incidental footprints, the space junk they left in various orbits, and indeed carefully crafted messages that were sent into deep space, all these items are left as witnesses for an anticipated future when other astronauts, human or not, will discover and make sense of them. This capacity of heritage to negotiate relations between present and future societies is what we call "heritage futures": managing our legacy in order to make an impact on the future.

In the space context, it is well understood that any future audiences, human or non-human, will not think and interpret their world as we do. At face value, the

Glossary

ALSEP—Apollo Lunar Surface Experiments Package (Apollo Missions 12–17)

APOH—A Profile of Humanity (See Appendix)

ASREM—A Simple Response to an Elemental Message (2016 interstellar transmission)

CE—Common Era (equivalent to *Anno Domini*, or AD, year designator)

EM—Electromagnetic (radiation)

ESA—European Space Agency

ETI—Extra Terrestrial-Intelligence

EVA—Extravehicular Activity (anything outside of habitable segments of spacecraft)

IR—Infrared (electromagnetic frequencies)

ISS—International Space Station

Lincos—Lingua Cosmica (artificial language, developed by Hans Freudenthal)

METI—Messaging to Extra-Terrestrial Intelligence (also known as "Active SETI")

NASA—National Aeronautics and Space Administration

NASDA—National Space Development Agency of Japan (JAXA precursor)

OST—The Outer Space Treaty/Treaty on Principles Governing the Activities of States in the Exploration and Use of Outer Space, including the Moon and Other Celestial Bodies (1967)

S-3—Surveyor 3 (pathfinding Lunar spacecraft)

SETI—Search for Extra-Terrestrial Intelligence

SETILO—Search for Extra-Terrestrial Intelligence Like Ourselves

VGR—Voyager Golden Record (1977 photographic message on Voyager 1 and 2)

Appendix

A Profile of Humanity: The Cultural Signature of Earth's Inhabitants Beyond the Atmosphere

PAUL E. QUAST

The underlying emphasis for the original *A Profile of Humanity* (APOH) catalogue (Quast, 2018a) was to document a broad range of purposeful cultural activities that, collectively, contribute to an emerging exoatmospheric archaeological record, mainly within the material domains of electromagnetic "interstellar transmissions," and also slower passive (physical) messages. However, the theoretic foundations principally stem from two moral complexities raised across three papers (Billingham and Benford, 2011; Brin, 2013; Harrison, 2014). On the one hand, there are several fundamental ethical concerns associated with sending electromagnetic messages into space on behalf of present-day humanity, of which, the perennial controversy surrounding Messaging Extra-Terrestrial Intelligence (METI) practices using de novo transmission strategies best illustrates the existentialism apparent in these debates (Zaitsev, 2008a; Musso, 2012; Brin, 2014; Korbitz, 2014; Gertz, 2016a; 2016b). Many transmissions may likely be benign, but a precautionary position is rightly advocated by opponents to these activities. The other hand presents us with a more pragmatic, yet very human, affair that should continue to occupy our minds regardless of how the METI philosophies or similar extraplanetary "contamination" discussions unfold: *How do we ensure those unborn generations, who do not readily "have a say" over our messaging activities, or other techno-cultural legacies, are adequately informed about the inherited consequences of such decisions committed on their behalf?*—often amusingly referred to as the "unordered pizza" argument. *"Remembering the Conversation"* (Quast, 2022), or more simply put, ensuring adequate custodianship over copies of these materials transmitted on behalf of Earth for future scientific scrutiny (for instance, as a referral guide in the event an ETI structures their response around this content) is yet another crucial facet only recently identified.

There have been some assessments performed by the SETI and METI communities (Vakoch, 2009; Zaitsev, 2012; Dumas, 2015) who responsibly document intelligible,

Original publication: Quast, P.E. (2018) A profile of humanity: The cultural signature of Earth's inhabitants beyond the atmosphere. *International Journal of Astrobiology*, 1–21.
https://doi.org/10.1017/S1473550418000290
Revision history: Received: 18 April 2018 (IJA)/Revised: 13 June 2018 (IJA)/Accepted: 21 June 2018 (IJA)/Published: 15 August 2018 (IJA)/Revised (US Eng.): 9 September 2022 (McFarland)/Published: March 2024 (McFarland)
Keywords: future archaeology, time capsules, long-term communication strategies, deep time messages, SETI, active SETI, METI, data storage, eternal memory archives
Paul E. Quast: info@beyondtheearth.org

electromagnetic envoys that intend to initiate diplomatic relations on behalf of Earth's populace, usually over protracted intervals of time. However, these few records are not very comprehensive, nor do they take into full account the other non-extraterrestrial directed messages which also contribute to this still growing technosignature. There are many differing motives that presently contribute to this intentional celestial property of our planetary system. As a sample platter: our desires to create secure "eternal memory" libraries to preserve information beyond our terrestrial environment (Guzman *et al.*, 2015); rational communication attempts with extraterrestrial intelligences (Zaitsev, 2006); expressions within "SpaceArt" (Paglen 2012); mission outreach initiatives (Sutherland, 2015); techno-colonialist propaganda objects as ideological claims in the "higher frontier" (Reeves, 1994); questionable "lifeboat" projects to conserve libraries of life beyond Earth using DNA or datasets; and also symbolic gestures devised to impart some profound heuristic about our observed position in the universe (Schulze-Makuch, 2016). Many of these diverse activities clearly do not represent equivalent degrees of "plausible risk" as those frequently levied against METI signals, but they do possess other degrees of significance for our enduring exoatmospheric archaeological record, alongside our increasing "bio-footprints," and other residual legacies we may yet uncover. Technological activities seemingly far outpace our ethical foresight.

The APOH catalogue was initially founded as the first step in facilitating proper documentation for messaging projects, and vital custodianship of such "essential" material legacies; it was established as a resource to enable posterity to commit informed stewardship decisions on behalf of their inherited circumstances or, more than likely, contribute insights for historical examinations of their ancestor's material culture practices. However, documenting METI activities is not this catalogue's sole function. In addition, and as part of our preliminary planetary protection protocols, the index also records purposeful messaging acts that now contribute to known "bio-footprints" off world, ensuring that potential contamination sites and events are suitably documented until intergovernmental space research authorities can then archive these legacies, alongside other messaging projects. The variegated range of cultural materials now representing the diverse populations of Earth is vast, counterintuitive, *and exponentially increasing*. While this index is a first step towards documenting these initiatives that represent Earth from afar, it is acknowledged that a counterpart library of contents, and full transmission sequences, will eventually be required as a crucial evidentiary guide for researching and communicating our expanding profile beyond the Earth.

Layout Notes

The original catalogue was informally compiled and presented at the 2018 UK SETI Research Network symposium in Oxford University before being published in the *International Journal of Astrobiology* (Quast, 2018a). This second edition has been consolidated and expanded as part of a much broader survey for the publication Speaking *Beyond Earth: Perspectives on Messaging Across Deep Space and Cosmic Time*. This volume aims to continue to chronicle and now examine the lengths we take in fabricating mnemonic artefacts of material culture which varyingly attempt to ensure our civilization is contemplated throughout time and space—by distant, spatiotemporal observers that we may not distinctively recognize as "human."

The below catalogue is presented as a work-in-progress directory to document the celestial legacy of our civilization's material culture deposits beyond Earth's atmosphere. As a departure from the initial version of this index, the revised edition attempts to re-contextualize the previously established categories to correspond with the taxonomic approach discussed in Chapter 1 of this volume— re-presenting some of the initial messaging projects under alternative "motivation" headings, while extensively building upon the previous quantity of recorded materials. Despite these changes, it is still anticipated that the established categories and overall framework for this directory will be updated over time as sustained research, peer-reviewed contributions, and future initiatives elucidate the anthropogenic signature emanating from Earth on behalf of all terrestrial life. The reader is encouraged to add to this index, to keep track of the many artefacts of material culture that will soon "Speak for Earth," and corroborate these revised details with future publications of this APOH index.

As a departure from the initial catalogue, this version now documents some short-term artefacts that will likely decay into Earth's atmosphere, while still maintaining a traditional emphasis of outlining messages with moderate to protracted lifespans. Despite the relevance of microbial contamination in this exoatmospheric record (i.e., as dense living libraries documenting billions of years of phylogenetic evolution, mutations, and genetic mistakes), this catalogue only documents *known examples* of these purposeful "bio-footprints" launched as part of intentional messages or, in some instances, sites contaminated with biological materials (e.g., Apollo lunar sites). As with the first edition, this index does not feature mission-oriented infrastructure (e.g., satellite communications, physical probes, launch vehicles, scientific payloads, etc.), or technosignatures such as the radiosphere (i.e., the unintentional, electromagnetic "leakage"). Some cancelled and planned initiatives have also been included to document these activities for communal review and insights into our increasing aspirations to leave our marks across other astronomical bodies. Furthermore, "Moonbounce" transmissions are not included alongside Celestis memorial flights due to the ephemerality of these contents. Some fields are marked with "--------" to signify lacunae, and all activity dates are now listed in CE notation.

List of Contents

METI Interstellar Radio Messages

This category is largely defined by scholars within the SETI community who classify the below Active SETI transmission events under a number of set criteria including relevant selection of signal target(s), adequate signal properties necessary to transverse the interstellar medium in order to be detected by an extra-terrestrial receiving array, appropriate modulation techniques that could theoretically be accessible for an Extra-Terrestrial Intelligence (ETI), and also selective content which denotes ideologies that should be comprehendible for extra solar denizens. Furthermore, these messages also score high on the San Marino Scale (Almár and Shuch, 2006)—a metric range used to assess the risks associated with deliberate transmissions from Earth to other civilizations.

Note: "Arrival" field denotes the estimated CE date for each signal to physically reach their stated target(s)—based upon widely-accepted, parallax measurements. This does not, however, delineate whether the signal contents will arrive in a legible manner or as recognizable packets of energy per unit of receiver surface area.

Initiative Name	Transmitted	Organizer(s)	Targeted Object(s)	Arrival	Transmitted Content(s) and Transmission Parameters
Arecibo Message (SNACI, 1975; Sagan, 1978; Goldsmith and Owen, 2001; Grinspoon, 2003; Atri et al., 2011)	16 Nov 1974	Frank Drake, Richard Isaacman, Linda May, James C.G. Walker, Carl Sagan	Towards NGC 6205/Messier 13, Hercules	26974	Numbers (base unit of 10), a selection of chemical elements, DNA formulas, number of DNA nucleotides, shape of double helix, human graphic, planets diagram and transmitting telescope—sent via binary phase shift keying using cardinality of 1,679 bits (a defined semi-prime number). Transmission: Arecibo Observatory 305m radio telescope, 2,380 MHz and modulated by shift of 10 Hz, 1,000 kW.

Initiative Name	Transmitted	Organizer(s)	Targeted Object(s)	Arrival	Transmitted Content(s) and Transmission Parameters
Greetings to Altair (Personal correspondence with Shin-ya Narusawa—collaborator on the JAXA METI Experiments.; Pink, 2008)	15 Aug 1983	Hisashi Hirabayashi, Masaki Morimoto	HD 187642, Aquila	Jul 1999	13 images depicting: (1) definitions of numbers, signs, operations and elements, (2) structure of the Solar System, (3) DNA nucleotides, (4) structure and replication of DNA, (5) protobionts, (6) jellyfish with dimensions, (7) fish with dimensions, (8) amphibian emerging from water to land with dimensions, (9) vertebrate organism with dimensions, (10) primate with dimensions, (11) family of humans, gene number and world population, (12) woman's face with depiction of raised [hello] arm gesture, (13) interpretive key [total number of "bits," wavelength of transmission frequency and radius of dish] and molecular formula for ethanol alongside English/Japanese words for "toast/cheers"—binary-encoded, each comprised of 71 × 71 mosaic. Transmission: Stanford 46m radio telescope "The Dish," 423 MHz.
NASDA METI Messages (Personal correspondence with Shin-ya Narusawa—collaborator on the JAXA METI Experiments.)	~22 Aug 1995 ~20 Aug 1996 ~12 Aug 1997 ~18 Aug 1998	National Science Development Agency of Japan (NASDA); Cosmic College for Students, Institute of Space and Astronautical Science (ISAS)	1995 Libra constellation 1996 --------- 1997 HD 116658, Virgo 1998 ---------	2147-2361 --------- 2247 ---------	(1995) 11 × 11 mosaic image detailing parents holding child's hands, 2111.539295 MHz. (1996) 11 × 11 mosaic image of an "alien" and an earthling. (1997) 11 × 11 mosaic images of (a) a human smiling (b) a rice dumpling and tea. (1998) 11 × 11 mosaic image depicting 1+1=2 in icons and corresponding Arabic integers. (All) Transmissions: 64m Usuda Deep Space Center antenna.

Initiative Name	Transmitted	Organizer(s)	Targeted Object(s)	Arrival	Transmitted Content(s) and Transmission Parameters
Cosmic Call 1 (Vakoch, 2009; Dumas, 2010; Zaitsev, 2011; Zaitsev and Ignatov, 1999; Oberhaus, 2019)	24 May 1999 30 Jun 1999 30 Jun 1999 1 Jul 1999	Team Encounter/ Alexander Zaitsev	HD 186408, Cygnus HD 190406, Sagitta HD 178428, Sagitta HD 190360, Cygnus	Nov 2069 Feb 2057 Oct 2067 Apr 2051	Four scientific messages followed by an archive of public messages. Scientific: (1) A noise-resistant exponent alphabet detailing numbers, mathematical concepts, physical units, chemical elements, physical, biological, astronomical units and other concepts visualized within 23 graphical pages (each 127×127 mosaic); (2) a mathematical language detailing the specifications of Team Encounter's proposed lightsail spacecraft; (3) messages from the Team Encounter staff; (4) copies of the Arecibo Message. Public: over 43,000 messages submitted by participating public—sent via binary phase shift keying. Transmissions: RT-70 Yevpatoria, 5010.024 MHz, 148, 152, 152, 152 kW.
Teen Age Message (Zaitsev, 2002a, 2002b; Zaitsev, 2008b; Zaitsev, 2016)	29 Aug 2001 3 Sep 2001 3 Sep 2001 3 Sep 2001 4 Sep 2001 4 Sep 2001	Alexander Zaitsev/ Yevpatoria RT-70 Radio Telescope Observatory, Education Department of Moscow	HD 197076, Delphinus HD 95128, Ursa Major HD 50692, Gemini HD 126053, Virgo HD 76151, Hydra HD 193664, Draco	Feb 2070 Jul 2047 Dec 2057 Jan 2059 May 2057 Jan 2059	Three sectioned transmission: (1) a monochromatic radio wave with doppler correction for the Earth's rotation and motion around the Sun. (2) A 15-minute concert performed on a Theremin musical instrument (7 songs in total). (3) TAM logo, texts of "Greeting" to ETI and the Bilingual Image Glossary of basic terrestrial concepts written in both Russian and English—digital information sent via binary phase shift keying and Theremin audio signals using single side band (SSB) modulation. Transmissions: RT-70 Yevpatoria, 5010 MHz, 150 kW (HD197076) and 96 kW (all other).

Initiative Name	Transmitted	Organizer(s)	Targeted Object(s)	Arrival	Transmitted Content(s) and Transmission Parameters
Cosmic Call 2 (Dumas, 2010; Dumas and Dutil, 2010; Dominus, 2015; Chorost, 2016; Dumas and Dutil, 2016; Braastad and Zaitsev, 2003; Zaitsev, 2003; Oberhaus, 2019)	6 Jul 2003 6 Jul 2003 6 Jul 2003 6 Jul 2003 6 Jul 2003	Team Encounter/ Alexander Zaitsev	Hip 4872, Cassiopeia HD 245409, Orion HD 75732, Cancer HD 10307, Andromeda HD 95128, Ursa Major	Apr 2036 Aug 2040 May 2044 Sep 2044 May 2049	Five scientific messages followed by an archive of public messages. Scientific: (1) A (re-evaluated) noise-resistant exponent alphabet detailing numbers, mathematical concepts, physical units, chemical elements, physical, biological, astronomical units and other concepts visualized on a single, long graphical page (mosaic is 127 pixels wide); (2) a copies of the Arecibo message; (3) the Bilingual Image Glossary (BIG) message of 12 binary images (101 × 101 lines); (4) messages from the Team Encounter staff; (5) a mathematical language detailing the specifications of Team Encounter's proposed lightsail spacecraft. Public: the "Ella" chatbot, audio files including "Starman" by David Bowie, a copy of "HellotoETI" website, drawings by Ukrainian schoolchildren, music and photographs of KFT band, 282 flags of the world, "Extraterrestrial Culture Day" bill adopted by New Mexico and over 90,000 messages submitted by participating public (text, still images, audio—sent in binary phase shift keying. (All) Transmissions: RT-70 Yevpatoria, 5010.024 MHz, 150 kW.
A Message from Earth (Atkinson, 2008; BBC, 2008; Kiss, 2008; Moore, 2008)	9 Oct 2008	RDF Digital/ Bebo, Alexander Zaitsev	Hip 74995, Libra	Feb 2029	501 text messages, photographs and drawings concerning participants own lives, ambitions, views of world peace and planet Earth along with images of celebrities, notable landmarks, each contained within a 347 × 347-pixel mosaic—sent in binary phase shift keying. Transmission: RT-70 Yevpatoria, 5010.024 MHz, 150 kW.

Initiative Name	Transmitted	Organizer(s)	Targeted Object(s)	Arrival	Transmitted Content(s) and Transmission Parameters
Lone Signal (Byrd, 2013; Gohring, 2013; Kramer, 2013; Pickard, 2014; Busch, 2013)	??? ???	Pierre Fabre and several entrepreneurs	Crab Pulsar Vela Pulsar	? ?	Default string of 0s (test signal and targets for equipment), 49m Madley Earth Station.
	??? ??? ??? 18 Jun 2013 19 Jun 2013 20 Jun 2013 21 Jun 2013 22 Jun 2013 23 Jun 2013 24 Jun 2013 8 Aug 2013 18 Aug 2013 29 Aug 2013		Cassiopeia A Geminga PSR J0437–471 HD 119850, Boötes " " " " " " " " " "	? ? ? Jul 2030 " " " " " " " " " "	? ? ? Continuous wave/hailing component (a binary encoding system, which uses octal intermediary to represent numbers, mathematical operators and other symbols) and 8,201 (144 character) messages submitted by the public on a separate frequency. Transmissions: Jamesburg 30m Earth Station, 200 kHz (Public), 6.720472513690 GHz (Hailing–Wavelength 44.60883626698388).
JAXA METI Experiments (Personal correspondence with Shin-ya Narusawa— collaborator on the JAXA METI Experiments.; Dumas, 2015; JAXA Space Education Center, 2013; 2014)	22 Sep 2013 23 Aug 2014	JAXA Space Camp, Shin-ya Narusawa (collaborator), Usuda Deep Space Center	HD 75732, Cancer HD 75732, Cancer	2053 2054	(2013) 11 × 11 mosaic image detailing parents holding a child's hands. (2014) A single image (potentially 11 × 11-pixel mosaic) of the Sun and a human. (All) Transmissions: 64m Usuda Deep Space Center antenna, 20 kW.

Initiative Name	Transmitted	Organizer(s)	Targeted Object(s)	Arrival	Transmitted Content(s) and Transmission Parameters
Sónar Calling GJ 273b (Personal correspondence with Alan Penny–METI International Vice President— at the UK SETI Research Network Conference 2018, University of Oxford.; Sónar, 2017; Vakoch, 2017)	16 Oct 2017 17 Oct 2017 18 Oct 2017 14 May 2018 15 May 2018 16 May 2018	Sónar Music Festival, Institute of Space Studies of Catalonia, METI International	GJ273b/Hip 36208, Canis Minor	(16–18 Oct) ~ Mar 2030 (14–16 May) ~ Oct 2031	16–18. A "hello" file consisting of (repeated) sequenced prime numbers along with an introduction to 8 bit "byte" encoding, a "tutorial" file demonstrating an 8-bit mathematical concept primers and physical concepts along with (16 October) seven 10-second music compositions, (17 October) six 10-second music compositions, (18 October) six 10-second music compositions. 14–16. A "hello" file consisting of (repeated) sequenced prime numbers along with an introduction to 8 bit "byte" encoding, a bitmap "tutorial" file derived from Cosmic Call messages that demonstrated mathematical concepts, chemical formulas and human anatomy as small graphics along with (14 May) five 10-second music compositions, (15 May) ten 10-second music compositions, (16 May) five 10-second music compositions from the Sónar Music Festival artists—transmitted 3 times a day, over three consecutive days. (All) Transmissions: Tromsø EISCAT 32 m dish, 930 MHz (16, 17, 18) and 929.0 MHz (14, 15, 16), 1.5 MW.
Pending					
Stihia Beyond Message (METI International, 2022; Stihia Beyond, 2022; Sparkes, 2022)	4 Oct 2022 (SEEMINGLY CANCELLED)	METI International, Stihia Music Festival, Goonhilly Earth Station, Ulugh Beg Astronomical Institute	TRAPPIST-1, Aquarius K2–18b/EPIC 201912552, Leo	~2063 ~2146	Counting and mathematic tutorial, Periodic Table of Elements, corresponding atomic depictions, stellar atlas of Mirzo-Ulugbek, and 15-second music samples—sent in binary using four different phases of same frequency (i.e., quadrature phase keying). Transmission: Goonhilly Satellite Earth Station 25.9 m "Arthur" dish.

Initiative Name	Transmitted	Organizer(s)	Targeted Object(s)	Arrival	Transmitted Content(s) and Transmission Parameters
A Beacon in the Galaxy (Jiang *et al.*, 2022)	~2023 (date unconfirmed)	---------	---------	---------	Revised versions of Cosmic Call bitmaps. (Intended) FAST—500 m Aperture Spherical Radio Telescope and Allen Telescope Array.
Message to the Milky Way (Diamond Sky Productions, 2013)	---------	Diamond Sky Productions/ Carolyn Porco, "The Day the Earth Smiled"	---------	---------	Public images taken on July 19, 2013, that "Describe us and our home planet…" and also "One element of this message will be a musical contribution from a member of the public." (Intended) Transmission: Arecibo Observatory 305m radio telescope.
Interstellar Beacon/Backup Earth (Kitchen, 2017)	---------	William Kitchen, Paul Shuch, John Spencer, Daniel Batcheldor, Armin T. Ellis, Nova Spivack, Ethan Siegel	---------	---------	An interstellar Rosetta stone encapsulating language and communication strategies, an interstellar time capsule detailing natural and anthropological history including the art and science of humanity, an interstellar "Noah's Ark" containing the genome sequences of numerous organisms and instructions for recreating humans "Human Nursery"—transmitted using laser pulses.

Outreach, Educational and Symbolic Transmissions

This category details signals that are not regarded (by the majority of the mainstream SETI community) as serious communication attempts with ETI civilizations, but rather as symbolic gestures, outreach activities, or educational opportunities. Some METI transmissions are listed within this category based upon the extent of criticism from SETI scholars under the criteria choice of target(s), signal properties, appropriate transmission equipment used, accessibility of contents, lack of encoding methods employed within the signal design, etc.

Initiative Name	Transmitted	Organizer(s)	Targeted Object(s)	Arrival	Transmitted Content(s) and Transmission Parameters
Morse Message (Valentine, 2011; Kotel'nikov Institute of Radio-engineering and Electronics, 2008)	19 Nov 1962 24 Nov 1962	Soviet Union Radioastronomers	Venus (passed by planet and now travelling towards HD 131336, Libra)	4122	(19 November) "MIR" (peace/world), (24 November "LENIN" (i.e., Vladimir Lenin) and "SSSR" (abbreviation for Soviet Union)—using Morse Code. Transmissions: Yevpatoria Pluton-M transmitter, 769 MHz and modulated by shift of 62.5 Hz, 50 kW.
Discovery; Calling All Aliens (Harrison, 2007; Dumas, 2015)	2005	Discovery Channel (Canada)	---------	---------	---------
Across the Universe (NASA Content Administrator, 2008; NBC News, 2008)	4 Feb 2008	Martin Lewis, NASA	HD 8890, Ursa Minor	2442	"Across the Universe" song by The Beatles along with message "Send my love to the aliens. All the best, Paul." by Paul McCartney. Transmission: Madrid Deep Space Communications Complex 70m antenna (DSS-63), 18 kW.
Hello from Earth (IT News, 2009; Leonard, 2009; Osborne, 2009)	28 Aug 2009	Wilson da Silva/Australia National Science Week/NASA	Hip 74995, Libra	2029	25,878 text messages from the public (each 160 characters in length)—encoded in binary format. Transmission: Canberra Deep Space Communications Complex 70m antenna (DSS-43).
Wow! Reply (National Geographic, 2012)	15 Aug 2012 15 Aug 2012 15 Aug 2012	National Geographic Channel, Campfire and Arecibo Observatory	HD 54351, Gemini HD 50692, Gemini HD 75732, Cancer	2162 2069 2053	20,000 Twitter messages with "training headers" and celebrity videos—sent in binary phase shift keying. Transmissions: Arecibo Observatory 305m radio telescope, 2380 MHz, 1 MW.
Toronto Science Fair METI Experiment [Canada] (Dumas, 2015)	2013	Toronto Science Fair, Algonquin Radio Observatory	HD 10700, Cetus Kepler-62, Lyra	2025 3213	"~100 words on video." Transmission: Algonquin Radio Observatory.

Initiative Name	Transmitted	Organizer(s)	Targeted Object(s)	Arrival	Transmitted Content(s) and Transmission Parameters
A Simple Response to an Elemental Message (McCracken, 2016; Schulze-Makuch, 2016; Scuka, 2016; Quast, 2017)	10 Oct 2016	Paul Quast, ESA, University of Edinburgh, UKATC	HD 8890, Ursa Minor	2450	3,775 ecological messages/poems/ statements submitted by the public in 146 countries, 81 historical quotes, and 70 photographs of Earth / humankind including several copies of the Arecibo Message—all encoded in binary format. Transmission: ESA 35m Cebreros Deep Space Ground Station, 7168.0089310 MHz, 20 kW.
Stephen Hawking Memorial Broadcast (European Space Agency, 2018; Hawking Foundation, 2018)	15 Jun 2018	ESA, The Stephen Hawking Foundation, Vangelis	1A 0620–00, Monoceros	5475	A six-minute message drawn from a Stephen Hawking speech (about preserving the planet) set against a specially written musical piece composed by Vangelis. Transmission: ESA 35m Cebreros Deep Space Ground Station, 20 kW.
Message to Proxima Centauri B (The Western Australian, 2018)	27 Nov 2018	Kevin Vinsen, International Centre for Radio Astronomy Research, ESA	Hip 70890 b, Centaurus	Feb 2023	Images, sounds, text. Transmission: ESA 35m New Norcia Deep Space Ground Station, 20 kW.
Roddenberry 100th Anniversary (NASA SCaN, 2021)	19 Aug 2021	Misc. NASA and DSN staff	HD 26976, Eridanus	2038	1976 short diversity speech by Gene Roddenberry. Goldstone Deep Space Communications Complex 34m antenna (DSS-13), 416.7 MHz, 20 kW.
Pending					
Portrait of Humanity (Portrait of Humanity, 2019; Cowan, 2020)	2020?	1854 Media, Magnum Photos, "Sent into Space"	---------	---------	200 Photographs in binary

Space Mission Outreach (Transmissions to Probes)

This category encompasses transmissions that are intended as public engagement activities for specific space missions or indirect (non-command function) signals sent specifically to spacecraft.

Initiative Name	Transmitted	Organizer(s)	Targeted Object(s)	Arrival	Transmitted Content(s) and Transmission Parameters
Wake-up command to Magellan probe (Reeves, 1994)	~17–30 Aug 1988	NASA/Jet Propulsion Laboratory	Magellan	~ 14.5 minutes (to Venus)	This powerful signal was a "wake-up" command sent to the Magellan probe in order to correct a "wobble" affecting the communication antenna. Transmission: Goldstone Deep Space Communications Complex 70m antenna (DSS-14), 350 kW.
Wake-up Rosetta (European Space Agency, 2013; 2014)	2014	ESA	Rosetta/67P (i.e., Churyumov-Gerasimenko comet)	~27 minutes	10 "wake-up" videos submitted by the public for the end of the Rosetta spacecraft hibernation sequence. Transmission: (unspecified facility) ESA Deep Space Ground Station, 20 kW.
#Goodbye Rosetta (May 2016; Miller, 2016)	2016	ESA, heritage-Futures, Sarah May	Rosetta/67P (i.e., Churyumov-Gerasimenko comet)	~27 minutes	Pictures, sentiments in various languages, video links and other media submitted by the global public through Twitter. Transmission: (unspecified facility) ESA Deep Space Ground Station, 20 kW.
Message to Voyager (NASA, 2017; Borgerding, 2017)	5 Sep 2017	NASA	Voyager 1	~20 hours	Single "We offer friendship across the stars. You are not alone." text message from #MessageToVoyager campaign. Transmission: Madrid Deep Space Communications Complex 70m antenna (DSS-63).
Beyond Pluto; The Ultima Thule Flyby (Johns Hopkins Applied Physics Laboratory, 2018)	1 Jan 2019	NASA, Johns Hopkins University	New Horizons	~6 hours	Unicode "greetings" (choice of "Go New Horizons! Go NASA!" / "Hello Ultima Thule!" / "Exploration Rocks!" / "Keep on Exploring!" / "Happy 2019!" / "Ultima is Far Out!") alongside the submitter's name. Johns Hopkins University Applied Physics Laboratory; SCF 60-Foot System, ~2kW.
Pending					
Voyager's Final Message (Voyager's Final Message, 2020)	---------	"Community"	Voyager 1	~21 hours	1,000-character (max) Unicode message.

Cultural Expression and Advertisement Messages

Category details transmissions which are commercial in nature (i.e., advertisement campaigns), or poetic gestures that are predominately orientated as expressions of cultural articulation (or "SpaceArt"). Signals within this category are also identified based upon criticism and criteria posed by the SETI community.

Initiative Name	Transmitted	Organizer(s)	Targeted Object(s)	Arrival	Transmitted Content(s) and Transmission Parameters
Poetica Vaginal (Van Damme *et al.*, 2009; Marshall, 2010; López del Rincón, 2015)	1986	Joe Davis/ Massachusetts Institute of Technology	HD 22049, Eridanus HD 10700, Cetus 2 other star systems (?)	(1) 1996 (2) 1998 (3/4) ------	A series of full-power test transmissions of vaginal contraction sounds (translated into text, music and phonetic speech) from ballet dancers. The U.S. Air Force prevented the official, "million Watt" transmission occurring. Transmissions: MIT's Millstone Hill Radar.
Message from Human Beings to the Universe– Nançay Message (Dumas, 2015; Malloy, 2016)	26 Jan 1987	Jean-Marc Philippe	Milky Way; Sagittarius A*	27000–29000	10,500 messages. Transmission: Nançay radio telescope.
Cosmic Connexion (Luxorion 2006; Zaitsev, 2012)	30 Sep 2006	Jean-Jacques Beineix/ Cargo Films, CNES, ARTE	HD 222404, Cepheus	2051	Animated (nude) presenters depicting humankind through images, sound, film, animations and cartoons within a short video file. Transmission: Issus-Aussaguel Station radio telescope.
Dorito Advertisement (Barras, 2008; Highfield, 2008; Space Daily, 2008)	12 Jun 2008	Doritos/ University of Leicester/ EISCAT	HD 95128, Ursa Major	2050	30 second video file encoded in binary. Transmission: EISCAT Svalbard radar, ~500 MHz.
Logo of Zhitomir City (Журнал Житомира, 2009)	~15 Sep 2009	Zhitomir City Council	---------	---------	The official logo of Zhitomir City. Transmission: RT-70 Yevpatoria.
RuBisCo Stars Message (Chandler, 2009; Gilster, 2009; Davis and Hofmans, 2010)	7 Nov 2009 7 Nov 2009 7 Nov 2009	Joe Davis	L1159–16, Aries SO J025300.5+165258, Aries HD 20630, Cetus	2024 2022 2039	DNA sequence of RuBisCo photosynthesis protein—encoded in binary (C=00 T=01 A=10 and G=11) text file and also performed as a sound piece along with secondary message "I am the Riddle of Life Know Yourself." Transmissions: Arecibo Observatory 305m radio telescope.

Initiative Name	Transmitted	Organizer(s)	Targeted Object(s)	Arrival	Transmitted Content(s) and Transmission Parameters
Break the Eerie Silence (Jones, 2010; Southorn, 2010)	12 Mar 2010	Penguin UK, Sent Forever, National Science and Engineering Week, The Big Bang	M42, Orion	~3334	~1,000 messages. Transmission: Goonhilly Satellite Earth Station 32m antenna. See also "Sent Forever" organization in the "Short-Range, Commercial Transmissions" category.
Message to Qo'nos (Reuters, 2010; U–The Opera, 2010)	18 Apr 2010	Pool Worldwide/ CAMRAS	HD 124897, Boötes (i.e., 'Qo'noS' location)	2047	An invitation (in Klingon) to attend an Earth-based performance of Klingon Opera. Transmission: Dwingeloo Radio Observatory 25m radio telescope.
Cogito (De Paulis, 2017; 2018)	(1) 26 Nov 2014 (2) Ongoing transmissions	Daniela de Paulis, Dwingeloo Radio Observatory, CAMRAS	(1) Titan (26 Nov 2014) (2) Subsequent messages— no specific, purposeful targets are defined	(1) ~79 minutes (2) ----- ----	Human (and other organism) brain waves are recorded in real-time by laboratory-grade electroencephalograph (EEG) before being converted into sound for instant transmission. Transmissions: Dwingeloo Radio Observatory 25m telescope.

Short-Range, Commercial Transmissions

This category details commercial entities that transmit public content into space for a fee, or as part of local outreach projects. Contents of signals placed within this category should not be decipherable past $1 \leq$ pc (1- 3ly) from Earth and possess an eclectic range of themes that may not be properly encoded for any of the above applications.

Transmitter Organization	Initiative Designation	Transmitted	Organizer(s)	Targeted Object(s)	Transmitted Content(s) and Transmission Parameters
Deep Space Communication Network (Personal correspondence with Jim Lewis— managing director of DSCN.; Lewis, 2017)	Live from Australia Festival	18 Jan 2015	Aphids	---------	Transmissions (for all in DSCN subset): "high-powered klystron amplifiers connected by a travelling wave-guide to a 5m parabolic dish," 6000–6250 MHz.
DSCN	Sam Klemke (Closer Productions, 2015; Salce, 2015)	13 Aug 2014	Closer Productions	---------	A film about "35 years of Sam Klemke living on Earth."
DSCN	Destination Selfie	12 Jun 2014	Nathaniel Stern	---------	---------

Transmitter Organization	Initiative Designation	Transmitted	Organizer(s)	Targeted Object(s)	Transmitted Content(s) and Transmission Parameters
DSCN	A Perfect Day (Browne, 2013)	22 Jan 2013	Susanna Browne	---------	A song sent via "high frequency radio transmission into space."
DSCN	Tweets in Space (Chakelian, 2012; Katz, 2012; Scharf, 2012a, 2012b; Tweets in Space, 2012)	28 Nov 2012	Scott Kildall, Nathaniel Stern	GJ 667Cc, Scorpius	1,500 tweets from global public via "analogue and digital signals."
DSCN	Space messages from Germany	26 Jan 2010	Cenyo Incorporation	---------	---------
DSCN	Live band performance	10 Jun 2010	Mercury Records Limited	---------	Audio files from live band performances.
DSCN	Text messages from England	24 Dec 2009	REaD Group	---------	Text messages (submitted as part of an advertising campaign).
DSCN	Text messages– Part 2	9 Oct 2009	Beeby Clark + Meyler	---------	Text messages (submitted as part of an advertising campaign).
DSCN	Text messages– Part 1	9 Oct 2009	Beeby Clark + Meyler	---------	Text messages (submitted as part of an advertising campaign).
DSCN	The Space Show 21	19 Aug 2009	David M. Livingston	---------	Internet radio talk show about space commerce and exploration.
DSCN	The Space Show 20	19 May 2009	David M. Livingston	---------	Internet radio talk show about space commerce and exploration.
DSCN	Deep space transmission	15 May 2009	Kapwani Kiwanga	---------	---------
DSCN	Eduardo Kac drawing	9 Mar 2009	Eduardo Kac	---------	Eduardo Kac Ph.D. drawing.
DSCN	Deep space Transmission	27 Jan 2009	Charles H. Andrews	---------	---------
DSCN	The Day the Earth Stood Still (Overbye, 2008; Rense, 2008)	12 Dec 2008	Twentieth Century–Fox	HD 128620, Centaurus	*The Day the Earth Stood Still* (film–2008 edition).
DSCN	The Space Show 19	12 Dec 2008	David M. Livingston	---------	Internet radio talk show about space commerce and exploration.

Transmitter Organization	Initiative Designation	Transmitted	Organizer(s)	Targeted Object(s)	Transmitted Content(s) and Transmission Parameters
DSCN	The Space Show 18	6 Oct 2008	David M. Livingston	---------	Internet radio talk show about space commerce and exploration.
DSCN	Deep space transmission	5 Sep 2008	Sui Genesis, Inc.	---------	---------
DSCN	The Space Show 17	Sep 2008	David M. Livingston	---------	Internet radio talk show about space commerce and exploration.
DSCN	The Space Show 16	19 Aug 2008	David M. Livingston	---------	Internet radio talk show about space commerce and exploration.
DSCN	Ophiuchus Improvisation	Sep 2008	Paul Amlehn, Robert Fripp	---------	"Queer Reflection Harmonic Minor" musical composition.
DSCN	Yelling at the Stars (Aphids, 2008; Real Time, 2008)	7 Apr 2008 /31 May 2008	Willoh S. Weiland, Aphids Residencies and Mentoring Scheme, Sidney Myer Music Bowl/Next Wave Festival	---------	40-minute audio/visual performance.
DSCN	The Space Show 15	Dec 2007	David M. Livingston	---------	Internet radio talk show about space commerce and exploration.
DSCN	The Space Show 14	Nov 2007	David M. Livingston	---------	Internet radio talk show about space commerce and exploration.
DSCN	The Space Show 13	Sep 2007	David M. Livingston	---------	Internet radio talk show about space commerce and exploration.
DSCN	The Space Show 12	Jul 2007	David M. Livingston	---------	Internet radio talk show about space commerce and exploration.
DSCN	The Space Show 11	Apr 2007	David M. Livingston	---------	Internet radio talk show about space commerce and exploration.
DSCN	The Space Show 10	Mar 2007	David M. Livingston	---------	Internet radio talk show about space commerce and exploration.
DSCN	The Space Show 9	Dec 2006	David M. Livingston	---------	Internet radio talk show about space commerce and exploration.
DSCN	The Space Show 8	Sep 2006	David M. Livingston	---------	Internet radio talk show about space commerce and exploration.
DSCN	The Space Show 7	Aug 2006	David M. Livingston	---------	Internet radio talk show about space commerce and exploration.

Transmitter Organization	Initiative Designation	Transmitted	Organizer(s)	Targeted Object(s)	Transmitted Content(s) and Transmission Parameters
DSCN	The Space Show 6	Jul 2006	David M. Livingston	---------	Internet radio talk show about space commerce and exploration.
DSCN	Romanian Gymnastic	Jul 2006	Romanian Gymnastics	---------	Video compilation of Romanian gymnastics.
DSCN	The Space Show 5	Apr 2006	David M. Livingston	---------	Internet radio talk show about space commerce and exploration.
DSCN	The Space Show 4	Mar 2006	David M. Livingston	---------	Internet radio talk show about space commerce and exploration.
DSCN	Audi engine photo	28 Feb 2006	---------	---------	A photograph of an Audi car engine.
DSCN	The Space Show 3	Feb 2006	David M. Livingston	---------	Internet radio talk show about space commerce and exploration.
DSCN	The Space Show 2	Dec 2005	David M. Livingston	---------	Internet radio talk show about space commerce and exploration.
DSCN	The Space Show 1	Dec 2005	David M. Livingston	---------	Internet radio talk show about space commerce and exploration.
DSCN	Birthday photos	24 Jul 2005	---------	---------	Birthday photographs.
DSCN	Craigslist Messages 2 (MacMillan, 2005; Naubaum, 2005; Than, 2005)	15 Mar 2005	Jim Buckmaster	---------	Video message from Craig Newmark, clip from the documentary 24 Hours on Craigslist, and "hundreds of thousands" of postings.
DSCN	Craigslist Messages 1	11 Mar 2005	Jim Buckmaster	---------	138,000 public messages/ads/postings from Craigslist forum.
DSCN	The Orange County register messages to deep space	---------	Residents of Orange County, California.	---------	Public messages from residents of Orange County, California.
DSCN	Deep space whale song	---------	---------	---------	Encoded whale song audio.
DSCN	Black Eyed Soul in deep space	---------	---------	---------	---------

Transmitter Organization	Initiative Designation	Transmitted	Organizer(s)	Targeted Object(s)	Transmitted Content(s) and Transmission Parameters
Talk2ETs (MacMillan, 2005; Talk2ETs, 2015)	2006–Present	Private, Commercial Enterprise	Eric Knight/ Civilian Space eXploration Team	None Designated	Public messages, sentiments (exact nature of content is unverifiable) etc., sent as text or recorded voice calls (from phone). Transmissions: 6m parabolic reflector and "very high power +500 Watt block-up converters."
Sent Forever (Thomason, 2008; Sent Forever, 2009)	2009–(?)	Private, Commercial Enterprise	Stephanie Baillache, Chris Thomason	None Designated	Public messages, statements, sentiments, images, music etc. Transmissions: Goonhilly Satellite Earth Station 32m Antenna.
Pan Galactic Email Station (Personal correspondence with Blackrock Castle Observatory faculty— facilitators of the Pan Galactic Email Station; Blackrock Castle Observatory, 2017)	2010–Present	Blackrock Castle Observatory, Cork	Blackrock Castle Observatory, Cork	None Designated	Public messages, statements, sentiments, images, music etc. Transmissions: Elfordstown Earthstation 32m radio telescope.
SpaceSpeak (Martin, 2020)	2013–Present	Private, Commercial Enterprise	SpaceSpeak Services, LLC.	None Designated	Public messages, statements, images, audio etc. Transmissions: undefined "Mobile transmitter," described by third parties as a limited "2.45 GHz ISM-band transmitter and a parabolic dish."
SpaceSpeak	Fun Kids signal (SpaceSpeak, 2022)	21 Feb 2022	Fun Kids radio station, SpaceSpeak Services, LLC.	None Designated	27-minute radio program featuring nearly 1000 children's voices.

Human Time Capsules and External Memory Initiatives

This category details informative, "eternal memory" initiatives and physical time capsule elements that may be largely accessible for future human agents. These initiatives provide a modest to elaborative quantity of intelligible information (assertion is predicated upon discovery by future humans with similar cognitive and morphological functions), with some objects capable of broadly compensating for anticipated shifts in cultural memory on Earth over protracted intervals of time.

Mission; Item Title	Date	COSPAR ID	Current Location	Organizer(s)	Content(s) Description and Carrier Medium(s)
LAGEOS; Time Capsule (O'Donnell and Worrell, 1976)	7 May 1976	1976–039A	Medium Earth Orbit	NASA	Binary arithmetic counting scheme, graphic of Earth's orbit around the Sun and 3 Mollweide projections of Earth's changing tectonic activity (Permian-age/Pangaea map, present day configuration and future projection in 8.4 million years)—etched on 10 × 18cm stainless steel sheets.
Cassini; "A Portrait of Humanity" (Benford, 1999; Lomberg, 2004)	ORIG. 1997	---------	Titan's surface (PROJECT CANCELLED)	Jon Lomberg, Gregory Benford, NASA	Stereoscopic photograph of a multi-ethnic grouping of individuals, 6-axis view of Earth, Huygens lander and Cassini probe, images of planets with symbols, a large image of Saturn, photographs of the Big Dipper, M51 and Hercules cluster and a scaled map of the Solar System depicting planets, symbols and probe/lander trajectories—etched onto a diamond disk wafer.
The Orbiting Unification Ring Satellite (O.U.R.S.) (O.U.R.S.) (Woods, 1992)	ORIG. 2000	---------	Low Earth Orbit (SEEMINGLY CANCELLED)	Arthur Woods/O.U.R.S. Foundation	A large orbiting ring sculpture with an electronic digital record of Earth's diverse cultures (in year 2000): pictures, sounds and text—stored on optical laser discs.
KEO Satellite (Butler, 1998; Ashraf, 2003; Wayne, 2011; KEO, 2013)	ORIG. 2001	---------	Helio-Geocentric Return Orbit (SEEMINGLY CANCELLED)	Jean-Marc Philippe/KEO Ltd.	4 page–long messages from each person on Earth (estimate was for 6 billion population during original launch), a diamond that encases samples of human blood, air, sea water and earth, DNA and the human genome graphics engraved onto a diamond face, an astronomical clock that shows the current rotation rates of several pulsars, photographs of people from all cultures and a contemporary encyclopedia of current human knowledge—encoded onto glass, radiation-resistant DVDs and enclosed within a hollow 1 m diameter sphere.
Encounter 2001 Spacecraft (Astronet, 1998)	ORIG. 2001	---------	Beyond Jupiter/Interstellar Space (PROJECT CANCELLED)	Charles Chafer/Jim Spellman/Encounter 2001/Celestis, AeroAstro	"Human hair from... 4.5 million people worldwide" along with pictures and small messages.

Mission; Item Title	Date	COSPAR ID	Current Location	Organizer(s)	Content(s) Description and Carrier Medium(s)
Rosetta Project (Rogers, 2017; The Long Now Foundation, 2009)	2 Mar 2004	2004–006A	Deep Space	ESA, Long Now Foundation	1,500 languages detailing the Universal Declaration of Human Rights spread across 13,000 pages with redundancy material and a basic Swadesh vocabulary list—micro-etched onto a pure nickel disc. A stylus is included.
Mars Phoenix Lander; Visions of Mars DVD (Lomberg, 2010; Planetary Society, 2008)	4 Apr 2007	2007–034A	Mars–Vastitas Borealis	NASA/JPL, University of Arizona, Planetary Society	Multimedia collection of literature/art; H.G. Wells' *The War of the Worlds* (with Orson Welles radio broadcast), Percival Lowell's *Mars as the Abode of Life*, Ray Bradbury's *The Martian Chronicles*, Kim Stanley Robinson's *Green Mars,* etc., messages to future Martian inhabitants and list of name (See below "Send Your Name into Space" category)—on special silica glass DVD.
International Space Station; Immortality Drive (Coyle, 2008; Reid, 2009)	12 Oct 2008	1998–067A (item later placed on board ISS)	Low Earth Orbit	Richard Garriott	DNA sequences of 8 humans (Stephen Hawking, Stephen Colbert, Jo Garcia, Richard Garriott, Tracy Hickman, Laura Hickman, Matt Morgan, Lance Armstrong) and digital copy of the book *George's Secret Key to the Universe*, on microchip.
EchoStar XVI; The Last Pictures (Paglen, 2012; Creative Time, 2017; O' Grady, 2017)	20 Nov 2012	2012–065A	Geostationary Orbit (Longitude: 61.5° W)	Trevor Paglen, Creative Time	100 micro-etched, black and white photographs—on silicon wafer disc, encased in gold-plated aluminum case.
NanoRosetta; Human Genome Project (Personal correspondence with Bruce Ha—Nanorosetta/Stamper Technology; Svec, 2013)	ORIG. ~2013	----------	Moon (SEEMINGLY CANCELLED)	Carnegie Mellon University, Nanorosetta, Astrobotic	3.2 billion base pairs from human genome—microscopically-etched onto 5 CD-sized nickel discs.
Falcon Heavy; Isaac Asimov's Foundation Trilogy (Spivack, 2016; 2016b; Britt, 2018; The Arch Mission, 2018)	6 Feb 2018	2018–017A	On trajectory to near-Ceres/asteroid belt orbit	Arch Mission Foundation, SpaceX	A copy of Isaac Asimov's Foundation trilogy (the novels *Foundation, Foundation and Empire,* and *Second Foundation*)—written onto a "5D quartz laser storage device."

Mission; Item Title	Date	COSPAR ID	Current Location	Organizer(s)	Content(s) Description and Carrier Medium(s)
Haiyang 2B; The Orbital Library (Freeman, 2018)	24 Oct 2018	2018–081A/B	Low Earth Orbit	Arch Mission Foundation, SpaceChain	"A copy of Wikipedia."
Beresheet; The Lunar Library (Mathewson, Space, 2018; Cohen, 2019; Holmes, 2019; Spivack, 2019; Powell, 2019)	22 Feb 2019	2019–009B	Moon–Mare Serenitatis (LIKELY DESTROYED UPON IMPACT)	SpaceIL, Israel Aerospace Industries, Arch Mission Foundation	Disc 1: Concise primer guide, accessibility instructions, about the organization feature, about SpaceIL feature, Israeli time capsule, historical context about human archiving and essays about the future. Disc 2: additional primer and sponsor details. Disc 3: Expanded Pictionary, device specifications, Rosetta project, Memory of Mankind archive and advisor archives. Disc 4: Wikipedia vital articles, public and private collections. Digital contents including "about" digital layers, English Wikipedia, Project Gutenberg (~25,000 books), Internet Archive, Wearable Rosetta, Panlex Project, CIA World Factbook, Type and Format collections, featured authors and private vault collections—30 million pages on 4 discs of Nickel nanofiche.
Lunar Mission One; Digital Memory Boxes (Griffin, 2014; Li, 2014; Wall, 2014; Gitlin, 2015)	ORIG 2024	---------	Moon (PROJECT CANCELLED)	Lunar Missions Trust, David Iron	An encyclopedia of Earth's biosphere, history of human civilization, a public archive, a private archive consisting of millions of digital memory boxes (contents presently undecided) with a separate "Footsteps on the Moon" image repository—"~100 Terabytes" of information.
Pending					
Project Mora (Graviton, 2018)	~ 2020	---------	Awaiting launch details	Graviton Industries, Astrobotic	Thousands of letters, sound files, sketches, paintings and (computer) coding files to be stored within a "digital museum" hexadecagon capsule.
Astrobotic Peregrine lander; Arch 2020 (Grush, 2018)	Stated 2020	---------	Awaiting launch details	The Arch Mission Foundation, Astrobotic, United Launch Alliance	25–50 million pages from the English edition of Wikipedia encyclopedia along with a copy of the "Rosetta Project" comprised of 1,500 Languages and additional details—micro-etched onto thirty-two 1.7 × 1.7cm square nickel sheets.
Moon Ark (Press Trust of India, 2015; Zhorov, 2016; MoonArk, 2017; Studio for Creative Inquiry, 2017)	Stated 2021	---------	Awaiting launch details	Carnegie Mellon University, Astrobotic	Hundreds of images, poems, musical compositions, nano-objects (including solid 18k gold icosahedrons, chondritic meteorite fragments, diamond samples and carbon 60 fullerenes), mechanisms and other earthly samples—contained within four independent 2" × 2" diameter chambers (labelled; earth, metasphere, moon and ether respectively according to their concept and contents).

Mission; Item Title	Date	COSPAR ID	Current Location	Organizer(s)	Content(s) Description and Carrier Medium(s)
The Earthling Project (Earthling Project, 2021)	Stated 2021	---------	Awaiting launch details	Earthling Project, SETI Institute, Astrobotic, Felipe Pérez Santiago, Arch Mission Foundation	Voice extracts and music compositions submitted by the public—likely rendered onto a nanofiche disk or "5D quartz laser storage device," launched to the moon on board of Astrobotic's Peregrine.
Sanctuary (Besson *et al.*, 2021)	Stated 2021	---------	Awaiting launch details (intended; Moon–Taurus–Littrow valley)	Sanctuary Team, Planetary Transportation Systems (PTS)	Time capsule of information, stories, illustrations, literature, genome sequences and art, micro-etched onto 17 sapphire disks, each 90 mm, and comprised of approximately 2 billion pixels per surface space.
The Beyond the Earth "Companion Guide to Earth" library (Quast, 2018)	Stated 2024–2025	---------	Awaiting project details	The Beyond the Earth Foundation	Spherical introductory guides composed of two hemispheres: (1) "North"—intuitive magnification/directionality cues, an exosemiotic, ideographic guide with corroborating Lincos lexicon, an atlas of terrestrial/celestial cultural vault locations, the complete *Ramazzottius varieornatus* genome, human anatomy and cognition markers, sample biota library and tree of life diagrams and instructions for collaborative tasks. (2) "South"—curated cultural content from the plurality of ethnic/indigenous heritages, linguistics Rosetta guide, population statistics and audio tracks with spectrograms—all information is micro-etched onto a series of 16 double-sided discs and enclosed within 32 mm capsules. All stated contents are presently under revision, in line with the foundation's *After the Horizon* catalogues.
Time Capsule to Mars (Explore Mars, 2014; Holler, 2014; UKSEDS, 2015)	------ ---	---------	Awaiting project details	Emily Briere, Iulia Jivanescu, David Rokeach, MIT Space Propulsion Lab and Explore Mars Inc.	Digital messages and photographs from the public—on "potentially quartz storage technology."
The Human Document Project (Manz, 2010; Elwenspoek, 2011; Human Document Project, 2017)	------ ---	---------	Awaiting project details	Human Document project consortium	A document on key aspects of contemporary human culture—no specific medium is presently decided.

Mission; Item Title	Date	COSPAR ID	Current Location	Organizer(s)	Content(s) Description and Carrier Medium(s)
Mars One; The HELENA payload (Richards, 2014)	------ ---	---------	Awaiting project details	Andre Van Vulpen, Angus Tavner, David Blair, Josh Richards and Mars One	"Content submitted by the public via social media during National Science Week 2015"—on a radiation-hardened DVD.
MOM on the MOON (Puli Space, 2016; Kunze, 2018)	------ ---	---------	Awaiting project details	Memory of Mankind, Puli Space	The Memory of Mankind token (denoting location of this underground vault in Hallstatt, Austria) and up to six tablets of ceramic microfilm.
PTScientists; ALINA Wikipedia Archive (Foust, 2016; Coldewey, 2016)	------ ---	---------	Awaiting project details	PTScientists, Spaceflight Industries, Audi	A large extract of the Wikipedia encyclopedia—etched onto ceramic data discs.

Passive METI Initiatives

Artefacts within this category are intended as "Rosetta stones" for intercepting ETI and are capable of providing a comprehendible (and in some cases limited) account of life on Earth for another intelligent civilization which may not share our sensory perceptions, morphology, genetic heredity, ontogenic or phylogenetic traits, mutually experienced environment, or cognitive capabilities.

Mission and Item Title	Date	COSPAR ID	Current Location	Organizer(s)	Content(s) Description and Carrier Medium(s)
Pioneer 10 and 11; Plaques (Sagan et al., 1972)	2 Mar 1972, 6 Apr 1973	1972-012A, 1973-019A	Interstellar Space (both)	NASA/ARC	Diagrams of humankind (prototypical male/female models), the spacecraft, our Solar System, hydrogen line and pulsar map—on gold anodized, aluminum plaques.
Voyager 1 and 2; Golden Records (Sagan, 1978)	5 Sep 1977, 20 Aug 1977	1977-084A, 1977-076A	Interstellar Space (both)	NASA, Cornell University Committee (Chaired by Carl Sagan)	A collection of 116 images (depicting Solar System, planets, humans, animals of Earth, architecture, physical constants, food, etc.), a variety of natural sounds of Earth (wind, thunder, animal calls, etc.), an eclectic musical selection representing numerous cultures and eras, spoken greetings in 55 ancient and modern languages, human sounds (footsteps, laughter, EEG of human brain activity, etc.), "per aspera ad astra" phrase in Morse code and printed messages from the 1977 U.S. President and U.N. Secretary General—on gold anodized, copper mother LP records, wrapped in the U.S. flag and enclosed within aluminum covers.

Mission and Item Title	Date	COSPAR ID	Current Location	Organizer(s)	Content(s) Description and Carrier Medium(s)
New Horizons; One Earth Message (New Horizons Message Initiative, 2015; Washburn, 2015; Lomberg, 2016; Shanks, 2016)	---------	2006–001A	Interstellar Space (New Horizons probe). (SEEMINGLY CANCELLED)	Galaxy Garden Enterprises LLC/Jon Lomberg	A rich, crowd-sourced encyclopedia of digital pictures, audio files, information about animals, people, places, history/world events and primer strategies for decoding along with the potential inclusion of software and 3 dimensional files—all submitted by the public and selected via communal voting before uplink to the internal memory of the New Horizons probe (with the option of periodically updating this "Golden Record 2.0" repository).
Pending					
The Tree of Life (Boucher, 2020)	---------	---------	--------- (intended for Proxima Centauri b)	Julia Christensen, NASA/JPL	"Songs" of trees (i.e., trees outfitted with devices allowing them to communicate growth information with the satellite) exchanged with Cubesat during transit. Material is intended to be beamed down to exoplanet surface.

Life as a Message Within Intentional "Bio-Footprints"

The category below documents known instances of microbial contamination on other astronomical bodies resulting from human activities, and also instances of using microbial life alongside other organisms as elements within messaging projects as either a data storage device, or as a dense library documenting billions of years of irreducible computation within phylogenetic evolution, mutation, genetic mistakes and biological adaptation to our specific ecosystems (in addition to serving as a hereditary blueprint for prior organisms).

Mission and Item Title	Date	COSPAR ID	Current Location	Organizer(s)	Content(s) Description and Carrier Medium(s)
Apollo 11 Excreted Astronaut Waste	16 Jul 1969	1969–059C	Moon–Sea of Tranquility	Neil A. Armstrong, Edwin E. Aldrin	Bags of feces matter, urine, vomit, food waste and other forms of waste which likely contain microbial life—largely stored within equipment bays of the Apollo descent stage.
Apollo 12 Excreted Astronaut Waste	14 Nov 1969	1969–099C	Moon–Mare Cognitum	Charles Conrad, Alan L. Bean	Bags of feces matter, urine, vomit, food waste and other forms of waste which likely contain microbial life—largely stored within equipment bays of the Apollo descent stage.
Apollo 14 Excreted Astronaut Waste	31 Jan 1971	1971–008C	Moon–Fra Mauro Base	Alan B. Shepard, Edgar D. Mitchell	Bags of feces matter, urine, vomit, food waste and other forms of waste which likely contain microbial life—largely stored within equipment bays of the Apollo descent stage.
Apollo 15 Excreted Astronaut Waste	26 Jul 1971	1971–063C	Moon–Hadley Rille	David R. Scott, James B. Irwin	Bags of feces matter, urine, vomit, food waste and other forms of waste which likely contain microbial life—largely stored within equipment bays of the Apollo descent stage.

Mission and Item Title	Date	COSPAR ID	Current Location	Organizer(s)	Content(s) Description and Carrier Medium(s)
Apollo 16 Excreted Astronaut Waste	16 Apr 1972	1972–031C	Moon–Descartes Highlands	John W. Young, Charles M. Duke	Bags of feces matter, urine, vomit, food waste and other forms of waste which likely contain microbial life—largely stored within equipment bays of the Apollo descent stage.
Apollo 17 Excreted Astronaut Waste	7 Dec 1972	1972–096C	Moon–Taurus-Littrow	Eugene A. Cernan, Harrison H. Schmitt	Bags of feces matter, urine, vomit, food waste and other forms of waste which likely contain microbial life—largely stored within equipment bays of the Apollo descent stage.
KEO Satellite (Butler, 1998; Ashraf, 2003; Wayne, 2011; KEO, 2013)	ORIG. 2001	- - - - - - - - -	Helio-Geocentric Return Orbit (SEEMINGLY CANCELLED)	Jean-Marc Philippe/ KEO Ltd.	A diamond that encases samples of human blood, air, sea water and earth, DNA and the human genome graphics engraved onto a diamond face.
Encounter 2001 Spacecraft (Astronet, 1998)	ORIG. 2001	- - - - - - - - -	Beyond Jupiter/ Interstellar Space (PROJECT CANCELLED)	Charles Chafer/Jim Spellman/ Encounter 2001/ Celestis, AeroAstro	"Human hair from... 4.5 million people worldwide" (apparently 6 strands per person).
Beresheet; The Lunar Library (Mathewson, Space, 2018; Cohen, 2019; Holmes, 2019; Spivack, 2019; Powell, 2019)	22 Feb 2019	2019–009B	Moon–Mare Serenitatis (LIKELY DESTROYED UPON IMPACT)	SpaceIL, Israel Aerospace Industries, Arch Mission Foundation	Desiccated Tardigrades and around 100 million cells from 25 people and other organisms encased within amber.
Lunar Mission One; Digital Memory Boxes (Griffin, 2014; Li, 2014; Wall, 2014; Gitlin, 2015)	ORIG 2024	- - - - - - - - -	Moon (PROJECT CANCELLED)	Lunar Missions Trust, David Iron	DNA from human hairs.
Pending					
LifeShip "Ark" (Harris, 2019; Dormehl, 2019)	Stated 2021	- - - - - - - - -	Awaiting launch details	LifeShip, Ben Haldeman, The Arch Mission Foundation	An "Ark" containing crowd-contributed submitted samples of human DNA alongside biota and food crop DNA—all desiccated and encased within artificial amber compiled into a 120 mm nanofiche storage device.

Space Mission Publicity and Outreach Initiatives

Initiatives within this category are identified as part of publicity and educational activities for a featured space mission and also by the esoteric content included aboard these

spacecraft. Most of the items within this category have been organized through outreach activities within numerous space agencies or have been the result of collaborative engagement between organizations.

Mission and Item Title	Date	COSPAR ID	Current Location	Organizer(s)	Content(s) Description and Carrier Medium(s)
Huygens: Music2Titan European Space Agency, 2004a)	15 Oct 1997	1997-061C	Titan–Xanadu region	ESA, Julien Civange, Louis Haéri	4 pop songs ("Lalala," "Bald James Dean," "Hot Time" and "No Love")—on CD-ROM (same CD as signatures—see 'Send your name into space' category below).
Eugene Shoemaker remains (McKinnon, 2017; Portalist, 2018)	7 Jan 1998	1998-001A	Moon—Shoemaker crater (DESTROYED-SURFACE IMPACT)	NASA	Aluminum, vacuum-sealed urn containing Eugene Shoemaker ashes—all wrapped in brass foil inscribed with a verse from Shakespeare's *Romeo and Juliet*.
Fedsat; CD "Timecapsule" (Australian Academy of Science, 2004; Gorman, 2005)	14 Dec 2002	2002-056B	Low Earth Orbit	Cooperative Research Centre for Satellite Systems, Australia	Paul Kelly and Kev Carmody song "From Little Things, Big Things Grow" and recorded (on March 2000) statements from several hundred Australian school children—recorded onto a nickel "master" disk.
Mars Express; Red Encounter (Red Encounter, 2003)	2 Jun 2003	2003-022A	Areocentric orbit around Mars	Ferrari, ESA	Sample of Ferrari red paint "Rosso Corsa"—contained in a glass "FRED" sphere sealed into a PMMA block.
MER Mars Exploration Rovers (Spirit and Opportunity); MarsDial (Lomberg, 2010; Boyle, 2012)	10 Jun 2003 7 Jul 2003	2003-027A, 2003-032A	Mars–Gusev Crater and Meridiani Planum	NASA/ Cornell University/ Planetary Society/Jon Lomberg	A Martian sundial (used to calibrate the Pancam on rovers as well as educational purposes) consisting of etched drawings/lettering, an additional inscription and the word "Mars" in 24 languages—on an aluminum plate with anodized metal surfaces in black, gold, color along with a silicon rubber compound.
MER A Mars Exploration Rovers (Spirit); Space Shuttle Colombia Plaque (NASA, 2004b)	10 Jun 2003	2003-027A	Mars–Gusev Crater	NASA	6-inch memorial plaque featuring the names of the Colombia Space Shuttles astronauts along with the NASA logo, U.S. and Israeli flags—mounted on back of rover's high gain antenna.
Beagle 2/ Mars Express; Blur call-sign and Damien Hirst painting (Beagles 2 website, 2004; Sutherland, 2015)	2 Jun 2003	2003-022C	Mars surface (LANDER MEMORY QUESTIONABLE)	British Space Agency, ESA	A Blur (the music band) call-sign to test communication channel (stored in internal memory) and Damien Hirst spot painting as a calibration target plate for cameras/spectrometers (physical painting on lander's surface).

Mission and Item Title	Date	COSPAR ID	Current Location	Organizer(s)	Content(s) Description and Carrier Medium(s)
New Horizons; Mementos (Collect Space, 2008; The Editors of Sky and Telescope, 2015)	19 Jan 2006	2006–001A	Interstellar Space	NASA	Florida/Maryland quarter coins, 2 United States flags, CD ROMs with photographs of mission team and names (see below category), piece of SpaceShipOne, 1991 U.S. postal stamp and 1 oz of Clyde Tombaugh ashes.
TerraSAR-X; Weltraum Visitor Sculpture (Weltraum Kunst, 2003–2016)	15 Jun 2007	2007–026A	Sun-synchronous orbit	Ragnhild Becker, Gunar Seitz, German Aerospace Centre (DLR), EADS Astrium	3-dimensional sculpture affixed to satellite hull.
Curiosity (Mars Explorational Rover); Graphics and Marsdial (Redazione, 2012; Shiner, 2013)	26 Nov 2011	2011–070A	Mars–Gale Crater	NASA	Graphic of Leonardo da Vinci, excerpt from his "Codex on the Flight of Birds" along with some essays, drawings and list of names (see below "Send Your Name into Space" category). A Martian sundial (left over from the Mars Explorational rover) consisting of etched drawings/lettering, an additional inscription and the word "Mars" in 24 languages—on an aluminum plate with anodized metal surfaces in black, gold, color along with a silicon rubber compound.
Juno; Galileo Plaque and Lego Figurines (Collect Space, 2011; NASA Content Administrator, 2011)	5 Aug 2011	2011–040A	Orbiting Jovian System (2021–end of mission deorbit)	NASA/ JPL, Lego, Italian Space Agency	Graphic of Galileo Galilei with hand-written paragraph (concerning Jupiter observations) and signature—on a 71 × 51 mm flight-grade aluminum plaque. Lego figurines of Juno, Galileo and Jupiter in aluminum included aboard craft.
Planet Labs AiR Program (Planet Labs, 2013)	2014–Present	2014-XXXAA	Low Earth Orbit	Planet Labs, Forest Stearns	Multi-chromatic illustrations resulting from artist in residence program at Planet Labs—all laser etched onto satellite hull and appendages.
CHEOPS: European Kids Drawings (Campbell, 2015; European Space Agency, 2016)	18 Dec 2019	2019–092B	Sun-synchronous orbit (around Earth)	University of Bern, ESA	3,000 children's drawings—engraved onto two metal plaques.

Mission and Item Title	Date	COSPAR ID	Current Location	Organizer(s)	Content(s) Description and Carrier Medium(s)
Lucy; The Lucy Plaque (NASA Lucy Mission, 2021)	16 Oct 2021	2021–093A	Heliocentric Orbit—intersecting Earth to Jupiter-trojan orbital regions	NASA, Lucy Mission Team	Time capsule featuring position of planets, Earth continents, spacecraft properties and defined "words of advice, joy, wisdom, inspiration," and other similar messages "from prominent members of our society" (about 20 quotes by poets, physicists, musicians, authors, activists, and writers)—inscribed onto a gold anodized, metal plate.
Pending					
'To Space, From Earth' (Mellor, 2020)	Stated 2022	2022-XXXA	Undefined region(s) of space	Beyond-Earth collective	Space Art DNA Capsule that will "Encode artworks into silicon-based synthetic DNA."

"Send Your Name into Space" Initiatives

This category details the popular "send your name into space" outreach initiatives that are conducted by space agencies for either the global public, or select groups of individuals/organizations (e.g., The Planetary Society).

Space Mission	Date	COSPAR ID	Current Location	Organizer(s)	Content(s) Description and Carrier Medium(s)
Viking 1 and Viking 2 (National Aeronautics and Space Administration, 1978; Benford, 1999)	20 Aug 1975, 9 Sep 1975	1975–075C, 1975–083C	Mars–Chryse Planitia and Utopia Planitia	NASA	Microdot of signatures from thousands of people (administrators, science teams, flight teams, camera technicians, support personnel, analysts and interns) who contributed towards the development of the Viking landers.
Mars Pathfinder (Planetary Society, 2017)	4 Dec 1996	1996–068A	Mars–Ares Vallis region	NASA/JPL/ Planetary Society	100,000 names (originally collected for the failed Mars '96 mission)—on microchip.
Cassini; "Send Your Signature to Saturn" (Murrill, 1997; NASA, 2004a; Zeluck, 1996)	15 Oct 1997	1997–061A	Saturn (DESTROYED-DE-ORBITED INTO SATURN)	NASA/JPL/ ESA/ASI/ Planetary Society	616,420 handwritten signatures (including Christiaan Huygens and Giovanni Cassini signatures)—on DVD disc.
Huygens CD (Cassini)/"Messages on Titan" (European Space Agency, 2004b)	15 Oct 1997	1997–061C	Titan–Xanadu region	ESA/ASI/ NASA	85,000 signatures (along with some texts, drawings and musical compositions)—on CD-ROM.

Space Mission	Date	COSPAR ID	Current Location	Organizer(s)	Content(s) Description and Carrier Medium(s)
Mars Polar Lander [Mission Failed] (Ainsworth, 1998)	3 Jan 1999	1999–001A	Mars–Planum Australe (DESTROYED–SURFACE IMPACT)	NASA/JPL	~1,000,000 names of kids from across the globe—on CD ROM.
Stardust (NExT); "Send your name to a comet" (NASA, 1999; Collect Space, 2014; Planetary Society, 2017)	7 Feb 1999	1999–003A	~312,000,000 km in deep space	NASA/JPL/ Planetary Society	Electronically-etched (~1,000,000) names of public and 58,214 Vietnam veterans memorial names—on 10.16 cm silicon chips.
Hayabusa; "Let's Fly to Meet Your 'Star Prince'" (Shujiro, 2003; Matogawa, 2005; Reddy, 2005)	9 May 2003	2003–019A	Deep space near 25143 Itokawa–List placed aboard 'Minilander' (FAILED TO LAND)	JAXA/Sawai Shujiro	880,000 signatures engraved onto a 10 cm aluminum sphere "target marker."
MER Mars Exploration Rovers (Spirit and Opportunity) (Planetary Society, 2017)	10 Jun 2003 7 Jul 2003	2003–027A, 2003–032A	Mars–Gusev Crater and Meridiani Planum	NASA/ Planetary Society	4,000,000 names apiece along with Stephen Little artwork "Monochrome (for Mars)"—on silica glass DVDs.
Deep Impact "send your name to a comet" (Carey, 2005)	12 Jan 2005	2005–001A	~431,000,000 km from Earth (DESTROYED—IMPACTED COMET)	NASA	625,000 names—on CD ROM.
Saparmurat Niyazov "Rukhnama" and 2 Flags (BBC News, 2005; Kalder, 2013)	23 Aug 2005	(Launched with) 2005–031A	Low Earth Orbit	Saparmurat Niyazov Roscosmos (?)/JAXA (?)	A copy of the book Rukhnama written by former Turkmenistan president Saparmurat Niyazov along with a Turkman flag and presidential standard.
New Horizons (Griggs, 2015; Johns Hopkins Applied Physics Laboratory, 2016; Planetary Society, 2017)	19 Jan 2006	2006–001A	Interstellar Space	NASA	434,738 names—on CD ROM.
Mars Phoenix Lander (Planetary Society, 2017)	4 Aug 2007	2007–034A	Mars–Vastitas Borealis	NASA/JPL, University of Arizona, Planetary Society	250,000 names—on DVD (same DVD as the "Visions of Mars DVD" archive—see above "Human Time Capsules and Eternal Memory Initiatives" category).
Dawn (NASA, 2007)	27 Sep 2007	2007–043A	Ceres (asteroid belt)	NASA	~365,000 names—on microchip.

Space Mission	Date	COSPAR ID	Current Location	Organizer(s)	Content(s) Description and Carrier Medium(s)
Kaguya (SELENE) (Kaplan, 2007)	14 Sep 2007	2007–039A	Moon–Gill crater	JAXA/ Planetary Society	412,627 names—printed on 280 × 160 mm aluminum sheet.
Kepler Space Observatory (NASA, 2008; Zimmer, 2009)	7 Mar 2009	2009–011A	(Earth-trailing) Heliocentric Orbit	NASA	~60,000 names—on DVD.
Lunar Reconnaissance Orbiter (LRO) (Jenner, 2009)	18 Jun 2009	2009–031A	Moon–Eccentric polar orbit	NASA	~1,600,000 names—on microchip.
Akatsuki (PLANET-C) (JAXA, 2009a; 2009b)	20 May 2010	2010–020D	Elliptical orbit around Venus	JAXA/ Planetary Society	260,214 names and messages (with 2 figurines)—on ~90 aluminum plates. "Hatsume Miku" manga character, character voice and "chibi" rendering of same character etched onto an additional 3 aluminum plates.
IKAROS (Par, 2010; Planetary Society, 2010)	20 May 2010	2010–020E	Heliocentric orbit ~110,000,000 km from Earth	JAXA/ Planetary Society	63,248 names and messages—on aluminum plates (stored in 3/4 of IKAROS' square-shaped sail corners). Content collected by JAXA only. 89,000 names and messages— separate names stored upon the Planetary Society's silica glass mini-DVD.
Curiosity (Mars Explorational Rover) (NASA, 2010)	26 Nov 2011	2011–070A	Mars–Gale Crater	NASA	1,246,445 names—etched onto 2 silicon microchips.
Mars Maven (University of Colorado, 2012; NBC News, 2014)	18 Nov 2013	2013–063A	Areocentric elliptic orbit around Mars	NASA/ Planetary Society	100,000 names, 377 student artworks and 1,000 haiku poems— on DVD.
Hayabusa 2 (Planetary Society, 2014; Yoshikawa et al., 2015)	3 Dec 2014	2014–076A	En-route to asteroid 162173 Ryugu	JAXA/ Planetary Society	~400,000 names, messages, illustrations and photographs— etched on a target marker (participants names only) and also on internal memory chip of spacecraft.
OSIRIS-REx (NASA, 2014; Lalwani, 2016; Planetary Society, 2016a; Planetary Society, 2017)	8 Sep 2016	2016–055A	En-route to asteroid 101955 Bennu	NASA/ Planetary Society	442,000 names and "We the Explorers" artworks—on silicon chip.
Mars Insight (Greicius, 2017)	5 May 2018	2018–042A	Mars–Elysium Planitia	NASA	2,429,807 names—on two microchips.
Parker Solar Probe; Hot Ticket (NASA, 2018)	12 Aug 2018	2018–065A	En-route to heliocentric orbit	NASA	1,137,202 names—on an SD card attached to a plaque with epigraph and photographs dedicated to Eugene N. Parker.

Space Mission	Date	COSPAR ID	Current Location	Organizer(s)	Content(s) Description and Carrier Medium(s)
LightSail 2 (#SelfieToSpace) (Davis, 2016; Planetary Society, 2016b)	25 Jun 2019	2019–036AC	Low Earth Orbit	Planetary Society	A mini-DVD containing a Planetary Society member roster, a list of Kickstarter contributors, and names and images from the Society's "Selfies to Space" campaign.
Mars 2020; Perseverance rover (NASA Content Administrator, 2019; NASA, 2020; Evans, 2020)	30 July 2020	2020–052A	Mars–Jezero Crater	NASA	10,932,295 names and 155 student essays—on three silicon microchips along with a commemorative placard with Mars and Earth graphic featuring the words "Explore as One" in Morse Code. Additionally, a 3 × 5-inch aluminum plaque depicting Earth being supported by Rod of Asclepius symbol as a tribute to healthcare workers, SHERLOC markers made from spacesuit materials, a "returned" Martian meteorite sample, a Mars rover "family portrait," MarsDial calibration plate along with evolution graphics, and inscription, and message "Joy of Discovery" in several languages.
Pending					
Team Indus; Millions2 Moon Movement (Analytics India, 2017)	Stated 2020	---------	Awaiting launch details	Team Indus (Bangladore)	Micro-engraved names—on an aluminum sheet.
AstroGrams (Atkinson, 2019)	Stated 2020	---------	Awaiting launch details	AstroGrams, Charles Duke	Personalized messages submitted by the general public—engraved on aluminum plaques.

Space Race Pseudo-Colonial Deposits

This category documents artefacts such as medals and pennants, SpaceArt objects, mementoes, novelty effigies, and other votive material culture deposits that were dispatched into outer space by early Space Age nations, mostly as pseudo-colonialist markers or commemorative applications.

Mission and Item Title	Date	COSPAR ID	Location	Organizer(s)	Content(s) Description
Luna 1 and 2 probes and boosters; USSR Pennants (Mitchell, 2004; Reeves, 1994)	2 Jan 1959 12 Sep 1959	1959–012A 1959–014A	Heliocentric Orbit Moon–Mare Imbrium	USSR	Pentagonal elements with USSR state seal with USSR on one side and "month" 1959 date (all in Cyrillic) on other side—minted on titanium with thermoresistant polysiloxane enamels. All pentagonal elements were arranged into a sphere with an internal explosive charge for scattering pentagons in all directions. Two capsules placed on Luna 2 probe (likely vaporized), one on the Luna 2 last stage rocket phase. Luna 1 potential possessed the same configuration with 2 spheres on probe alone.

Mission and Item Title	Date	COSPAR ID	Location	Organizer(s)	Content(s) Description
Venera 1; Earth-Venus Pennant (Mitchell, 2004)	12 Feb 1961	1961-003A	Heliocentric orbit	USSR	Disc pennant depicting graphic of the inner Solar System (with 1961 configuration of planetary orbits), mission name, launch date and "USSR" (all in Cyrillic) with USSR state seal on other side, enclosed within a metallic globe of the Earth (featuring blue-tinted oceans and gold-tinted continents)—enclosed within a protective shell of stainless-steel pentagonal elements, each inscribed with "Earth Venus 1961."
TRAAC: "For A Space Prober" Poem (Landsberg and Mieghem, 1972; Leonard, 2017)	15 Nov 1961	1961-031B	Low Earth Orbit	Thomas G. Bergin, Johns Hopkins University; Applied Physics Lab, US Navy	Poem "For A Space Prober" written by Thomas G. Bergin along with launch team names—inscribed upon the spacecraft's instrumentation panel.
Zond 1; Earth-Venus Pennant (Mitchell, 2004)	2 Apr 1964	1964-016D	Heliocentric orbit	USSR	Disc pennant depicting graphic of the inner Solar System (with 1964 configuration of planetary orbits), mission name, launch date and "USSR" (all in Cyrillic) with USSR state seal on other side, enclosed within a metallic globe of the Earth (featuring blue-tinted oceans and gold-tinted continents)—enclosed within a protective shell of stainless-steel pentagonal elements, each inscribed with "Earth Venus 1964."
Venera 3; Earth-Venus Pennant (Mitchell, 2004)	16 Nov 1965	1965-092A	Venus—surface	USSR	Disc pennant depicting graphic of the inner Solar System (with 1965 configuration of planetary orbits), mission name, launch date and "USSR" (all in Cyrillic) with USSR state seal on other side, enclosed within a metallic globe of the Earth (featuring blue-tinted oceans and gold-tinted continents)—enclosed within a protective shell of stainless-steel pentagonal elements, each inscribed with "Earth Venus 1965."
Luna 9; Pennants (Mitchell, 2004)	31 Jan 1966	1966-006A	Moon–Mare Imbrium	USSR	Singular pennants containing USSR state seal, USSR, and the landing date "January-1966" (in Cyrillic). Luna 3, 8 and 13 likely carried similar, singular pennants.
Luna 10; Pennants and "The Internationale" anthem (Mitchell, 2004; Reeves, 1994)	31 Mar 1966	1966-027A	Moon—inclined orbit at 71.9°	USSR	Pennants containing USSR state seal, USSR and date (in Cyrillic). "The Internationale" left-wing anthem transmitted to Earth from the Luna 10 orbiter on 8 April 1966.
Luna 11 and 12; Pennants (Mitchell, 2004)	24 Aug 1966 22 Oct 1966	1966-078A 1966-094A	Moon–Impacted Moon–Impacted	USSR	Pennants containing graphical depictions of Kremlin clock tower and space probe in Lunar orbit with USSR state seal, USSR and date (in Cyrillic) on the other side.

Mission and Item Title	Date	COSPAR ID	Location	Organizer(s)	Content(s) Description
Venera 4; Pennants (Mitchell, 2004)	12 Jun 1967	1967–058A	Venus–Eisila region	USSR	Pennants containing USSR state seal, USSR and date (in Cyrillic).
Venera 5 and 6; Pennants (Mitchell, 2004)	5 Jan 1969 10 Jan 1969	1969–001A, 1969–002A	Venus—surface Venus—surface	USSR	Square/pentagonal pennants containing USSR state seal, USSR and date (in Cyrillic), bas-relief of Lenin and graphical depictions of spacecraft in Solar System. Dates and names on each pennant vary per probe.
Apollo 11; Goodwill Messages (NASA, 1969; Pearlman, 2007; Rahman, 2007)	16 Jul 1969	1969–059C	Moon–Sea of Tranquility	NASA	Statements of goodwill from 73 (1969) world leaders—engraved on a silicon disc and enclosed within an aluminum case.
Apollo 11; Gold Olive Branch and commemorative bag (Minnesotastan, 2010; McKinnon, 2017)	16 Jul 1969	1969–059C	Moon–Sea of Tranquility	NASA	A gold (presumably cast gold) olive branch—a traditional symbol of peace—Apollo 1 mission patch and a diamond studded astronaut pin (both objects commemorating the mission team) along with Soviet Union medals (unofficially included to commemorate Yuri Gagarin and Vladimir Komarov).
Apollo 11; Lunar Plaque (Johnson, 2008; Smithsonian Institution, 2018)	16 Jul 1969	1969–059C	Moon–Sea of Tranquility	NASA/Jack Kinzler	Two hemisphere maps of Earth, an inscription and astronauts' signatures—engraved on stainless steel.
Apollo 12; Moon Museum (Landes, 2016; Museum of Modern Art, 2018)	14 Nov 1969	1969–099C	Moon–Mare Cognitum	NASA/Jack Kinzler	Six miniature artworks by Robert Rauschenberg, David Novros, John Chamberlain, Claes Oldenburg, Forrest Myers and Andy Warhol—on a small, ceramic wafer potentially attached to leg of Apollo 12 lander.
Apollo 12; Lunar Plaque	14 Nov 1969	1969–099C	Moon–Mare Cognitum	NASA/Jack Kinzler	Inscription and astronauts' signatures—engraved on stainless steel.
Apollo 12; Astronaut Wings (NASA History Program Office, 2012)	14 Nov 1969	1969–099C	Moon–Mare Cognitum	Alan Bean	Silver astronaut pin and naval aviator wings badge placed on Moon in remembrance of Clifton C. Williams.
Venera 7; Pennants (Mitchell, 2004)	17 Aug 1970	1970–060A	Venus—surface	USSR	Square/pentagonal pennants containing USSR state seal, USSR and date (in Cyrillic), bas-relief of Lenin with commemorative inscription (100 years since birth) and graphical depictions of spacecraft in Solar System.

Mission and Item Title	Date	COSPAR ID	Location	Organizer(s)	Content(s) Description
Luna 16; Pennants (Mitchell, 2004)	12 Sep 1970	1970–072A	Moon–Mare Fecunditatis	USSR	Pennants containing graphical depictions of Luna 16 lander, a caricature of the spacecraft's launch trajectory originating from USSR, the USSR state seal, USSR, date and "Earth-Moon-Earth" (in Cyrillic).
Luna 17/ Lunokhod 1; Pennants (Mitchell, 2004)	10 Nov 1970	1970–095A	Moon–Sea of Rains	USSR	Pennants containing graphical depictions of Lenin, Luna 17 lander, Lunokhod 1 rover, a caricature of the spacecraft's launch trajectory originating from USSR, the USSR state seal, USSR and date (in Cyrillic).
Apollo 14; Lunar Plaque	31 Jan 1971	1971–008C	Moon–Fra Mauro Base	NASA/Jack Kinzler	Two hemisphere maps of Earth, an inscription and astronauts' signatures—engraved on stainless steel.
Apollo 14; Bible and Microfilm of Genesis verse (Noble, 1997)	31 Jan 1971	1971–008C	Moon–Fra Mauro Base	Edgar Mitchell	Printed copy of common King James Bible along with a microfilm printing of a verse from Genesis.
Mars 2 and 3; Pennants (Mitchell, 2004)	19 May 1971 28 May 1971	1971–045D, 1971–049D	Mars—likely vaporized Mars–Ptolemaeus Crater	USSR	Pennants containing graphical depictions of each Mars landers, a graphic detailing the inner Solar System planets and probe home world, mission communication infrastructure of the spacecraft, the USSR state seal, USSR and date (in Cyrillic).
Apollo 15; Lunar Plaque	26 Jul 1971	1971–063C	Moon–Hadley Rille	NASA/Jack Kinzler	Two hemisphere maps of Earth, an inscription and astronauts' signatures—engraved on stainless steel.
Apollo 15; Fallen Astronaut (Smithsonian Air and Space Museum, 2017)	26 Jul 1971	1971–063C	Moon–Hadley Rille	David Scott and Paul Van Hoeydonck	8.5 cm aluminum figurine in spacesuit.
Apollo 15; $2 Bills (Jefferson Space Museum, 2019; McKinnon, 2017; Portalist, 2018)	26 Jul 1971	1971–063C	Moon—Hadley Rille	Dave Scott and Jim Irwin	Stack of one hundred $2 currency bills.
Apollo 15; Bible (McKinnon, 2017; Portalist, 2018)	26 Jul 1971	1971–063C	Moon- Hadley Rille	Jim Irwin	Printed copy of common King James Bible with red cover.
Apollo 15; Jim Irwin object cache (McKinnon, 2017; Wilkinson, 2013; Jones, 1996; Irwin, 1973)	26 Jul 1971	1971–063C	Moon–Hadley Rille	Jim Irwin	Silver medallions containing the fingerprints of Jim Irwin's wife and children along with a photograph of a person (photograph of an unrelated stranger—apparently father of J.B. Irwin). A four-leaf clover and sliver of lava from Devil Lake, Oregon, were also included.

Mission and Item Title	Date	COSPAR ID	Location	Organizer(s)	Content(s) Description
Luna 20; Pennants (in descent stage) (Mitchell, 2004)	14 Feb 1972	1972–007A	Moon–Apollonius Highlands	USSR	Pennants containing graphical depictions of Luna 20 lander, a caricature of the spacecraft's launch trajectory originating from USSR, the USSR state seal, USSR, date and "Earth-Moon-Earth" (in Cyrillic). Luna 15 likely possessed similar pennants.
Venera 8; Pennants (Mitchell, 2004)	27 Mar 1972	1972–021A	Venus–Vasilisa region	USSR	Square/pentagonal pennants containing USSR state seal, USSR and date (in Cyrillic), bas-relief of Lenin and graphical depictions of spacecraft in Solar System. Dates and names on each pennant vary per probe.
Apollo 16; Lunar Plaque	16 Apr 1972	1972–031C	Moon–Descartes Highlands	NASA/Jack Kinzler	Two hemisphere maps of Earth, an inscription and astronauts' signatures—engraved on stainless steel.
Apollo 16; Family Photograph (Orwig, 2015; McKinnon, 2017)	16 Apr 1972	1972–031C	Moon–Descartes Highlands	Charles Duke and Family	Printed 3 × 4-inch photograph of the astronaut Charles Duke and his family with text on back "This is the family of astronaut Charlie Duke from planet Earth who landed on the moon on April 20, 1972."
Apollo 17; Lunar Plaque	7 Dec 1972	1972–096C	Moon–Taurus-Littrow	NASA/Jack Kinzler	Two hemisphere maps of Earth, lunar landing site map, an inscription and astronauts' signatures—engraved on stainless steel.
Luna 21/ Lunokhod 2; Pennants (Mitchell, 2004)	8 Jan 1973	1973–001A	Moon–Le Monnier Crater	USSR	Pennants containing graphical depictions of Lenin, Luna 21 lander, Lunokhod 2 rover, a caricature of the spacecraft's launch trajectory originating from USSR, the USSR state seal, USSR and date (in Cyrillic).
Venera 9 and 10; Pennants (Mitchell, 2004)	8 Jun 1975 14 Jun 1975	1975–050D, 1975–054D	Venus–Beta Regio Venus–Beta Regio	USSR	Square/pentagonal pennants containing USSR state seal, USSR and date (in Cyrillic), bas-relief of Lenin and graphical depictions of spacecraft in Solar System. Dates and names on each pennant vary per probe.
Luna 24; Pennants (in descent stage) (Mitchell, 2004)	9 Aug 1976	1976–081A	Moon–Mare Crisium	USSR	Pennants containing graphical depictions of Luna 24 lander, a caricature of the spacecraft's home country USSR, the USSR state seal, USSR, date and "Earth-Moon-Earth" (in Cyrillic). Luna 23 likely possessed similar pennants.
Venera 11 and 12; Pennants (Mitchell, 2004)	9 Sep 1978 14 Sep 1978	1978–084D, 1978–086C	Venus–Phoebe Regio Venus—surface	USSR	Square/pentagonal pennants containing USSR state seal, USSR and date (in Cyrillic), bas-relief of Lenin and graphical depictions of spacecraft in Solar System. Dates and names on each pennant vary per probe.
Venera 13 and 14; Pennants (Mitchell, 2004)	30 Oct 1981 4 Nov 1981	1981–106D, 1981–110D	Venus–Phoebe Regio Venus–Phoebe Regio	USSR	Square/pentagonal pennants containing USSR state seal, USSR and date (in Cyrillic), bas-relief of Lenin and graphical depictions of spacecraft in Solar System. Dates and names on each pennant vary per probe.

Mission and Item Title	Date	COSPAR ID	Location	Organizer(s)	Content(s) Description
Venera 15 and 16; Pennants (Mitchell, 2004)	2 Jun 1983 7 Jun 1983	1983–053A, 1983–054A	Cytherocentric orbit Cytherocentric orbit	USSR	Square pennants containing USSR state seal, USSR and date (in Cyrillic) and graphical depictions of the spacecraft in orbit and also the Kremlin clock tower. Dates and names on each pennant vary per probe.
Vega 1 and 2; Pennants (Mitchell, 2004)	15 Dec 1984 21 Dec 1984	1984–125E, 1984–128E	Venus–Aphrodite Terra Venus–Aphrodite Terra	USSR	Square/pentagonal pennants containing USSR state seal, USSR and date (in Cyrillic), text "Interkosmos CNES" from joint mission planners, graphical depictions of spacecraft/lander from the surface of Venus and also a rendering of the mission's weather balloon in Venus's atmosphere. Dates and names on each pennant vary per probe.
Phobos 1 or 2; Asaph Hall inscription (Reeves, 1994)	7 Jul 1988 12 Jul 1988	1988–058A or 1988–059A	Interplanetary space or Martian/Phobos orbit	USSR	Aluminum plaque containing a photographic etching duplicating a page from Asaph Hall's telescope logbook (from the night he discovered Phobos with a 26-inch refractor telescope) along with inscription "USSR Phobos Mission, 1988."

Reflector Satellites and Larger Objects-Artefacts

Entire satellites placed into Earth orbit or deep space for non-scientific, military, navigation, or telecommunication purposes.

Mission and Item Title	Date	COSPAR ID	Current Location	Organizer(s)	Content(s) Description and Carrier Medium(s)
Mayak (Byrd, 2017; Brown, 2017)	14 Jul 2017	2017–042F	Low Earth Orbit (failed to deploy)	"Your Sector of Space" group, Moscow State University	A 3U CubeSat that would deploy four triangular reflectors, x4 m2 each, which form a tetrahedral shape.
Humanity Star (Grush, 2018; King, 2018; McGowan, 2018)	21 Jan 2018	2018–010F	Decayed into Earth's atmosphere; 22 Mar)	Peter Beck/ Rocket Lab (US)	A geodesic sphere made from carbon fiber with 65 highly reflective panels, designed to produce flares visible from Earth's surface.
Falcon Heavy; Tesla Roadster and "Starman" (Leonard, 2018; Malkin, 2018; Rein *et al.*, 2018)	6 Feb 2018	2018–017A	En-route to near-Ceres/ asteroid belt orbit	Space X/Elon Musk	(Cherry-Red) Tesla Roadster electric sports car and a space-suited mannequin strapped into the driver's seat with David Bowie's "Space Oddity" song looping on the car radio (along with a sign stating; "Don't Panic"). See also "Falcon Heavy; Isaac Asimov's Foundation Trilogy" in "Human Time Capsules and Eternal Memory Initiatives" category.
G-Satellite (Olympic News, 2020)	7 Mar 2020	2020–016	Low Earth Orbit	Tokyo 2020 Olympic Games	Satellite with an exterior gold (Earth-facing) surface with slogan "G-Satellite Tokyo 2020" along with 2 miniature anime characters "Mobile Suit Gundam" and "Char's Zaku" with partial articulate movement capabilities.

ACKNOWLEDGMENTS

    The author would like to thank Carl Walker and Mark McCaughrean (both ESTEC) for providing an intricate account of European Space Agency activities that have contributed to the cultural signature of Earth, alongside Shin-ya Narusawa (JAXA Messages), Alexander Zaitsev (Cosmic Call 1 and 2, Teen Age Message), Jim Lewis (Deep Space Communications Network), Jacob Haqq-Misra (Lone Signal), Jon Lomberg (Voyager 1 and 2 Records, One Earth Message, and other projects), Alan Penny (METI International), Daniela de Paulis (Cogito), and also the staff at the Blackrock Castle Observatory (Cork) for facilitating access to their message contents and records for this catalogue. The author would also like to thank David Brin, Gregory Benford, and Duncan Forgan for their invaluable input and guiding comments to maintain the accuracy of this index. Finally, the author would also like to acknowledge the SETI/METI communities for their sustained investigations to classify off-world transmission events. This catalogue would have taken far longer to compile had it not been for their sustained efforts and intriguing analysis performed across a myriad of disciplines. There is much more work yet to be done.

REFERENCES

Ainsworth, D. (1998) Mars '98 payload integrates as scientists view first close-ups of strange, layered polar terrain. NASA. Retrieved on 4 September 2017, available at http://www.jpl.nasa.gov/releases/98/m98integ.html.

Almár I., and Shuch, P.H. (2006) The San Marino Scale: A new analytical tool for assessing transmission risk. *Acta Astronautica*, Vol. 60, Issue 1, pp. 57–59.

Analytics India (2017) Claim your moonshot, send your name to moon via Team Indus Moon Mission. *Analytics India Magazine*. Retrieved on 19 July 2018, available at https://analyticsindiamag.com/claim-moonshot-send-name-moon-via-team-indus-moon-mission/.

Aphids (2008) Yelling at the Stars. AphIds Archive (online). Retrieved on 27 September 2017, available at https://aphids.net/projects/yelling-at-stars/.

The Arch Mission (2018) The Arch Mission. The Arch Mission website. Retrieved on 7 February 2018, Available at https://www.archmission.com/.

Ashraf, S.F. (2003) Once upon a time, 50,000 years ago…. Rediff (online). Retrieved on 1 September 2016, available at http://www.rediff.com/news/2002/oct/15spec.htm.

AstroGrams (2019) AstroGrams™: When nothing on Earth will do…. Create an original "Artifact" to send into space! Astrograms website. Retrieved on 3 January 2020, available at https://astrograms.com/Default.php.

Astronet (1998) Spacecraft with human hair and DNA planned for interstellar flight. Astronet website. Retrieved on 16 February 2018, available at https://carlkop.home.xs4all.nl/humanha.html.

Atkinson, N. (2008) Messages from Earth beamed to alien world. *Universe Today; Space and Astronomy News*. Retrieved on 5 January 2017, available at https://www.universetoday.com/19335/messages-from-earthbeamed-to-alien-world/.

Atkinson, N. (2020) New project headed by Apollo's Charlie Duke to send messages to space. *Universe Today: Space and Astronomy News*. Retrieved on 3 January 2020, available at https://www.universetoday.com/144304/new-project-headed-by-apollos-charlie-duke-to-send-messages-to-space/.

Atri, D., DeMarines, J. and Haqq-Misra, J. (2011) A protocol for messaging to extraterrestrial intelligence. *Acta Astronautica*, Vol. 27, pp. 165–169.

Australian Academy of Science (2004) FedSat satellite, launch and deployment, 14 December 2002: Report to the Committee for Space Research. Australian National Committee for Space Science publications. Retrieved on 21 July 2018, available at https://www.science.org.au/files/userfiles/support/submissions/2004/nc-space-cospar2004.pdf.

Barras, C. (2008) First space ad targets hungry aliens. *New Scientist* (online). Retrieved on 5 January 2017, available at https://www.newscientist.com/article/dn14130-first-space-ad-targets-hungry-aliens/?feedId=onlinenews_rss20.

BBC News (2005) Turkmen book "blasted into space." *BBC News* (online). Retrieved on 14 May 2016, available at http://news.bbc.co.uk/1/hi/world/asia-pacific/4190148.stm.

BBC News (2008) Is anyone listening out there? *BBC News*: Science and Environment. Retrieved on 29 August 2017, available at http://news.bbc.co.uk/2/hi/science/nature/7660449.stm.

Beagles 2 (2004) Blur call sign. Beagle 2 website. Retrieved on 19 August 2017, available at http://www.beagle2.com/resources/blursignal.htm.

Benford, G. (1999) *Deep Time: How Humanity Communicates Across Millennia*. New York: Avon Books. (ISBN: 0380975378).

Besson, N., Faiveley, B., Freese, M., Krzywinski, M., Benson, M., Lehoucq, R., Pietriga, E., Uzan, J.P., and Steyer, S. (2021) Sanctuary: Redefining new frontiers—press packet. Sanctuary website. Retrieved on 7 September 2021, available at sanctuary.press.kit.pdf (sanctuaryproject.eu).

Billingham, J., and Benford, J. (2011) Costs and difficulties of large-scale "messaging," and the need for international debate on potential risks. *Journal of the British Interplanetary Society*, Vol. 67, p. 22 (2014).

Blackrock Castle Observatory (2017) Pan Galactic Email Station. Retrieved on16 September 2016, Available at https://www.bco.ie/whats-here/cosmos-atthe-castle/pangalactic/.

Borgerding, K. (2017) NASA beamed a tweet into space in honour of Voyager 1's 40th anniversary. Recode.net. Retrieved on 1 November 2018, available at https://www.recode.net/2017/9/5/16234078/nasa-beaming-tweet-space-voyager-1-40th-anniversary.

Boucher, B. (2020) This artist wanted to teach aliens about life on Earth. So, she teamed up with NASA scientists to send an artwork into outer space. Artnet website. Retrieved on 20 December 2020, available at https://news.artnet.com/art-world/julia-christensen-aliens-artwork-space-1929514.

Boyle, R. (2012) How a sundial lets curiosity see Mars in living color. *Popular Science*. Retrieved on 5 November 2017, available at https://www.popsci.com/science/article/2012-08/how-mars-rover-curiositys-sundial-will-help-rover-see-mars-living-color.

Braastad, R., and Zaitsev, A.L. (2003) Synthesis and transmission of Cosmic Call 2003 interstellar radio message. Kotel'nikov Institute of Radio-engineering and Electronics. Retrieved on 30 September 2016, available at http://www.cplire.ru/html/ra&sr/irm/CosmicCall-2003/index.html.

Brin, D. (2013) Shouting at the cosmos. David Brin website. Retrieved on 23 August 2016, available at https://www.davidbrin.com/nonfiction/shouldsetitransmit.html.

Brin, D. (2014) The search for extra-terrestrial intelligence (SETI) and whether to send "messages" (METI): A case for conversation, patience and due diligence. *Journal of the British Interplanetary Society*, Vol. 67, pp. 8–16.

Britt, R. (2018) SpaceX launches Isaac Asimov's "Foundation" books into deep space. Inverse.com. Retrieved on 7 February 2018, available at https://www.inverse.com/article/41025-space-x-tesla-spaceman-asimov-foundation-archbooks-falcon.

Brown, M.J.I. (2017) Russia's Mayak satellite: Crowd-funded cosmic pest or welcome nightly visitor? *The Conversation* (online). Retrieved on 7 August 2017, available at http://theconversation.com/russias-mayak-satellite-crowd-funded-cosmic-pest-or-welcome-nightly-visitor-81322

Browne, S. (2013) A perfect day. Susanna Browne website. Retrieved on 16 November 2017, available at http://susannabrowne.com/A-Perfect-Day.

Busch, M.W. (2013) Lone Signal and Jamesburg Earth Station Technologies: Technical Setup. Retrieved on 14 September 2016, available at http://www.webcitation.org/6Ik4qwRQ5?url=https://s3.amazonaws.com/lonesignal-prod-web/Lone%2BSignal%2BTechnical%2BSetup_06042013.pdf.

Butler, D. (1998) Space 'time capsule' could send a message to the future. *Nature*, Vol. 391, p. 112. Retrieved on 14 September 2017, available at https://www.nature.com/articles/34252.

Byrd, D. (2013) Got something to say to an alien? Lone Signal can beam your message. EarthSky. Retrieved on 1 November 2016, available at http://earthsky.org/space/got-something-to-say-to-an-alien-lone-signal-wants-to-beam-your-message.

Byrd, D. (2017) Have you seen the Mayak satellite? Earthsky. Retrieved on 08 May 2018, available at https://earthsky.org/space/mayak-bright-russian-satellite-july-august-2017

Campbell, H. (2015) Kids, send your art into space on the Cheops satellite. Space 2.0.com. Retrieved on 13 March 2017, available at http://www.science20.com/kids_send_your_art_into_space_on_the_cheops_satellite-155368.

Carey, B. (2005) 625,000 names to be vaporized in Deep Impact. Space.com. Retrieved on 17 February 2017, available at https://www.space.com/1255-625-000-names-vaporized-deep-impact.html.

Chakelian, A. (2012) Tweets in space: Contacting E.T., 140 characters at a time. *Time Magazine* (online). Retrieved on 16 August 2017, available at http://newsfeed.time.com/2012/05/10/tweets-in-space-contacting-e-t-140-characters-at-a-time/.

Chandler, D.L. (2009) ET: Check your voicemail. *MIT News* (online). Retrieved on 2 October 2017, available at http://news.mit.edu/2009/sketch-rubisco.

Chilton, M. (2016) 100 weird objects sent into space. *The Telegraph*. Retrieved on 27 September 2016, available at http://www.telegraph.co.uk/music/news/100-weird-objects-in-space/.

Chorost, M. (2016) How a couple of guys built the most ambitious alien outreach project ever. *Smithsonian Air and Space Magazine* (online). Retrieved on 7 June 2017, Available at http://www.smithsonianmag.com/science-nature/how-couple-guys-built-most-ambitious-alien-outreach-projectever-180960473/.

Closer Productions (2015) Sam Klemke's Time Machine. Closer Productions. Retrieved on 19 October 2017, Available at http://closerproductions.com.au/films/sam-klemkes-time-machine.

Cohen, B. (2019) What's in the time capsule Israel is leaving on the moon? From The Grapevine (online). Retrieved on 22 February 2019, available at https://www.fromthegrapevine.com/innovation/israel-moon-launch-time-capsule-what-is-inside-spaceil-beresheet.

Coldewey, D. (2016) To the Moon! Lunar XPRIZE team looks to send Wikipedia into space aboard homemade rover. TechCrunch. Retrieved on 20 April 2018, available at https://techcrunch.com/2016/04/21/to-the-moon-lunar-xprize-team-looks-to-send-wikipedia-into-space-aboard-homemade-rover/.

Collect Space (2008) To Pluto, with postage: Nine mementos fly with NASA's first mission to the last planet. CollectSpace.com. Retrieved on 3 January 2017, available at http://www.collectspace.com/news/news-102808a.html.

Collect Space (2011) LEGO figures flying on NASA Jupiter probe. CollectSpace.com. Retrieved on 17 January 2017, Available at http://www.collectspace.com/news/news-080411a.html.

Collect Space (2014) Your name in space: NASA asteroid probe latest mission to fly names. Collect Space.com. Retrieved on 18 February 2017, available at http://www.collectspace.com/news/news-011514a-send-your-name-space.html.

Cowan, K. (2020) Hundreds of photographs to be sent into space to broadcast a message of peace and unity from humankind. Creative Boom (online). Retrieved on 26 January 2020, available at https://www.creativeboom.com/resources/200-photographs-to-be-sent-into-space-to-broadcast-a-message-of-peace-and-unity-from-humankind/.

Coyle, J. (2008) Stephen Colbert to have his DNA sent to space. Welt (online). Retrieved on 7 June 2017, available at https://www.welt.de/english-news/article2411553/Stephen-Colbert-to-have-his-DNA-sent-to-space.html.

Creative Time (2017) Trevor Paglen: The Last Pictures. Creative Time website. Retrieved on 2 January 2017, available at http://creativetime.org/projects/the-last-pictures/.

Davis, J. (2016) Selfies, messages and names delivered for LightSail 2 flight. The Planetary Society. Retrieved on 17 April 2017, available at http://www.planetary.org/blogs/jason-davis/2016/20160328-mini-DVD-delivered.html.

Davis, J., and Hofmans, D. (2010) RuBisCo stars and the riddle of life. Proceedings from the Astrobiology Science Conference 2010. Retrieved on 11 October 2019, available at https://www.lpi.usra.edu/meetings/abscicon2010/pdf/5370.pdf.

Debczak, M. (2017) Artist plans to launch a giant sculpture into orbit. MentalFloss.com. Retrieved on 2 October 2017, available at http://mentalfloss.com/article/504849/artist-plans-launch-giant-sculpture-orbit.

De Paulis, D. (2017) Cogito. Daniela de Paulis website. Retrieved on 12 June 2018, available at http://www.danieladepaulis.com/?p=307.

De Paulis, D. (2018) Cogito. Cogito website. Retrieved on 12 June 2018, available at http://www.cogitoinspace.org/.

Diamond Sky Productions (2013) Message to the Milky Way. Diamond Sky Productions. Retrieved on 14 August 2017, Available at http://diamondskyproductions.com/recent/index.php#mmw.

Dominus, M. (2015) A message to the aliens (introduction). The Universe of Discourse (blog). Retrieved on 18 August 2017, available at http://blog.plover.com/aliens/dd/intro.html.

Dormehl, L. (2019) This biotech startup wants to put your DNA in a vault on the moon. Digital Trends (online). Retrieved on 28 February 2020, available at https://www.digitaltrends.com/cool-tech/lifeship-dna-archive-on-the-moon/.

Dumas, S. (2010) The 1999 and 2003 messages explained. Retrieved on 1 February 2017, available at http://www.plover.com/misc/Dumas-Dutil/messages.pdf.

Dumas, S. (2015) Message to extra-terrestrial intelligence—a historical perspective. Researchgate. Available at https://www.researchgate.net/publication/281036518_Message_to_Extra-Terrestrial_Intelligence_-_a_historical_perspective.

Dumas, S., and Dutil, Y. (2010) The Evpatoria Messages. SETI League; Contact in Context. Retrieved on 23 April 2017, available at https://www.plover.com/misc/Dumas-Dutil/evpatoria07.pdf.

Dumas, S., and Dutil, Y. (2016) Annotated cosmic call primer. *Smithsonian Air and Space Magazine* (online). Retrieved on 14 October 2016, available at https://www.smithsonianmag.com/science-nature/annotated-cosmic-callprimer-180960566/.

The Earthling Project (2021) The Earthling Project. Earthling Project website. Retrieved on 30 January 2021, available at https://earthlingproject.com/our-mission/.

The Editors of Sky and Telescope (2015) "Behind the scenes" with New Horizons. *Sky and Telescope Magazine* (online). Retrieved on 18 January 2017, available at http://www.skyandtelescope.com/sky-and-telescope-magazine/beyond-the-printed-page/behind-the-scenes-with-new-horizons-05142015/.

Elwenspoek, M.C. (2011) Long-time data storage: Relevant time scales. *Challenges*, Vol. 2, Issue 1, pp 19–36.

European Space Agency (2004a) Rock'n'Rolling for Titan. European Space Agency: Our Activities. Retrieved on 5 November 2017, available at http://www.esa.int/Our_Activities/Space_Science/Cassini-Huygens/Rock_n_roll_heading_for_Titan.

European Space Agency (2004b) CD-ROM attached to huygens. European Space Agency: Cassini-Huygens. Retrieved on 4 January 2017, available at http://sci.esa.int/cassini-huygens/26564-cd-rom-attached-to-huygens/.

European Space Agency (2013) Wake up, Rosetta! European Space Agency: Our Activities. Retrieved on 6 April 2015, available at http://www.esa.int/Our_Activities/Space_Science/Rosetta/Wake_up_Rosetta.

European Space Agency (2014) The competition winners who helped us wake up Rosetta! European Space Agency: Rosetta. Retrieved on 12 November 2017, available at http://www.esa.int/Our_Activities/Space_Science/Rosetta/The_competition_winners_who_helped_us_wake_up_Rosetta.

European Space Agency (2016) Three thousand drawings to fly into space on CHEOPS. European Space Agency: CHEOPS. Retrieved on 5 April 2017, available at http://sci.esa.int/cheops/57659-three-thousand-drawings-to-flyinto-space-on-cheops/.

European Space Agency (2018) ESA honoured to take part in Hawking tribute. European Space Agency: Art and Culture in Space. Retrieved on 15 June 2018, available at https://www.esa.int/About_Us/Art_Culture_in_Space/ESA_honoured_to_take_part_in_Hawking_tribute.

Evans, K. (2021) Have you found all the "Easter eggs" Perseverance smuggled to Mars yet? IFLScience website. Retrieved on 1 March 2021, available at https://www.iflscience.com/space/have-you-found-all-the-easter-eggs-perseverance-smuggled-to-mars-yet/.

Explore Mars (2014) Time capsule to Mars. Time Capsule to Mars website. Retrieved on 13 September 2017, available at http://www.timecapsuletomars.com/.

Foust, J. (2016) German X Prize team announces launch contract. SpaceNews (online). Retrieved on 20 April 2018, available at http://spacenews.com/german-x-prize-team-announces-launch-contract/.

Freeman, D. (2018) The Arch Mission Foundation and SpaceChain create Orbital Library™, first archive in space. SpaceChain website. Retrieved on 6 February 2019, available at https://spacechain.com/the-arch-mission-foundation-and-spacechain-create-orbital-library/.

Gertz, J. (2016a) Reviewing METI: A critical analysis of the arguments. *Journal of the British Interplanetary Society (JBIS)*, Vol. 69, pp. 31–36 (arXiv: 1605.05663).

Gertz, J. (2016b) Post-detection SETI protocols and METI: The time has come to regulate them both. *Journal of the British Interplanetary Society*, Vol. 69, pp. 263–270 (JBIS Refcode: 2016.69.263).

Gilster, P. (2009) "RuBisCo Stars" and the riddle of life. Centauri-Dreams (online). Retrieved on 19 May 2017, available at https://www.centauri-dreams.org/2009/11/18/%e2%80%9crubisco-stars%e2%80%9d-and-the-riddle-of-life/.

Gitlin, J.M. (2015) Building an archive on the Moon (and doing science, too). Ars Technica. Retrieved on 15 March 2017, available at https://arstechnica.com/science/2015/12/lunar-mission-one-is-crowdsourcing-a-trip-to-the-moon/.

Gohring, N. (2013) Lone Signal aims to send "hello!" tweets to extraterrestrials. CNN Money. Retrieved on 28 May 2017, available at http://www.webcitation.org/6Hk9s9TD4?url=http://money.cnn.com/2013/06/17/technology/enterprise/lone-signal/index.html.

Goldsmith, D., and Owen, T.C. (2001) *The Search for Life in the Universe*. Herndon, VA: University Science Books (ISBN: 1891389165).

Gorman, A. (2005) The archaeology of orbital space. In: Australian Space Science Conference 2005, pp. 338–357. Melbourne: RMIT University.

Graviton (2018) Culture across time and space. Project Mora website. Retrieved on 20 September 2018, available at https://projectmora.com/index.

Greicius, T. (2017) More than 2.4 million names are going to Mars. NASA. Retrieved on 14 September 2017, available at https://www.nasa.gov/feature/jpl/more-than-24-million-names-are-going-to-mars.

Griffin, A. (2014) Lunar Mission One to crowdfund moon drilling project by offering space time capsules on Kickstarter. *Independent* (online). Retrieved on 4 February 2017, Available at http://www.independent.co.uk/news/science/lunar-mission-one-to-crowdfund-moon-drilling-project-byoffering-space-time-capsules-on-kickstarter-9869561.html.

Griggs, M.B. (2015) How to see if your name is going to pluto on The New Horizons spacecraft. *Popular Science*. Retrieved on 23 August 2017, available at https://www.popsci.com/check-and-see-if-your-name-going-pluto-new-horizons-spacecraft.

Grinspoon, D. (2003) *Lonely Planets: The Natural Philosophy of Alien Life*. New York: HarperCollins Publishers Inc. (ISBN: 9780060185404).

Grush, L. (2018) Rocket Lab secretly launched a disco ball satellite on its latest test flight. The Verge (online). Retrieved on 3 March 2018, available at https://www.theverge.com/2018/1/24/16926426/rocket-lab-humanity-star-secret-satellite-electron-test-launch.

Grush, L. (2018) This non-profit plans to send millions of Wikipedia pages to the Moon—printed on tiny metal sheets. The Verge (online). Retrieved on 20 May 2018, available at https://www.theverge.com/2018/5/15/17353194/lunar-library-wikipedia-moon-arch-foundation-astrobotic-spacex.

Guzman, M., Welch, C., and Hein, A.M. (2015) Eternal memory: Long-duration storage concepts for space. Conference Paper: 66th International Astronautical Congress, Jerusalem, Israel.

Harris, M. (2019) This startup wants to stash your DNA on the Moon: LifeShip says it will offer lunar backup storage of human DNA in "artificial amber." IEEE Spectrum. Retrieved on 28 February 2019, available at https://spectrum.ieee.org/tech-talk/aerospace/space-flight/this-startup-wants-to-stash-your-dna-on-the-surface-of-the-moon.

Harrison, A.A. (2007) *Starstruck: Cosmic Visions in Science, Religion and Folklore*. Oxford, NY: Berghahn Books (ISBN: 9781845452865). Retrieved on 14 January 2016, available at https://books.google.co.uk/

books?id=ORNcQUBAEjUC&pg=PA52&lpg=PA52&dq=Discovery;+Calling+All+Aliens+transmission&source=bl&ots=O-oOYahJ7-&sig=L2tyDjQXEYyIus8hIUnLiSfhyMk&hl=en&sa=X&ved=0ahUKEwjtspWpgqfVAhWEmLQKHZ1UDmAQ6AEIJDAB#v=onepage&q=Discovery%3B%20Calling%20All%20Aliens%20transmission&f=false.

Harrison, A.A. (2014) Speaking for Earth: Projecting cultural values across deep space and time. In: Vakoch, D.A. (ed) *Archaeology, Anthropology and Interstellar Communication*. Washington: The NASA History Series (ISBN: 9781501081729).

Hawking Foundation (2018) The Stephen Hawking tribute CD. Stephen Hawking interment (website). Retrieved on 16 June 2018, available at https://www.stephenhawkinginterment.com/thecd/.

Highfield, R. (2008) UK astronomers to broadcast adverts to aliens. *The Telegraph* (online). Retrieved on 2 February 2017, available at http://www.telegraph.co.uk/news/science/science-news/3335306/UK-astronomers-to-broadcast-adverts-to-aliens.html.

Holler, W. (2014) Student led time capsule to Mars project most ambitious crowd-funded campaign in history. Explore Mars website. Retrieved on 7 February 2017, available at https://www.exploremars.org/student-led-time-capsule-to-mars-project-most-ambitious-crowd-funded-campaign-in-history.

Holmes, O. (2019) Israel to launch first privately funded Moon mission. *The Guardian* (online). Retrieved on 22 February 2019, available at https://www.theguardian.com/science/2019/feb/20/israel-to-launch-first-privately-funded-moon-mission.

Human Document Project (2017) The Human Document Project (HUDOC). Retrieved on 4 September 2017, available at http://hudoc2017.manucodiata.org/.

Irwin, J. (1973) *To Rule the Night: The Discovery Voyage of Astronaut Jim Irwin*. Philadelphia: A.J. Holman Co. (ISBN: 978-0879810245).

IT News (2009) NASA to text message interplanetary cousins. IT News.com. Retrieved on 15 August 2016, available at https://www.itnews.com.au/news/nasa-to-text-message-interplanetary-cousins-152737.

JAXA (2009a) Venus climate orbiter "AKATSUKI" (PLANET-C) message campaign. Japan Aerospace Exploration Agency (JAXA). Retrieved on 5 February 2016, available at http://global.jaxa.jp/press/2009/10/20091023_akatsuki_campaign_e.html.

JAXA (2009b) We will deliver your message to the brightest star Venus—"AKATSUKI" message campaign—Deadline for accepting messages extended. Japan Aerospace Exploration Agency (JAXA). Retrieved on 15 March 2016, available at http://global.jaxa.jp/projects/sat/planet_c/topics.html.

JAXA Space Education Center (2013) Usuda Space Camp year 2013 (Japanese). Retrieved on 26 November 2017, available at http://edu.jaxa.jp/education/participation/cosmic_advanced/archive/2013/saku_udsc1/.

JAXA Space Education Center (2014) Usuda Space Camp year 2014 (Japanese). Retrieved on 26 November 2017, available at http://edu.jaxa.jp/education/participation/cosmic_advanced/archive/2014/udsc_2/.

Jefferson Space Museum (2019) Apollo 15: Flown $2 bill. Jefferson Space Museum (online). Retrieved on 20 July 2019, available at http://www.jeffersonspacemuseum.com/apollo-15

Jenner, L. (2009) 1.6 million names to the Moon. NASA. Retrieved on 16 February 2017, available at https://www.nasa.gov/mission_pages/LRO/multimedia/million_names.html.

Jiang, J.H., Li, H., Chong, M., Jin, Q., Rosen, P.E., Jiang, X., Fahy, K.A., Taylor, S.F., Kong, Z., Hah, J., and Zhu, Z-H. (2022) A beacon in the galaxy: Updated Arecibo Message for potential FAST and SETI projects. Cornell University Library (arXiv: 2203.04288). Retrieved on 30 March 2022, available at https://arxiv.org/abs/2203.04288.

Johns Hopkins Applied Physics Laboratory (2016) Search by name and Last Name. Pluto.JHUAPL.edu. Retrieved on 16 September 2017, available at http://pluto.jhuapl.edu/Mission/Communications/Search-Name.php.

Johns Hopkins Applied Physics Laboratory (2018) "Beam" your greetings to New Horizons. Pluto.JHUAPL.edu. Retrieved on 9 December 2018, available at http://pluto.jhuapl.edu/News-Center/News-Article.php?page=20181204.

Johnson, S.L. (2008) Red, white and blue: U.S. flag at home on the Moon. *Houston History Magazine*, Vol. 6, p. 60.

Jones, E.M. (1996) Transcript from Apollo 15 lunar mission. Apollo 15 Lunar Surface Journal (online). Retrieved on 20 July 2019, available at https://www.hq.nasa.gov/alsj/a15/a15.clsout3.html#1670255.

Jones, T. (2010) How would you break the "Eerie Silence"—winners! Zoonomian (blog): Communicate Science. Retrieved on 2 May 2017, available at https://communicatescience.com/zoonomian/2010/03/15/how-would-you-break-the-eerie-silence-winners/.

Журнал Житомира (2009) В Евпатории отправили эмблему города Житомира в космос на 70 лет (Ukrainian). Журнал Житомира (online). Retrieved on 14 May 2017, available at http://zhzh.info/news/2009-09-15-5762.

Kalder, D. (2013) Turkmenistan: Proudly maintaining the tradition of dictator literature. Publishing Perspectives.com. Retrieved on 14 October 2017, available at https://publishingperspectives.com/2013/12/turkmenistan-proudly-maintaining-the-tradition-of-dictator-literature/.

Kaplan, M. (2007) Send a New Year's message to the Moon on Japan's SELENE mission. The Planetary

Society. Retrieved on 17 August 2016, available at http://www.planetary.org/press-room/releases/ 2007/0111_Send_a_New_Years_Message_to_the_Moon.html.

Katz, L. (2012) Finally, a chance to tweet to aliens. Cnet. Retrieved on 29 October 2017, available at https:// www.cnet.com/news/finally-a-chance-to-tweet-to-aliens/.

KEO (2013) KEO: The satellite that carries the hopes of the world. KEO.org. Retrieved on 12 September 2015, available at http://www.keo.org/uk/pages/message.php.

King, B. (2018) Humanity Star: Bright idea or Dark Sky nemesis? *Sky and Telescope* magazine (online). Retrieved on 14 February 2018, available at https://www.skyandtelescope.com/observing/humanity-star/.

Kiss, J. (2008) Bebo tries to contact Earth-like planet. *The Guardian*. Retrieved on 4 August 2017, available at https://www.theguardian.com/media/2008/jul/29/bebo.digitalmedia.

Kitchen, W.J. (2017) The Interstellar Beacon. Retrieved on 12 June 2017, available at https://www.interstellarbeacon.org/

Korbitz, A. (2014) Towards understanding the Active SETI debate: Insights from risk communication and perception. *Acta Astronautica*, Vol. 105, Issue 2, pp. 517–520.

Kotel'nikov Institute of Radio-Engineering and Electronics (2008) MIR, LENIN, SSSR. Retrieved on 24 August 2015, available at http://www.cplire.ru/html/ra&sr/irm/MIR-LENIN-SSSR.html.

Kramer, M. (2013) New project will send your messages to aliens in deep space. Space.com. Retrieved on 13 June 2017, available at https://www.space.com/21528-alien-intelligence-messages-lone-signal.html.

Kunze, M. (2018) Memory of Mankind. Memory of Mankind website. Retrieved on 14 December 2017, available at https://www.memory-of-mankind.com/.

Lalwani, M. (2016) NASA wants to send your art on a round-trip to space. Engadget.com. Retrieved on 14 March 2017, available at https://www.engadget.com/2016/02/19/nasa-wants-to-send-your-art-on-a-round-trip-to-space/.

Landes, N. (2016) In 1969,Warhol, Rauschenberg, and Chamberlain sent their art to the Moon. Artsy (online). Retrieved on 13 June 2018, available at https://www.artsy.net/article/artsy-editorial-how-warhol-rauschenberg-and-chamberlain-smuggled-art-onto-the-moon.

Landsberg, H.E., and Mieghem, J.V. (1972) *Advances in Geophysics*, Vol 15. Cambridge, MA: Academic Press (ISBN: 9780080568430).

Leonard, D. (2009) Send ET a text message from Earth. Space.com. Retrieved on 24 August 2017, available at https://www.space.com/7140-send-text-message-earth.html.

Leonard, D. (2018) Is the Tesla roadster flying on the Falcon Heavy's maiden flight just space junk? Space.com. Retrieved on 7 February 2018, Available at https://www.space.com/39602-falcon-heavy-tesla-not-just-space-junk.html.

Leonard, P. (2017) Message to the gods: Space poetry that transcend human rivalries. Space.com. Retrieved on 7 February 2019, available at https://www.space.com/39144-space-poetry-transcends-human-rivalries.html.

Lewis, J. (2017) Deep Space Communications Network. Retrieved on 3 August 2017, available at http://deep-spacecom.net/view-transmission/.

Li, S. (2014) A time capsule on the Moon. *The Atlantic* (online). Retrieved on 27 February 2017, available at https://www.theatlantic.com/technology/archive/2014/11/a-time-capsule-on-the-moon/383052/.

Lomberg, J. (2004) A portrait of humanity. *Contact in Context* (online). Retrieved on 13 August 2015, available at www.jonlomberg.com/articles.html.

Lomberg, J. (2010) Artist Jon Lomberg. Jon Lomberg website. Retrieved on 3 January 2017, available at https://www.jonlomberg.com/profile.html.

Lomberg, J. (2016) One Earth Message. One Earth Message website. Retrieved on 8 March 2017, available at http://www.oneearthmessage.org.

The Long Now Foundation (2009) The Rosetta Project. The Long Now Foundation. Retrieved on 6 April 2017, available at www.rosettaproject.org.

López del Rincón, D. (2015) Bioarte: arte vida en la era de la biotechnología (Spanish). Gobierno de España: Ministerio de Cultura (ISBN: 9788446042464).

Luxorion (2006) Messages aux extraterrestres: Cosmic Connexion (French). Luxorion. Retrieved on 4 November 2017, available at http://www.astrosurf.com/luxorion/seti-messages.htm.

Macmillan, A. (2005) Tele-spamming our alien brethren. *Popular Science*, Vol. 266, Issue 98, p. 101. Retrieved on 6 September 2016, available at https://books.google.co.uk/books?id=Xfgb-rsXX-iMC&pg=PA98&lpg=PA98&dq=craigslist+transmission+into+space&source=bl&ots=FFic-GKiSyt&sig=A7ewbQSh0fpi7fWVU6cuYgqI1pc&hl=en&sa=X&ved=0ahUKEwjlhtS-uZD-VAhUJY1AKHfneCLgQ6AEINDAC#v=onepage&q=craigslist%20transmission%20into%20space&f=false.

Malkin, B. (2018) SpaceX oddity: How Elon Musk sent a car towards Mars. *The Guardian* (online). Retrieved on 7 February 2018, available at https://www.theguardian.com/science/2018/feb/07/space-oddity-elon-musk-spacex-car-mars-falcon-heavy.

Malloy, J. (2016) *Social Media Archaeology and Poetics*. Cambridge, MA: MIT Press (ISBN: 9780262034654).

Manz, A. (2010) The Human Document Project and Challenges. *Challenges*, Vol. 1, Issue 1, pp. 3–4 (doi: 10.3390/challe1010003).

Marshall, M. (2010) Earth calling: A short history of radio messages to ET. *New Scientist* (online). Retrieved on 16 October 2016, available at https://www.newscientist.com/article/dn18417-earth-calling-a-short-history-of-radiomessages-to-et/.

Martin, S. (2020) Alien contact: THIS is how YOU could send a message to aliens. *Express* (online). Retrieved on 16 August 2020, available at https://www.express.co.uk/news/science/1294657/alien-contact-message-universe-space-news-spacespeak-aliens.

Mathewson, S. (2018) World's first private Moon lander will carry Israeli time capsule. Space.com. Retrieved on 22 February 2019, available at https://www.space.com/42783-spaceil-private-moon-lander-israeli-time-capsule.html.

Matogawa, Y. (2005) Hayabusa: To 880,000 little princes and princesses. Japan Aerospace Exploration Agency (JAXA). Retrieved on 5 February 2016, available at http://www.isas.ac.jp/e/snews/2005/1130_tm.shtml.

May, S. (2016) #Goodbye Philae: Memorials and space futures. *Heritage-Futures* (online). Retrieved on 06 December 2019, available at https://heritage-futures.org/goodbyephilae-memorials-space-futures/.

McCracken, E. (2016) Artistic odyssey to send messages to stars. University of Edinburgh. Retrieved on 14 February 2016, available at http://www.ed.ac.uk/news/2016/starmessage-030216.

McGowan, M. (2018) "Space graffiti": Astronomers angry over launch of fake star into sky. *The Guardian* (online). Retrieved on 27 January 2018, available at https://www.theguardian.com/world/2018/jan/26/space-graffiti-astronomers-angry-over-launch-of-fake-star-into-sky.

McKinnon, M. (2017) Ashes, art, and other surprising things humans have left on the Moon. Outer Space (online). Retrieved on 20 July 2019, available at https://www.outerplaces.com/science/item/15706-ashes-art-and-other-surprising-things-humans-have-left-on-the-moon.

Mellor, C. (2020) Message in a space capsule: Art works preserved in DNA break free of the biosphere. Blocks and Files website. Retrieved on 11 November 2020, available at https://blocksandfiles.com/2020/11/06/beyond-earth-dna-data-storage/.

METI International (2022) Scientists and musicians will transmit radio messages to nearby star, attempting to make first contact and raise awareness about the climate crisis. METI International news feed. Retrieved on 13 April 2022, available at http://meti.org/en/blog/scientists-and-musicians-will-transmit-radio-messages-nearby-star-attempting-make-first-contact.

Miller, N. (2016) Philae Lander: Farewell messages to Philae as life support ends. *BBC News* (online). Retrieved on 12 December 2019, Available at https://www.bbc.co.uk/news/science-environment-36904368.

Minnesotastan (2010) Gold olive branch left on the Moon. Neatorama.com. Retrieved on 19 December 2016, available at http://www.neatorama.com/2010/11/10/gold-olive-branch-left-on-the-moon/.

Mitchell, D.P. (2004) Soviet spacecraft pennants. Mentallansdacape.com. Retrieved on 21 July 2018, available at http://mentallandscape.com/V_Pennants.htm.

MoonArk (2017) Moon ark: An epochal artifact designed to communicate forward across time and space. Moon Arts website. Retrieved on 18 August 2017, available at http://moonarts.org/.

Moore, M. (2008) Messages from Earth sent to distant planet by Bebo. *The Telegraph*. Retrieved on 14 June 2017, available at http://www.telegraph.co.uk/news/newstopics/howaboutthat/3166709/Messages-from-Earth-sentto-distant-planet-by-Bebo.html.

Murrill, M.B. (1997) Signatures from Earth board spacecraft to Saturn. NASA. Retrieved on 9 October 2017, available at https://saturn.jpl.nasa.gov/news/2157/signatures-from-earth-board-spacecraft-to-saturn/.

Museum of Modern Art (MOMA) (2018) Various artists with Andy Warhol, Claes Oldenburg, David Novros, Forrest Myers, Robert Rauschenberg, John Chamberlain—The Moon Museum 1969. Museum of Modern Art website. Retrieved on 13 June 2018, available at https://www.moma.org/collection/works/62272.

Musso, P. (2012) The problem of active SETI: An overview. *Acta Astronautica*, Vol. 78, pp. 43–54.

NASA (1969) Apollo 11 goodwill messages. NASA Press Kit for Apollo 11 Goodwill Messages. NASA. Retrieved on 16 January 2016, available at https://history.nasa.gov/ap11-35ann/goodwill/Apollo_11_material.pdf.

NASA (1978) *The Martian Landscape*. Washington, D.C: NASA, U.S. Government Printing Office (Catalogue No. 1.21:425).

NASA (1999) Stardust launch: Press Kit. NASA. Retrieved on 4 April 2017, available at https://www2.jpl.nasa.gov/files/misc/stardust.pdf.

NASA (2004a) DVD with signatures on way to Saturn. NASA. Retrieved on 9 October 2017, available at https://saturn.jpl.nasa.gov/news/2803/dvd-withsignatures-on-way-to-saturn/.

NASA (2004b) Spirit honors the crew of Space Shuttle Colombia. NASA. Retrieved on 13 July 2019, available at https://www.nasa.gov/missions/shuttle/f_marsplaque.html.

NASA (2007) Dawn Community: All aboard the Dawn Spacecraft. NASA. Retrieved on 14 May 2017, available at https://dawn.jpl.nasa.gov/DawnCommunity/.

NASA (2008) NASA Kepler Mission offers opportunity to send names into space. NASA. Retrieved on 14 June 2016, available at https://www.jpl.nasa.gov/news/news.php?release=2008-073.

NASA (2010) Send your name to Mars. Mars.NASA.gov. Retrieved on 14 July 2017, available at https://mars.nasa.gov/msl/participate/sendyourname/.

NASA (2014) NASA Invites public to send names on an asteroid mission and beyond. NASA. Retrieved on 17 October 2017, available at https://www.nasa.gov/press/2014/january/nasa-invites-public-to-send-names-on-an-asteroid-mission-and-beyond/.

NASA (2017) NASA beamed your #MessageToVoyager. NASA. Retrieved on 1 November 2018, available at https://voyager.jpl.nasa.gov/message/.

NASA (2018) More than 1.1 million names installed on NASA's Parker Solar Probe. NASA. Retrieved on 2 May 2018, available at https://www.nasa.gov/feature/goddard/2018/more-than-11-million-names-installed-on-nasa-s-parker-solar-probe.

NASA (2020) Healthcare workers to be honored on Mars. NASA. Retrieved 10 September 2020, available at https://www.jpl.nasa.gov/spaceimages/details.php?id=PIA23921.

NASA Content Administrator (2008) NASA beams Beatles' "Across the Universe" into space. NASA. Retrieved on 14 September 2016, available at https://www.nasa.gov/topics/universe/features/across_universe.html.

NASA Content Administrator (2011) Juno Jupiter Mission to carry plaque dedicated to Galileo. NASA. Retrieved on 17 February 2017, available at https://www.nasa.gov/mission_pages/juno/news/galileo20110803.html.

NASA Content Administrator (2019) NASA invites public to submit names to fly aboard next Mars Rover. NASA. Retrieved on 30 May 2019, available at https://www.nasa.gov/press-release/nasa-invites-public-to-submit-names-to-fly-aboard-next-mars-rover.

NASA History Program Office (2012) *Catalogue of Manmade Material on the Moon*. Washington, D.C.: NASA History Program Office. Retrieved on 17 May 2019, available at https://history.nasa.gov/FINAL%20Catalogue%20of%20Manmade%20Material%20on%20the%20Moon.pdf.

NASA Lucy Mission (2021) The Lucy Plaque. NASA Lucy Mission website. Retried on 18 October 2021, available at www.lucy.swri.edu/LucyPlaque.html.

NASA SCaN (2021) Transmission successful! NASA Twitter Account. Retrieved on 20 August 2021, available at https://twitter.com/NASASCaN/status/1428427580827373570.

National Geographic (2012) The Wow Signal: How we'll reply. Retrieved on 16 April 2017, available at http://web.archive.org/web/20120819043043/http://channel.nationalgeographic.com/channel/chasing-ufos/the-wow-signal-how-we-ll-reply/.

Naubaum, A. (2005) Intergalactic communications: Tele-spamming our alien brethren. *Popular Science* (online). Retrieved on 14 May 2017, available at http://www.popsci.com/military-aviation-space/article/2005-05/intergalactic-communications.

NBC News (2008) NASA beaming Beatles tune to the stars. Retrieved on 13 September 2016, available at http://www.nbcnews.com/id/22951001#.WWuzR-ko-Uk.

NBC News (2014) Maven and MOM orbiters close in on their moments of truth at Mars. NBC News. Retrieved on 3 March 2017, available at http://www.nbcnews.com/science/space/maven-mom-orbiters-close-their-momentstruth-mars-n207616.

New Horizons Message Initiative (2015) New Horizons Message Initiative. New Horizons Message.com. Retrieved on 12 March 2017, available at https://www.newhorizonsmessage.com/.

Noble, D.F. (1997) The religion of technology: The divinity of man and the spirit of invention. New York: Alfred A. Knopf (ISBN: 0679425640).

Oberhaus, D. (2019) *Extraterrestrial Languages*. Cambridge, MA: MIT Press (ISBN: 9780262043069).

O'Donnell, B., and Worrell, D. (1976) NASA press kit for Project LAGEOS. NASA. Retrieved on 16 October 2015.

Oglethorpe University (1990) International Time Capsule Society. Retrieved 12 March 2017, available at https://crypt.oglethorpe.edu/international-time-capsule-society/.

O'Grady, M. (2017) Art for a post-surveillance age. *The New York Times* (online). Retrieved on 29 August 2017, available at https://www.nytimes.com/2017/08/29/t-magazine/art/trevor-paglen.html.

Olympic News (2020) Tokyo 2020 "G-Satellite" carried to the International Space Station. Olympic News. Retrieved on 9 March 2020, available at https://www.olympic.org/news/tokyo-2020-g-satellite-carried-to-the-international-space-station

Orbital Reflector (2017) The Orbital Reflector. Orbital Reflector.com. Retrieved on 19 November 2017, available at http://orbitalreflector.com/.

Orwig, J. (2015) Apollo 16 astronaut explains hidden message behind the family portrait he left on the moon. *Independent* (online). Retrieved on 17 April 2016, available at http://www.independent.co.uk/news/science/apollo-16-astronaut-explains-hidden-message-behind-the-family-portraithe-left-on-the-moon-a6718111.html.

Osborne, D. (2009) Send a text message to ET via Hello From Earth. Geek.com. Retrieved on 17 August 2016, available at https://www.geek.com/news/send-a-text-message-to-et-via-hello-from-earth-871361/.

Overbye, D. (2008) One alien to another: A broadcast to the stars. *The New York Times* (online). Retrieved on 2 March 2017, available at http://www.nytimes.com/2008/12/12/science/space/12earth.html.

Paglen, T. (2012) *The Last Pictures*. Berkeley: University of California Press (ISBN: 9780520275003).

Par, R. (2010) IKAROS. Onward to Venus! Parman.blogspot (public blog). Retrieved on 2 November 2017, available at http://parman.blogspot.com/2010/05/jaxa-onward-to-venus-at-258-pm-pacific.html.

Pearlman, R.Z. (2007) The untold story: How one small disc carried one giant message for mankind. Space.com. Retrieved on 14 September 2017, available at https://www.space.com/4655-untold-story-small-disc-carried-giant-message-mankind.html.

Pickard, G. (2014) Was Lone Signal Tweets to Space project just a scam? Top Secret Writer (See article and comments). Retrieved on 14 August 2017, available at http://www.topsecretwriters.com/2014/02/was-lone-signal-tweets-to-space-project-just-a-scam/.

Pink Tentacle (2008) Alien e-mail reply to arrive in 2015? Retrieved on 24 March 2017, available at http://pinktentacle.com/2008/05/alien-e-mail-reply-to-arrive-in-2015/.

Planet Lab (2013) Using art to expand our understanding of life on Earth. Planet Labs website. Retrieved on 14 January 2019, available at https://www.planet.com/company/art/.

Planetary Society (2010) The IKAROS names disc. The Planetary Society. Retrieved on 14 February 2017, available at http://www.planetary.org/multimedia/space-images/spacecraft/ikaros_names_disc.html.

Planetary Society (2014) Messages from Earth: Hayabusa 2. The Planetary Society. Retrieved on 28 January 2017, available at http://www.planetary.org/get-involved/messages/hayabusa-2/.

Planetary Society (2016a) OSIRIS-Rex: Messages to Bennu! Retrieved on 2 November 2017, available at http://www.planetary.org/get-involved/messages/bennu/.

Planetary Society (2016b) Selfie to space. The Planetary Society. Retrieved on 16 January 2017, available at http://www.planetary.org/get-involved/messages/lightsail/.

Planetary Society (2017) Members' names flying to space. The Planetary Society. Retrieved on 2 November 2017, available at http://www.planetary.org/get-involved/messages/namesinspace.html.

Planetary Society (2018) Visions of Mars. Planetary Society. Retrieved on 16 January 2018, available at http://www.planetary.org/explore/projects/vom//.

Portalist Staff (2018) 15 of the weirdest things people have left on the Moon. *Portalist* (online). Retrieved on 20 July 2019, available at https://theportalist.com/ashes-art-and-other-surprising-things-humans-have-left-on-the-moon.

Portrait of Humanity (2019) Portrait of Humanity is a global initiative by 1854 Media, publisher of *British Journal of Photography*. Portrait of Humanity website. Retrieved on 26 January 2020, available at https://portraitofhumanity.co/about/.

Powell, C.S. (2019) A 30-million page library is heading to the Moon to help preserve human civilization. NBC news (online). Retrieved on 9 March 2019, available at https://www.nbcnews.com/mach/science/30-million-page-library-heading-moon-help-preserve-human-civilization-ncna977786.

Press Trust of India (2015) Rover to carry Earth's art, poetry, music to the Moon. *India Express* (online). Retrieved on 24 August 2017, available at https://indianexpress.com/article/trending/rover-to-carry-earths-art-poetry-music-to-the-moon/.

Puli Space (2016) MOM on the MOON. Puli Space website. Retrieved on 18 January 2018, available at http://pulispace.com/en/mom-on-the-moon.

Quast, P.E. (2017) A human perspective of Earth: An overview of dominant themes to emerge from global "A Simple Response..." messages (DOI: 10.13140/RG.2.2.36341.78563).

Quast, P.E. (2018a) A profile of humanity: The cultural signature of Earth's inhabitants beyond the atmosphere. *International Journal of Astrobiology*, Vol. 20, Issue 3, pp. 194–214 (DOI: 10.1017/S1473550418000290).

Quast, P.E. (2018b) Beyond the Earth; Schematics for "Companion Guide for Earth" archival elements residing within geosynchronous orbit (DOI: 10.13140/RG.2.2.14177.97127).

Quast, P.E. (2022) Remembering the conversation? Discussions about the challenges and next steps in establishing a long-term METI archive. In: *Proceedings from the UK SETI Research Network (UKSRN) Conference 2022*, University of Durham, United Kingdom, 7–8 July 2022.

Rahman, T. (2007) *We Came in Peace for All Mankind: The Untold Story of the Apollo 11 Silicon Disc*. Leawood, KS: Leathers Publishing (ISBN: 978-1585974412).

Red Encounter (2003) What is Red Encounter? RedEncounter.ESA.Net. Retrieved on 28 November 2017, available at http://redencounter.esa.int/.

Redazione, ASI (2012) Curiosity has landed on Mars. ASI- Agenzia Spaziale Italiana News. Retrieved on 21 February 2017, available at http://www.asi.it/en/news/curiosity-has-landed-mars.

Reddy. F. (2005) Hayabusa owns a piece of the rock. Astronomy.com. Retrieved on 11 June 2018, available at http://www.astronomy.com/newsobserving/news/2005/11/hayabusa%20owns%20a%20piece%20of%20the%20rock.

Reeves, R. (1994) *The Superpower Space Race: An Explosive Rivalry Through the Solar System*. New York: Plenum Publishing Corporation (ISBN: 9780306447686).

Reid, M. (2009) Immortality Drive. Gandt Blog. Retrieved on 6 April 2017, available at http://gandt.blogs. brynmawr.edu/2009/04/19/immortality-drive/.

Rein, H., Tamayo, D., and Vokrouhlicky, D. (2018) The random walk of cars and their collision probabilities with planets. *Aerospace*, Vol. 5, Issue 2, p. 57 (DOI: 10.3390/aerospace5020057).

Rense (2008) Fox to beam Day the Earth Stood Still into space. Rense.com. Retrieved on 17 April 2017, available at http://www.rense.com/general84/fox.htm.

Reuters (2010) Klingon opera prepares for interstellar debut. Reuters.com. Retrieved on 14 August 2017, available at http://www.reuters.com/article/us-klingons-idUSTRE6891EZ20100910.

Richards, J. (2014) HELENA—Oxygen production and art time capsule. Mars One Community (online). Retrieved on 10 November 2016, available at https://community.mars-one.com/projects/helena.

Rogers, S. (2017) This tiny little disk contains a microscopic archive of all languages in the world. Interesting Engineering (online). Retrieved on 19 June 2017, available at https://interestingengineering.com/ this-tiny-little-disk-contains-a-microscopic-archive-of-all-languages-in-the-world.

Sagan, C., Drake, F.D., Druyan, A., Ferris, T., Lomberg, J., and Salzman-Sagan, L. (1978) *Murmurs of Earth: The Voyager Interstellar Record*. New York: Ballantine Books (ISBN: 978-0345315366).

Sagan, C., Salzman-Sagan, L., and Drake, F. (1972) A message from earth. *Science* 175, pp. 881–884.

Saint-Gelais, R. (2014) Beyond linear B; the metasemiotic challenge of communication with extraterrestrial intelligence. In Vakoch, D.A. (ed.), *Archaeology, Anthropology, and Interstellar Communication*. Washington, D.C.: NASA History series, pp. 79–94 (ISBN: 9781501081729).

Salce, C. (2015) Sam Klemke's time machine review: A brutally honest time capsule. Nuke the Fridge. Retrieved on 19 October 2017, available at http://nukethefridge.com/sam-klemkes-time-machine-review-a-brutally-honest-time-capsule/.

Scharf, C.A. (2012a) Tweets in space are go—today! *Scientific American*. Retrieved on 23 September 2017, available at https://blogs.scientificamerican.com/life-unbounded/tweets-in-space-are-go-today/.

Scharf, C.A. (2012b) Tweets in space! *Scientific American*. Retrieved on 23 September 2017, available at https://blogs.scientificamerican.com/life-unbounded/tweets-in-space/.

Schulze-Makuch, D. (2016) Is beaming messages to other stars a wise idea? *Smithsonian Air and Space Magazine* (online). Retrieved on 9 October 2016, available at https://www.airspacemag.com/daily-planet/ beamingmessages-other-stars-wise-idea-180960723/?no-ist.

Scuka, D. (2016) A simple response. ESA Rocket Science Blog. Retrieved on 28 July 2016, available at http:// blogs.esa.int/rocketscience/2016/07/28/a-simple-response/.

Sent Forever (2009) Personal messages sent into space at the speed of light. Send2press.com. Retrieved on 2 May 2017, available at https://www.send2press.com/wire/2009-07-0716-002/.

Shanks, S. (2016) One Earth is working to keep the legacy alive. Planetarium (online). Retrieved on 14 June 2017, available at http://c.ymcdn.com/sites/www.ips-planetarium.org/resource/resmgr/Opportunities/ One_Earth_Message.pdf.

Shiner, L. (2013) First man on Mars: Leonardo da Vinci. *Smithsonian Air and Space Magazine* (online). Retrieved on 22 November 2016, available at https://www.airspacemag.com/daily-planet/ first-man-on-mars-leonardo-davinci-7012414/.

Shujiro, S. (2003) With the hopes of 880,000 people: Hayabusa target markers. Japan Aerospace Exploration Agency (JAXA). Retrieved on 5 February 2016, available at http://global.jaxa.jp/article/special/hay-abusa/sawai_e.html.

Smithsonian Air and Space Museum (2017) Sculpture, fallen astronaut. Smithsonian Air and Space Museum website. Retrieved on 3 July 2016, available at https://airandspace.si.edu/collection-objects/ sculpture-fallen-astronaut?object=nasm_A19860035000.

Smithsonian Institution (2018) Apollo 11 plaque (ID: S69-38749). Smithsonian National Air and Space Museum. Retrieved on 10 June 2018, available at https://airandspace.si.edu/multimedia-gallery/5515hjpg.

SNACI–The Staff at the National, A. and C. Ionosphere (1975) The Arecibo Message of November 1974. *Icarus*, Vol. 26, Issue 4, pp. 462–466 (DOI: 10.1016/0019-1035(75)90116-5).

Sónar (2017) Sónar celebrates 25 years of the festival by contacting intelligent extraterrestrial life. Sónar Music Festival Press Release (online). Retrieved on 29 March 2018, available at http://intranet.sonar.es/ mailing/1128/en.html.

Southorn, G. (2010) The eerie silence—winning messages! *Sky at Night Magazine* (online). Retrieved on 17 July 2017, available at http://www.skyatnightmagazine.com/forum/the-eerie-silence-winning-messages-t110073.html.

Space Daily (2008) Doritos makes history with world's first ET advert. Space Daily (online). Retrieved on 2 November 2016, available at http://www.spacedaily.com/reports/Doritos_Makes_History_With_ World_First_ET_Advert_999.html.

SpaceSpeak (2022) 2022 SpaceSpeak with fun kids the UK's children's radio station. SpaceSpeak website. Retrieved on 25 March 2022, available at https://spacespeak.com/RadioUK_FunKids.

Sparkes, M. (2022) Group that wants to contact aliens will transmit to TRAPPIST-1 system. *New Scientist* (Online). Retrieved on 13 April 2022, available at https://www.newscientist.com/ article/2315676-group-that-wants-to-contact-aliens-will-transmit-to-trappist-1-system/.

Spivack, N. (2016a) Let's put Wikipedia in space: The Arch Project. Nova Spivack (blog). Retrieved on 7 February 2018, available at http://www.novaspivack.com/uncategorized/wikipedia-in-space.

Spivack, N. (2016b) The ARCH Foundation. *Space Talk: The Next Generation*, Autumn/Winter 2016. Retrieved on 7 February 2018, available at https://drive.google.com/file/d/10-LXJ2ZKFpc0L6LKvD4nb AOlvg-6EXFm/view.

Spivack, N. (2019) The Lunar Library: Genesis Mission. Arch Mission website. Retrieved on 9 March 2019, available at https://www.archmission.org/lunar-library-overview.

Stihia Beyond (2022) Radio dispatch to deep space. Stihia Beyond website. Retrieved on 25 April 2022, available at http://www.stihia-beyond.org/.

Studio for Creative Inquiry (2017) MoonArk. Carnegie Mellon University. Retrieved on 15 September 2017, available at http://studioforcreativeinquiry.org/grants/moonark.

Sutherland, P. (2015) Has Britain's lost Mars probe been found after 11 Years? UK Space Agency mysteriously announces it will give "update" on missing Beagle spacecraft on Friday. *Daily Mail* (online). Retrieved on 4 June 2017, available at http://www.dailymail.co.uk/sciencetech/article-2907441/Missing-Mars-probe-Beagle-2-s-status-updated-UK-Space-Agency-Friday.html.

Svec, J. (2013) "BANG, ZOOM! Straight to the Moon!" Kickstarter.com. Retrieved on 14 February 2016, available at https://www.kickstarter.com/projects/87824834/nanorosetta-own-a-print-of-the-human-genome/posts/452964.

Talk2ETs (2015) Talk2ETs. Retrieved on 4 August 2017, available at http://www.talk2ets.com/.

Than, K. (2005) Craigslist gets beamed into space. CNN.com. Retrieved on 3 December 2016, available at http://edition.cnn.com/2005/TECH/space/03/23/craigslist.space/.

Thomason, C. (2008) Radio ensures memorable messages are sent forever. Press Dispensary website. Retrieved on 1 May 2017, available at https://pressdispensary.co.uk/releases/c991699/Radio-Ensures-Memorable-Messages-are-Sent-Forever.html.

Troup, C. (2008) Silloh S. Weiland and Yelling at the Stars. Real Time Arts (online). Retrieved on 4 November 2017, available at http://www.realtimearts.net/article/88/9233.

Tweets in Space (2012) Tweets in space. Tweets in Space website. Retrieved on 11 September 2017, available at http://tweetsinspace.org/.

U–The Opera (2010) Official invitation sent to Qo'nos. U–The Opera (online). Retrieved on 14 August 2017, available at http://www.u-theopera.org/official-invitation-sent-to-qo%E2%80%99nos.

UKSEDS (2015) "Time Capsule to Mars" aims for 2018 Mars landing. UKSEDS.org. Retrieved on 4 February 2017, available at http://ukseds.org/2015/02/tc2m/.

University of Colorado (2012) Maven: Send your name and message to Mars! University of Colorado, Bolder: Laboratory for Atmospheric and Space Physics. Retrieved on 6 July 2017, available at http://lasp.colorado.edu/maven/goingtomars/send-your-name/.

Vakoch, D.A. (2009) Asymmetry in active SETI: A case for transmissions from earth. *Acta Astronautica*, Vol. 68, pp. 476–488

Vakoch, D.A. (2017) The message we're sending to nearby aliens is no threat to Earth. *New Scientist* (online). Retrieved on 29 March 2018, available at https://www.newscientist.com/article/2153948-the-message-were-sending-tonearby-aliens-is-no-threat-to-earth/.

Valentine, G. (2011) An awkward history of our space transmissions. Gizmodo. Retrieved on 19 July 2017, available at https://gizmodo.com/5780084/an-awkward-history-of-our-space-transmissions.

Van Damme, C., Van Eeckhaut, M., Scherlippens, B., and Willems, S. (2009) *Look/Alike: Kunstenaarsprofielen en Artistiek Rollenspel in Hedendaagse Kunst*. *Gent*. Belgium: Academia Press (9789038215228).

Voyager's Final Message (2020) Your message to the Voyager spacecraft. Voyager's Last Message website. Retrieved on 14 December 2020, available at https://www.voyagersfinalmessage.com.

Wall, M. (2014) Private Moon mission aims to drill into lunar south pole by 2024. Space.com. Retrieved on 3 October 2016, available at https://www.space.com/27807-private-moon-mission-lunar-one.html.

Washburn, M. (2015) New Horizons One Earth Message. The Planetary Society. Retrieved on 22 June 2017, Available at http://www.planetary.org/blogs/guest-blogs/2015/0424-new-horizons-one-earth-message.html.

Wayne, G. (2011) A short history of long-term thinking, for our fifty thousand year time capsule. Motherboard. Retrieved on 16 September 2016, available at https://motherboard.vice.com/en_us/article/3dd7xk/ashort-history-of-long-term-thinking-for-our-50-000-year-time-capsule.

Weltraum Kunst (2003–2016) The project. Weltraum Kunst.de. Retrieved on 12 July 2016, available at http://www.weltraumkunst.de/inhalte/project.htm.

The Western Australian (2018) Phone home. *The Western Australian* newspaper, 29 November 2018.

Wilkinson, J. (2013) There's a four-leaf clover on the Moon! Moonzoo blog (online). Retrieved on 20 July 2019, available at https://moonzooblog.wordpress.com/2013/08/20/theres-a-four-leaf-clover-on-the-moon/.

Woods, A. (1992) OURS—The Orbiting Unification Ring Satellite: A global artwork in space for the year 2000. Conference Paper; 1st European Space Art Symposium, Montreusx, Switzerland. Retrieved on 20 January 2018, available at http://www.ours.ch/publications.php.

Yoshikawa, M., Hosoda, S., Sawada, H., Ogawa, N., Tsuda, Y., Kishi, A., Asakura, H., and Hayabusa2 Public

Outreach Team (2015) Public outreach of Hayabusa2 Mission. 46th Lunar And Planetary Science Conference [2015]. Retrieved on 14 July 2017, available at https://www.hou.usra.edu/meetings/lpsc2015/pdf/1644.pdf.

Zaitsev, A.L. (2002a) A teen-age message to the stars. SETI League. Retrieved on 16 September 2016, available at http://www.setileague.org/articles/tam.htm.

Zaitsev, A.L. (2002b) Design and implementation of the 1st Theremin Concert for Aliens. Kotel'nikov Institute of Radio-engineering and Electronics. Retrieved on 24 August 2015, available at http://www.cplire.ru/html/ra&sr/irm/Theremin-concert.html.

Zaitsev, A.L. (2003) Synthesis and transmission of Cosmic Call 2003 interstellar radio message. Kotel'nikov Institute of Radio-engineering and Electronics. Retrieved on 9 December 2018, available at http://www.cplire.ru/html/ra&sr/irm/CosmicCall-2003/index.html.

Zaitsev, A.L. (2006) Messaging to extra-terrestrial intelligence. Moscow, Russia: Cornell University Library (arXiv: physics/0610031).

Zaitsev, A.L. (2008a) Sending and searching for interstellar messages. *Science Direct*, Vol. 63, Issue 5-6, pp. 614–617 (DOI: 10.1016/j.actaastro.2008.05.014).

Zaitsev, A.L. (2008b) The first musical interstellar radio message. *Journal of Communications Technology*, Vol. 53, pp. 1107–1113.

Zaitsev, A.L. (2011) Rationale for METI. Cornell University Library (arXiv: 1105.0910 9 physics.gen-ph]). Retrieved on 16 December 2017, available at https://arxiv.org/abs/1105.0910.

Zaitsev, A.L. (2012) Classification of interstellar radio messages. *Acta Astronautica*, Vol. 78, pp. 16–19 (DOI: 10.1016/j.actaastro.2011.05.026).

Zaitsev, A.L. (2016) Messaging to ETI; Theremin's melodies. Messaging to ETI (Google website). Retrieved on 05 September 2018, available at https://sites.google.com/site/messagingtoeti/Home/theremin-s-melodies.

Zaitsev, A.L., and Ignatov, S.P. (1999) Broadcast for extra-terrestrial intelligence from Evpatoria Deep Space Centre. Kotel'nikov Institute of Radio-engineering and Electronics. Retrieved on 29 July 2015, available at http://www.cplire.ru/html/ra&sr/irm/report-1999.html.

Zeluck, S.R. (1996) "Send Your Messages to Saturn" draws huge response. JPL.NASA.gov. Retrieved on 11 June 2017, available at https://www.jpl.nasa.gov/releases/96/cdsign2.html.

Zhorov, I. (2016) An artistic time capsule prepares to hitch a ride to the Moon. National Public Radio (online). Retrieved on 6 December 2016, available at http://www.npr.org/2016/01/03/461795258/arts-capsule-tohitch-a-ride-to-the-moon-on-carnegie-mellon-s-rover.

Zimmer, T. (2009) The DVD on the Kepler Telescope. TorstenZimmer.com. Retrieved on 28 January 2017, available at http://www.torstenzimmer.com/pages/kepler.htm.

Bibliography

Almár, I., and Shuch, H.P. (2007) The San Marino Scale: A new analytical tool for assessing transmission risk. *Acta Astronautica*, Vol. 60, Issue 1, pp. 57–59 (DOI: 10.1016/j.actaastro.2006.04.012).

Anderson, M. (2016) Forever data in quartz: The quest for the immortal bit. IEEE Spectrum (online). Retrieved on 20 April 2016, available at https://spectrum.ieee.org/tech-talk/semiconductors/memory/forever-data-in-quartz-the-quest-for-the-immortal-bit.

Apollo Prayer League (2020) The Apollo Prayer League. Apollo Prayer League website. Retrieved on 20 April 2020, Available at www.apolloprayerleague.com/?The_First_Lunar_Bible___The_Apollo_Prayer_League.

Arbib, M.A. (2013) Evolving an extraterrestrial intelligence and its language-readiness. In: Dunér, D., Parthemore, J., Persson, E. and Holmberg, G. (eds.), *The History and Philosophy of Astrobiology: Perspectives on Extraterrestrial Life and the Human Mind*. Newcastle-upon-Tyne: Cambridge Scholars Publishing.

Bal, M. (1997) *Narratology: Introduction to the Theory of Narrative*. Toronto: University of Toronto Press (ISBN: 9780802078063).

Barclay, R.L., and Brooks, R. (2002) In situ preservation of historical spacecraft. *Journal of the British Interplanetary Society*, Vol. 55, Issue 5-6, pp. 173–181 (DOI: 10.1201/9781420084320-c37).

Barker, P. (1982) Omnilinguals. In: Smith, N.D. (ed.), *Philosophers Look at Science Fiction*. Chicago: Nelson-Hall (ISBN: 9780882298078).

Baylis, G.C., and Driver, J. (1995) One-sided edge assignment in Vision: 1. Figure-ground segmentation and attention to objects. *Current Directions in Psychological Science*, Vol. 4, Issue 5, pp. 140–146 (DOI: 10.1111/1467-8721.ep10772580).

BBC News (2005) Turkmen book "blasted into space." *BBC News* (online). Retrieved on 14 May 2016, available at http://news.bbc.co.uk/1/hi/world/asia-pacific/4190148.stm.

Benford, G. (1992) Saving the "library of life." *Proceedings of the Natural Academy of Sciences of the United States of America*, Vol. 89, Issue 22, pp. 11098–11101 (DOI: 10.1073/pnas.89.22.11098).

Benford, G. (1999) *Deep Time: How Humanity Communicates Across Millennia*. New York: Avon Book (ISBN: 0380975378).

Benford, J.N., and Benford, D.J. (2016) Power beaming leakage radiation as a SETI observable. *The Astrophysical Journal*, Vol. 825, Issue 2 (DOI: 10.3847/0004-637X/825/2/101).

Bennett, A.T.D., and Cuthill, I.C. (1994) Ultraviolet vision in birds: What is its function? *Vision Research*, Vol. 34, Issue 11, pp. 1471–1478 (DOI: 10.1016/0042-6989(94)90149-x).

Berger, J. (1973) *Ways of Seeing*. London: Penguin Books (ISBN: 9780563122449).

Bertamini, M., and Wagemans, J. (2013) Processing convexity and concavity along a 2-d contour: Figure-ground, structural shape, and attention. *Psychonomic Bulletin & Review*, Vol. 20, Issue 2, pp. 191–207 (DOI: 10.3758/s13423-012-0347-2).

Billing, L. (2017) Should humans colonize other planets? No. *Theology and Science*, Vol. 13, Issue 3, pp. 321–332 (DOI: 10.1080/14746700.2017.1335065).

Billingham, J., and Benford, J. (2011) Costs and difficulties of large-scale "messaging," and the need for international debate on potential risks. *Journal of the British Interplanetary Society*, Vol. 67, pp. 17–23 (JBIS Refcode: 2014.67.17).

Billings, L. (2007) Overview: Ideology, advocacy, and spaceflight—evolution of a cultural narrative. In: Dick, S.J. and Launius, R.D. (eds.), *Societal Impact of Spaceflight*. Washington: NASA History Series (SP-2007-4801).

Binford, L.R. (1962) Archaeology as anthropology. *American Antiquity*, Vol. 28, Issue 2, pp. 217–225 (DOI: 10.2307/278380).

Binford, L.R. (1978) Dimensional analysis of behavior and site structure: Learning from an Eskimo hunting stand. *American Antiquity*, Vol. 43, Issue 3, pp. 330–361 (DOI: 10.2307/279390).

Binford, L.R. (1982) *In Pursuit of the Past: Decoding the Archaeological Record*. Berkeley: University of California Press (ISBN: 9780520233393).

Bohm, D. (1980) *Wholeness and the Implicate Order*. London: Routledge (ISBN: 9780415289795).

Bohm, D. (1985) Fragmentation and wholeness in religion and science. *Zygon: Journal of Religion & Science*, Vol. 20, Issue 2, pp. 125–133 (DOI: 10.1111/j.1467-9744.1985.tb00587.x).

Boulding, K.E. (1965) Earth as a spaceship. Washington State University Committee on Space Sciences, 10 May 1965. Papers of Kenneth E. Boulding (Archives Box 38), University of Colorado. Retrieved on 15 May 2019, Available at https://bertaux.files.wordpress.com/2014/08/boulding-earth-as-spaceship-1965.pdf.

Bowker, G. (2008) *Memory Practices in the Sciences*. Cambridge, MA: MIT Press (ISBN: 9780262524896).

Bracewell, R.N. (1960) Communications from superior galactic communities. *Nature*, Vol. 186, Issue 4726, pp. 670–671 (DOI: 10.1038/186670a0).

Bradbury, R.J. (1997–2000) Matrioshka Brains. Retrieved on 24 August 2015, Available at https://www.gwern.net/docs/ai/1999-bradbury-matrioshkabrains.pdf.

Brin, D. (1983) The "Great Silence"—the controversy concerning extraterrestrial intelligent life. *Quarterly Journal of the Royal Astronomical Society*, Vol. 24, Issue 3, pp. 283–309. Retrieved on 22 March 2017, available at https://www.researchgate.net/publication/234496344_The_%27Great_Silence%27_The_Controversy_Concerning_Extraterrestrial_Intelligent_Life

Brin, D. (2014) The search for extraterrestrial intelligence (SETI) and whether to send "messages" (METI): A case for conversation, patience and due diligence. *Journal of the British Interplanetary Society*, Vol. 67, pp. 8–16 (JBIS Refcode: 2014.67.8).

Brin, D. (2019) The "Barn Door" Argument, The Precautionary Principle, and METI as "Prayer"—an appraisal of the top three rationalizations for "Active SETI." *Theology and Science*, Vol. 17, Issue 1, pp. 16–28 (DOI: 10.1080/14746700.2018.1557391).

Brunner, J. (Woodcott, K.) (1961) *I Speak for Earth*. London: Orion Publishing Group Ltd. (ISBN: 9780575101203).

Bryld, M., and Lykke, N. (2000) *Cosmodolphins: Feminist Cultural Studies of Technology, Animals and the Sacred*. New York: Zed Books (ISBN: 9781856498159).

Buchli, V. and Lucas, G. (2001) *Archaeologies of the Contemporary Past*. London: Routledge.

Burra Charter (2013) The Burra Charter: The Australia ICOMOS charter for places of cultural significance. Australia ICOMOS Incorporated International Council on Monuments and Sites. Retrieved on 14 May 2018, available at http://portal.iphan.gov.br/uploads/ckfinder/arquivos/The-Burra-Charter-2013-Adopted-31_10_2013.pdf.

Bury, J.B. (1932) *The Idea of Progress: An Inquiry into Its Origin and Growth*. Fairford, UK: Echo Library (ISBN: 9781406801088).

Cabrol, N.A. (2016) Alien mindscapes—Perspective on the search for extraterrestrial intelligence. *Astrobiology*, Vol. 16, Issue 9, pp. 661–676 (DOI: 10.1089/ast.2016.1536).

Capelotti, P.J. (1996) A conceptual model for aerospace archaeology: A case study from the Wellman Site, Virgohamna, Danskøya, Svalbard. PhD Thesis for the Department of Anthropology at Rutgers University. Ph.D. Dissertation (University Microfilms #9633681).

Capelotti, P.J. (2004) Space: The final [archaeological] frontier. *Archaeology*, Vol. 57, Issue 6. Retrieved on 14 July 2018, available at https://archive.archaeology.org/0411/etc/space.html.

Capelotti, P.J. (2009) Culture of Apollo: A catalog of manned exploration of the Moon. In: Darrin, A., and O'Leary, B.L. (eds.), *The Handbook of Space Engineering, Archaeology, and Heritage*. Boca Raton: CRC Press (ISBN: 9781420084313).

Capelotti, P.J. (2010) *The Human Archaeology of Space: Lunar, Planetary and Interstellar Relics of Exploration*. Jefferson, NC: McFarland (ISBN: 9780786458592).

Capova, K.A. (2008) The Voyager message: Messages on the Voyager probes. Masters dissertation for Department of Anthropology at the Charles University, Prague. Retrieved on 07 September 2017, available at https://is.cuni.cz/webapps/zzp/detail/64647/.

Capova, K.A. (2013a) The charming science of the other: The cultural analysis of the scientific search for life beyond earth. PhD Thesis for the Department of Anthropology at the Durham University (DOI: 10.13140/RG.2.2.15164.54406).

Capova, K.A. (2013b) The detection of extraterrestrial life: Are we ready? In: Vakoch, D.A. (ed.), *Astrobiology, History, and Society: Life Beyond Earth and the Impact of Discovery*. New York: Springer (ISBN: 9783642359835).

Capova, K.A. (2021) Introducing humans to the extraterrestrials: The pioneering missions of the Pioneer and Voyager probes. *Frontiers in Human Dynamics*, Vol. 3 (DOI: 10.3389/fhumd.2021.714616).

Carrasco, M., Ling, S., and Read, S. (2004) Attention alters appearance. *Nature Neuroscience*, Vol. 7, Issue 3, pp. 308–313 (DOI: 10.1038/nn1194).

Chadwick, J. (2008) *The Decipherment of Linear B*. Cambridge, UK: Cambridge University Press (ISBN: 9780521398305).

Chen, C. (2003) *Mapping Scientific Frontiers: The Quest for Knowledge Visualization*. London: Springer (ISBN: 9781447151272).

Chertok, B. (2006) *Rockets and People: Creating a Rocket Industry, Vol. II*. Washington, D.C.: The NASA History Series (ISBN: 0160766729).

Chomsky, N. (1999) *Language, Mind, and Politics*. Cambridge, UK: Polity Press (ISBN: 9780745618883).

Chomsky, N. (2000) Minimalist inquiries: The framework. In: Martin, R., Michaels, D., and Uriagereka, J. (eds.), *Step by Step: Essays on Minimalist Syntax in Honor of Howard Lasnik*. Cambridge, MA: MIT Press (ISBN: 9780262133616).

Chomsky, N., and Gliedman, J. (1983) Things no amount of learning can teach. *Omni Publications International* (magazine), Vol. 6, Issue 11 (Catalog ID: 0149–8711). Retrieved on 03 June 2018, Available at https://chomsky.info/198311__/.

Chua, D.K.L. (2007) Rioting with Stravinsky: A particular analysis of *The Rite of Spring*. *Music Analysis*, Vol. 26, Issue 1–2, pp. 59–109 (DOI: 10.1111/j.1468–2249.2007.00250.X).

Ćirković, M.M. (2018) *The Great Silence: Science and Philosophy of Fermi's Paradox*. Oxford, UK: Oxford University Press (ISBN: 9780199646302).

Clare, E.L., and Holderied, M.W. (2015) Acoustic shadows help gleaning bats find prey but may be defeated by prey acoustic camouflage on rough surfaces. *eLife sciences* (online). Retrieved on 16 January 2020, available at https://elifesciences.org/articles/07404.

Clark, N. (2005) Ex-orbitant Globality. *Theory, Culture and Society*, Vol. 22, Issue 5, pp. 165–185 (DOI: 10.1177/0263276405057198).

Clarke, A.C. (1951) *The Sentinel*. New York: HarperVoyager (ISBN: 9780586212042).

Clarke, A.C. (1962) *Profiles of the Future: An Inquiry into the Limits of the Possible*. London: Gollancz.

Clarke, A.C. (1964) *Profiles of the Future: An Inquiry into the Limits of the Possible*. New York: Bantam Books (ASIN: B0007DWSF0).

Clery, D. (2016) Here's who could win the $20 million XPrize for roving on the moon—but will any science get done? American Association for the Advancement of Science, *Sciencemag* (online). Retrieved on 02 October 2017, available at https://www.sciencemag.org/news/2016/12/heres-who-could-win-20-million-xprize-roving-moon-will-any-science-get-done.

Coates, M.M. (2003) Visual ecology and functional morphology of Cubozoa (Cnidaria). *Integrative and Comparative Biology*, Vol. 43, Issue 4, pp. 542–548 (DOI: 10.1093/icb/43.4.542).

Cocconi, G. and Morrison, P. (1959) Searching for interstellar communications. *Nature*, Vol. 184, Issue 4690, pp. 844–846 (DOI: 10.1038/184844a0).

Cockell, C. (2005) Planetary protection—A microbial ethics approach. *Space Policy*, Vol. 21, Issue 4, pp. 287–292 (DOI: 10.1016/j.spacepol.2005.08.003).

Cockell, C. (2018) *The Equations of Life: How Physics Shapes Evolution*. London: Atlantic Books (ISBN: 9781786493026).

Cockell, C., and Horneck, G. (2004) A planetary park system for Mars. *Space Policy*, Vol. 20, Issue 4, pp. 291–295 (10.1016/j.spacepol.2004.08.003).

Cockell, C.S., Santomartino, R., McMahon, S., Reekie, P., Alberti, S.J.M.M., Phillipson, T., and Russell, S. (2019) A laboratory for multi-century science. *Astronomy & Geophysics*, Vol. 60, Issue 6, pp. 26–28 (DOI: 10.1093/astrogeo/atz192).

Coes, M. (2011) *Breaking the Maya Code*. London: Thames & Hudson (ISBN: 9780500289556).

Cohen, J.B. and Kappauf, W.E. (1982) Metameric color stimuli, fundamental metamers, and Wyszecki's metameric blacks. *American Journal of Psychology*, Vol. 95, Issue 4, pp. 537–564.

Collin, S.P., Davies, W.L., Hart, N.S., and Hunt, D.M. (2009) The evolution of early vertebrate photoreceptors. *Philosophical Transactions of the Royal Society of London—B-Biological Sciences Series*, Vol. 365, Issue 1531, pp. 2925–2940 (DOI: 10.1098/rstb.2009.0099).

Comer, D.C., Chapman, B.D., and Comer, J.A. (2017) Detecting landscape disturbances at the Nazca Lines using SAR data collected from airborne and satellite platforms. *Geosciences*, Vol. 7, Issue 4 (DOI: 10.3390/geosciences7040106).

Cooper, T. (2010) *Longer Lasting Products: Alternatives to the Throwaway Society*. London: Routledge (ISBN: 9780566088087).

Cornips, L.E.A., and Corrigan, K.P. (2005) *Syntax and Variation: Reconciling the Biological and the Social, Vol. 265*. Amsterdam: John Benjamins Publishing (ISBN: 9781588116406).

Cosgrove, D. (1994) Contested global visions: One⊠world, whole⊠earth, and the Apollo space photographs. *Annals of the Association of American Geographers*, Vol. 84, Issue 2, pp. 270–294 (DOI: 10.1111/j.1467–8306.1994.tb01738.x).

Cott, H.B. (1940) *Adaptive Coloration in Animals*. Oxford, UK: Oxford University Press (ISBN: 9780416300505).

Cronin, T.W., Johnsen, S., Marshall, N.J., and Warrant, E.J. (2014) *Visual Ecology*. Princeton, NJ: Princeton (ISBN: 9780691151847).

Cross, I. (2001) Music, cognition, culture, and evolution. *Annals of the New York Academy of Sciences*, Vol. 930, Issue 1, pp. 28–42 (DOI: 10.1111/j.1749–6632.2001.tb05723.x).

Croteau, D., and Hoynes, W. (2003) *Media/Society: Industries, Images, and Audiences*. Thousand Oaks, CA: Pine Forge Press (ISBN: 0761987738).

Crouch, D., and Damjanov, K. (2015) Extra-planetary digital cultures. *M/C Journal* (online). Retrieved on 27 February 2017, available at http://journal.media-culture.org.au/index.php/mcjournal/article/view/1020.

Crowe, M.J. (2003) *The Extraterrestrial Life Debate, 1750–1900: The Idea of a Plurality of Worlds from Kant to Lowell*. New York: Dover Publications (ISBN: 9780486406756).

Crutzen, P.J., and Stoermer, E.F. (2000) The Anthropocene. IGBP Global Change Newsletter 41, pp. 17–18.

Cutting, J.E., and Vishton, P.M. (1995) Perceiving layout: The integration, relative dominance, and contextual use of different information about depth. In: Epstein, W. and Rogers, S. (eds.), *Perception of Space and Motion (Handbook of Perception and Cognition)*. New York: Academic Press (ISBN: 9780122405303).

Dacke M., Baird, E., Byrne, M., Scholtz, C.H., and Warrant, E.J. (2013) Dung beetles use the Milky Way for orientation. *Current Biology*, Vol. 23, Issue 4, pp. 149–150 (DOI: 10.1016/j.cub.2012.12.034).

Dacke, M., Nilsson, D.E., Scholtz, C.H., Byrne, M., and Warrant, E.J. (2003) Insect orientation to polarized moonlight. *Nature*, Vol. 424 (DOI: 10.1038/424033a).

Daniels, P.T., and Bright, W. (1996) *The World's Writing Systems*. Oxford: Oxford University Press (ISBN: 9780195079937).

Dantzig, T. (2007) *Number: The Language of Science*. New York: Plume—Penguin Group (ISBN: 978045 2288119).

David, B. (2017) *Cave Art*. London: Thames & Hudson (ISBN: 9780500204351).

David, L. (2018) Tesla roadster gets interplanetary ID. Space.com website. Retrieved on 16 May 2018, available at https://theconversation.com/a-sports-car-and-a-glitter-ball-are-now-in-space-what-does-that-say-about-us-as-humans-91156.

Davidson, I. (1992) There's no art—to find the mind's construction—in offence. *Cambridge Archaeological Journal*, Vol. 2, Issue 1, pp. 52–57.

Davidson, I. (2020) Marks, pictures and art: Their contribution to revolutions in communication. *Journal of Archaeological Method and Theory*, Vol. 27, Issue 3, pp. 745–770 (DOI: 10.1007/s10816-020-09472-9).

Davies, P.C.W. (2010) *The Eerie Silence: Renewing Our Search for Alien Intelligence*. Boston: Houghton Mifflin Harcourt (ISBN: 9780547133249).

Davis, J., Bisson-Filho, A., Kadyrov, D., De Kort, T.M., Biamonte, M.T., Thattai, M., Thutupalli, S., and Church, G.M. (2020) In vivo multi-dimensional information-keeping in Halobacterium salinarum. *BioRxiv* Preprint server for biology (DOI: 10.1101/2020.02.14.949925). Retrieved on 19 February 2020, available at https://www.biorxiv.org/content/10.1101/2020.02.14.949925v1.

Deavours, C.A. (1985) Extraterrestrial communication: a cryptologic perspective. In: Regis, E. (ed.), *Extraterrestrials: Science and Alien Intelligence*. Cambridge, UK: Cambridge University Press (ISBN: 0521262275).

Denning, K. (2006) Ten thousand revolutions: Conjectures about civilizations. *Acta Astronautica*, Vol. 68, pp. 381–388 (DOI: 10.1016/j.actaastro.2009.11.019).

Denning, K. (2011a) Is life what we make of it? *Philosophical Transactions of The Royal Society: A Mathematical Physical and Engineering Sciences*, Vol. 369, Issue 1936, pp. 669–678 (DOI: 10.1098/rsta.2010.0230).

Denning, K. (2011b) Unpacking the Great Transmission debate. In: Vakoch, D.A. (ed.), *Communication with Extraterrestrial Intelligence*. Albany, NY: State University of New York Press (ISBN: 9781438437941).

Denning, K. (2011c) "L" on Earth. In: Vakoch, D.A., and Harrison, A.A. (eds.), *Civilizations Beyond Earth: Extraterrestrial Life and Society*. Oxford, UK: Berghahn Press (ISBN: 9781782383154).

Denning, K. (2014) Learning to read: Interstellar message decipherment from archaeology and enthropology perspectives. In: Vakoch, D.A. (ed.), *Archaeology, Anthropology and Interstellar Communication*. Washington, D.C.: The NASA History series (ISBN: 9781501081729).

D'Errico, F., and Nowell, A. (2000) A new look at the Berekhat Ram figurine. Implications for the origins of symbolism. *Cambridge Archaeological Journal*, Vol. 10, Issue 1, pp. 123–167 (DOI: 10.1017/S0959774300000056).

De Valois, R.L., Smith, C.J., Kitai, S.T., and Karoly, A.J. (1958) Response of single cells in monkey lateral geniculate nucleus to monochromatic light. *Science*, Vol. 127, Issue 3292, pp. 238–239 (DOI: 10.1126/science.127.3292.238).

DeVito, C. (2011) On the universality of human mathematics. In: Vakoch, D.A. (ed.), *Communication with Extraterrestrial Intelligence*. Albany: State University of New York Press (ISBN: 9781438437934).

DeVito, C. (2013) *Science, SETI and Mathematics*. New York: Berghahn Books (ISBN: 9781782380696).

DeVito, C., and Oehrle, R. (1990) A language based on the fundamental facts of science. *Journal of the British Interplanetary Society*, Vol. 43, Issue 12, pp. 561–568.

Dick, S.J. (1984) *Plurality of Worlds: The Extraterrestrial Life Debate from Democritus to Kant*. Cambridge, UK: Cambridge University Press (ISBN: 9780521319850).

Dick, S.J. (1989) The concept of extra-terrestrial intelligence—An emerging cosmology? *Planetary Report*, Vol. 9, Issue 2, pp. 13–17.

Dick, S.J. (1998) *Life on Other Worlds: The 20th-Century Extraterrestrial Life Debate*. Cambridge, UK: Cambridge University Press (ISBN: 9780521799126).

Dick, S.J. (2003) Cultural evolution, the postbiological universe, and SETI. *International Journal of Astrobiology*, Vol. 2, Issue 1, pp. 65–74 (DOI: 10.1017/S147355040300137X).

Dor-Ziderman, Y., Lutz, A., and Goldstein, A. (2019) Prediction-based neural mechanisms for shielding the self from existential threat. *NeuroImage*, Vol. 202 (DOI: 10.1016/j.neuroimage.2019.116080).

Douglas, M. (1970) *Natural symbols: Explorations in cosmology*. London: Barrie and Rockliff.

Doyle, A.C. (1891) "A Scandal in Bohemia." London: George Newnes Ltd.

Doyle, L., McCowan, B., Johnston, S., and Hanser, S. (2011) Information theory, animal communication and the search for extraterrestrial intelligence. *Acta Astronautica*, Vol. 68, Issue 3–8, pp. 406–417 (DOI: 10.1016/j.actaastro.2009.11.018).

Dumas, S. (2015) Message to extra-terrestrial intelligence—a historical perspective. Researchgate. Retrieved on 14 August 2016, available at https://www.researchgate.net/publication/281036518_Message_to_ Extra-Terrestrial_Intelligence_-_a_historical_perspective.

Dunér, D. (2011) Cognitive foundations of interstellar communication. In: Vakoch, D.A. (ed.), *Communication with Extraterrestrial Intelligence*. Albany, NY: State University of New York Press (ISBN: 9781438437941).

Dunér, D. (2014) Interstellar intersubjectivity: The significance of shared cognition for communication, empathy, and altruism in space. In: Vakoch, D.A. (ed.), *Extraterrestrial Altruism*. New York: Springer (ISBN: 9783642377501).

Dunér, D. (2018) Semiotics of biosignatures. *Southern Semiotics Review*, Vol. 9, pp. 47–63 (DOI: 10.33234/ ssr.9.4).

Durrans, B. (1992) Posterity and paradox: Some uses of time capsules. In: Wallman, S. (ed.), *Contemporary Futures: Perspectives from Social Anthropology*. London: Routledge (ISBN: 9780415066631).

Eco, U. (1978) *A Theory of Semiotics*. Bloomington: Indiana University Press (ISBN: 9780253202178).

Elliott, J.R. (2011) A post-detection decipherment strategy. *Acta Astronautica*, Vol. 68, Issue 3-4, pp. 441– 444 (DOI: 10.1016/j.actaastro.2010.01.003).

Elliott, J.R., and Baxter, S. (2011) The DISC Quotient: A post-detection strategy. In: Vakoch, D.A. (ed.), *Communication with Extraterrestrial Intelligence*. Albany, NY: State University of New York Press (ISBN: 9781438437941).

Ellis, R.S. (1968) *Foundation Deposits in Ancient Mesopotamia*. New Haven, CT: Yale University Press (ISBN: 9780300004427).

Elwenspoek, M.C. (2011) Long-time data storage: Relevant time scales. *Challenges*, Vol. 2, Issue 1, pp. 19–36 (DOI: 10.3390/challe2010019).

ESOC (2020) Space Debris by the Numbers. European Space Agency; Space Debris. Retrieved on 03 September 2019, available at https://www.esa.int/Our_Activities/Operations/Space_Debris/ Space_debris_by_the_numbers.

Everett, D.L. (2005) Cultural constraints on grammar and cognition in Pirahã: Another look at the design features of human language. *Current Anthropology*, Vol. 46, Issue 4, pp. 621–634 (DOI: 10.1086/431525).

Ferris, T. (2017) *Voyagers in Space & Time*. Ozma Records Publication from: Voyager Golden Record 40th Anniversary Collection. Available at https://ozmarecords.com/products/voyager-golden-record-3xlp-box-set.

Fewer, G. (2002) Towards an LSMR & MSMR (Lunar & Martian Sites & Monuments Records): Recording planetary spacecraft landing sites as archaeological monuments of the future. In: Russel, M. (ed.), *Digging Holes in Popular Culture: Archaeology and Science Fiction*. Bournemouth, UK: Oxbow Books (ISBN: 1842170635).

Field, D.J., Hayes, A., and Hess, R.F. (1993) Contour integration by the human visual-system—evidence for a local association field. *Vision Research*, Vol. 33, Issue 2, pp. 173–193 (DOI: 10.1016/0042 6989(93)90156-q).

Finney, B.R. (1992) *From Sea to Space*. Palmerston North, New Zealand: Massey University (ISBN: 9780908665594).

Finney, B.R., and Bentley, J.H. (2014) A tale of two analogues: Learning at a distance from the ancient Greeks and Maya and the problem of deciphering extraterrestrial radio transmissions. In: Vakoch, D.A. (ed.), *Archaeology, Anthropology and Interstellar Communication*. Washington, D.C.: The NASA History series (ISBN: 9781501081729).

Fisk, A.S., Tam, S.K.E., Brown, L.A., Vyazovskiy, V.V., Bannerman, D.M., and Peirson, S.N. (2018) Light and cognition: Roles for circadian rhythms, sleep, and arousal. *Frontiers in Neurology*, Vol. 9, Issue 56 (DOI: 10.3389/fneur.2018.00056).

Fodor, J.A. (1983) *The Modularity of Mind*. Boston, MA: MIT Press (ISBN: 9780262060844).

Forbes, P. (2009) *Dazzled and Deceived: Mimicry and Camouflage*. New Haven, CT: Yale University Press (ISBN: 9780300178968).

Foucault, M. (2001) *The Order of Things: An Archaeology of the Human Sciences*. London: Routledge (ISBN: 9780415267373).

Franklin, S. (1995) Science as culture, cultures of science. *Annual Review of Anthropology*, Vol. 24, pp. 163– 184 (DOI: 10.1146/annurev.an.24.100195.001115).

Freeman, D. (2019) The Arch Mission Foundation and SpaceChain create Orbital Library™, first archive in space. SpaceChain website. Retrieved on 6 February 2019, available at https://spacechain.com/ the-arch-mission-foundation-and-spacechain-create-orbital-library/.

Freudenthal, H. (1960) *Lincos, Design of a Language for Cosmic Intercourse, Part 1*. Amsterdam: North-Holland Publishing Company (ASIN: B001KPS1DC).

Fuller, R.B. (1967) *Operating Manual for Spaceship Earth*. Baden, Switzerland: Lars Muller Publishers (ISBN: 9783037781265).

Garcia, X. (2016) Remaster the Golden Record. Science Friday website. Retrieved on 04 January 2020, Available at https://www.sciencefriday.com/educational-resources/remaster-the-golden-record/.

Gerhardstein, P., Tse, J., Dickerson, K., Hipp, D., and Moser, A. (2012) The human visual system uses a global closure mechanism. *Vision Research*, Vol. 71, pp. 18–27 (DOI: 10.1016/j.visres.2012.08.011).

Gertz, J. (2016a) Reviewing METI: A critical analysis of the arguments. *Journal of the British Interplanetary Society (JBIS)*, Vol. 69, pp. 31–36 (JBIS Refcode: 2016.69.31).

Gertz, J. (2016b) Post-detection SETI protocols & METI: The time has come to regulate them both. *Journal of the British Interplanetary Society*, Vol. 69, pp. 263–270 (JBIS Refcode: 2016.69.263).

Gibson, J.J. (1979) *The Ecological Approach to Visual Perception*. Boston: Houghton Mifflin (ISBN: 9780395270493).

Gibson, R. (2001) Lunar archaeology: The application of federal historic preservation law to the site where humans first set foot upon the Moon. Master's Thesis for the Department of Anthropology at the New Mexico State University.

Gillespie, C., and Vishwanath, D. (2019) A shape-level flanker facilitation effect in contour integration and the role of shape complexity. *Vision Research*, Vol. 158, pp. 221–236 (DOI: 10.1016/j.visres.2019.02.002).

Gindilis, L.M., and Gurvits, L.I. (2018) SETI in Russia, USSR and the post–Soviet space: A century of research. *Acta Astronautica*, Vol. 162, pp. 1–13 (DOI: 10.1016/j.actaastro.2019.04.030).

Glavin, D.P., Dworkin, J.P., Lupisella, M., Kminek, G., and Rummel, J.D. (2004) Biological contamination studies of lunar landing sites: Implications for future planetary protection and life detection on the Moon and Mars. *International Journal of Astrobiology*, Vol. 3, Issue 3, pp. 265–271 (DOI: 10.1017/S1473550404001958).

Goddard, R.H. (1920) Report to Smithsonian Institution concerning further developments of the rocket method of investigating space, March 1920. In: Goddard, E.C. (ed.), *The Papers of Robert H. Goddard*. New York: McGraw-Hill. Retrieved on 16 May 2017, available at https://siarchives.si.edu/history/featured-topics/stories/march-1920-report-concerning-further-developments-space-travel.

Gollisch, T., and Meister, M. (2010) Eye smarter than scientists believed: Neural computations in circuits of the retina. *Neuron*, Vol. 65, Issue 2, pp. 150–164 (DOI: 10.1016/j.neuron.2009.12.009).

Gordin, V.L. (1924) *Grammar of the logical language AO/ Грамматика логического языка АО*. Moscow: Self-published manuscript (ISBN: 789949868780).

Gordon, I.E. (2005) Theories of visual perception, third edition. London: Psychology Press (ISBN: 1841693839).

Goris, R.C. (2011) Infrared organs of snakes: An integral part of vision. *Journal of Herpetology*, Vol. 45, Issue 1, pp. 2–14 (DOI: https://doi.org/10.1670/10-238.1).

Gorman, A.C. (2001) The archaeology of body modification. The identification of symbolic behaviour through use wear and residues on flaked stone tools. Unpublished PhD thesis: University of New England, Armidale NSW.

Gorman, A.C. (2009a) The gravity of archaeology. *Archaeologies*, Vol. 5, Issue 2, pp. 344–359.

Gorman, A.C. (2009b) Beyond the space race: the significance of space sites in a new global context. In: Holtorf, C., and Piccini, A. (eds.), *Contemporary Archaeologies: Excavating Now*. Frankfurt: Peter Lang AG (ISBN: 9783631576373).

Gorman, A.C. (2009c) The cultural landscape of space. In: Darrin, A. and O'Leary, B.L. (eds.), *The Handbook of Space Engineering, Archaeology, and Heritage*. Boca Raton: CRC Press (ISBN: 9781420084313).

Gorman, A.C. (2013) Beyond the morning star: The real tale of the Voyagers' Aboriginal music. *The Conversation* (online). Retrieved on 04 January 2016, available at https://theconversation.com/beyond-the-morning-star-the-real-tale-of-the-voyagers-aboriginal-music-18288.

Gorman, A.C. (2014) The Anthropocene in the Solar System. *Journal of Contemporary Archaeology*, Vol. 1, Issue 1, pp. 89–132 (DOI: 10.1558/jca.v1i1.87).

Gorman, A.C. (2016) Tracking cable ties: Contemporary archaeology at a NASA satellite tracking station. In: Frederick, U.K. and Clarke, A. (eds.), *That Was Then, This Is Now: Contemporary Archaeology and Material Cultures in Australia*. Cambridge, UK: Cambridge Scholars Press (ISBN: 9781443885386).

Gorman, A.C. (2017a) Pale blue dot: Everyday material culture on the International Space Station. Day of Archaeology. Retrieved on 28 July 2017, available at https://www.dayofarchaeology.com/pale-blue-dot-everyday-material-culture-on-the-international-space-station/.

Gorman, A.C. (2017b) Not all space debris is junk—a comprehensive management strategy for culturally significant spacecraft. In: *68th International Astronautical Congress (IAC) conference*, Adelaide, Australia, 25–29 September 2017.

Gorman, A.C. (2018) A sports car and a glitter ball are now in space—what does that say about us as humans? *The Conversation* (online). Retrieved on 16 May 2018, Available at https://theconversation.com/a-sports-car-and-a-glitter-ball-are-now-in-space-what-does-that-say-about-us-as-humans-91156.

Gorman, A.C. (2019) *Dr. Space Junk vs The Universe: Archaeology and the Future*. Sydney: NewSouth Publishing (ISBN: 9781742236247).

Gorman, A.C., and O'Leary, B.L. (2007) An ideological vacuum: The Cold War in outer space. In: Schofield, J., and Cocroft, W. (eds.), *A Fearsome Heritage: Diverse Legacies of the Cold War*. Walnut Creek: Left Coast Press (ISBN: 9781598742589).

Grace, G.W. (1987) *The Linguistic Construction of Reality*. Abingdon: Routledge Kegan & Paul (ISBN: 9780709938866).

Grinspoon, D. (2016) The Golden Spike of Tranquility Base. *Sky and Telescope Magazine* (online). Retrieved on 19 August 2018, Available at https://www.skyandtelescope.com/astronomy-blogs/the-golden-spike-of-tranquility-base.

Grossberg, S. (1994) 3-D vision and figure-ground perception by visual cortex. *Perception & Psychophysics*, Vol. 55, pp. 48–120 (DOI: 10.3758/bf03206880).

Groys, B. (2018) *Russian Cosmism*. Cambridge, MA: MIT Press (ISBN: 9780262037433).

Guillochon, J., and Loeb, A. (2015) SETI via leakage from light sails in exoplanetary systems. *The Astrophysical Journal Letters*, Vol. 811, Issue 2, pp. 1–6 (DOI: 10.1088/2041-8205/811/2/L20).

Guzman, M., Welch, C., and Hein, A.M. (2016) Extremely long-duration storage concepts for space. *Acta Astronautica*, Vol. 130, Issue 1–2, pp. 128–136 (DOI: 10.1016/j.actaastro.2016.10.007).

Hanlon, M.L.D. (2019) The case for protecting the Apollo landing areas as heritage sites. *Astronomy* magazine (online). Retrieved on 13 March 2019, available at http://www.astronomy.com/news/2019/02/the-case-for-protecting-the-apollo-landing-areas-as-heritage-sites.

Haqq-Misra, J., Busch, M.W., Som, S.M., and Baum, S.D. (2013) The benefits and harm of transmitting into space. *Space Policy*, Vol. 29, Issue 1, pp. 40–48 (DOI: 10.1016/j.spacepol.2012.11.006).

Harding, S.G. (1991) *Whose science? Whose knowledge?: Thinking from women's lives*. Ithaca, NY: Cornell University Press (ISBN: 9780801497469).

Harding, S.G. (1992) After eurocentrism: Challenges for the philosophy of science. In: *PSA: Proceedings of the Biennial Meeting of the Philosophy of Science Association*, Vol. 1992, pp. 311–319 (ISSN: 02708647).

Harrison, A.A. (2014) Speaking for Earth: Projecting cultural values across deep space and time. In: Vakoch, D.A. (ed.), *Archaeology, Anthropology and Interstellar Communication*. Washington: The NASA History Series (ISBN: 9781501081729).

Harrison, R., DeSilvey, C., Holtorf, C., Macdonald, S., Bartolini, N., Breithoff, E., Fredheim, H., Lyons, A., May, S., Morgan, J., and Penrose, S. (2020) *Heritage Futures: Comparative Approaches to Natural and Cultural Heritage Practices*. London: UCL Press (ISBN: 9781787356016).

Hart, N.S., Bailes, H.J., Vorobyev, M., Marshall, N.J., and Collin, S.P. (2008) Visual ecology of the Australian lungfish (*Neoceratodus forsteri*). *BMC Ecology*, Vol. 8, Issue 21 (DOI:10.1186/1472-6785-8-21).

Hartline, H.K. (1938) The response of single optic nerve fibers of the vertebrate eye to illumination of the retina. *American Journal of Physiology*, Vol. 121, Issue 2, pp. 400–415 (DOI: 10.1152/ajplegacy.1938.121.2.400).

Haskins, E.V. (2007) Between archive and participation: Public memory in a digital age. *Rhetoric Society Quarterly*, Vol. 37, Issue. 4, pp. 401–422 (DOI: 10.1080/02773940601086794).

Hauer, B., and Kondrak, G. (2016) Decoding anagrammed texts written in an unknown language and script. *Transactions of the Association for Computational Linguistics*, Vol. 4, pp. 75–86. Retrieved on 03 April 2018, available at https://transacl.org/ojs/index.php/tacl/article/view/821.

Hawkins, G.S. (1983) *Mindsteps to the Cosmos*. New York: HarperCollins (ISBN: 9780060151560).

Hays, P.L., and Lutes, C.D. (2007) Towards a theory of spacepower. *Space Policy*, Vol. 23, Issue 4, pp. 206–209 (DOI: 10.1016/j.spacepol.2007.09.003).

Hegarty, P. (2006) Noise Music. *The Semiotic Review of Books*, Vol. 16, Issue 1–2, pp. 1–4.

Heidmann, J. (1992) *Extraterrestrial Intelligence*. Cambridge, UK: Cambridge University Press (ISBN: 0521585635).

Heidmann, J. (1993) A reply from Earth: Just send them the encyclopaedia. *Acta Astronautica*, Vol. 29, Issue 3, pp. 233–235 (DOI: 10.1016/0094-5765(93)90053-Y).

Heiken, G.H., Vaniman, D.T., and French, B.M. (1991) *Lunar Sourcebook: A User's Guide to the Moon*. Cambridge, UK: Cambridge University Press (ISBN: 0521334446).

Heiligenberg, W. (1973) Electrolocation of objects in the electric fish Eigenmannia (Rhamphichthyidae, Gymnotoidei). *Journal of Comparative Physiology*, Vol. 87, Issue 2, pp. 137–164 (DOI: 10.1007/BF01352158).

Helmreich, S. (2014) Remixing the Voyager interstellar record: Or, as extraterrestrials might listen. *Journal of Sonic Studies*, Vol. 8 (online). Retrieved on 08 January 2020, available at https://www.researchcatalogue.net/view/109536/109537/0/0.

Herkert, J.R. (2011) Ethical challenges of emerging technologies. In: Marchant, G.E., Allenby, B.R., and Herkert, J.R. (eds.), *The Growing Gap Between Emerging Technologies and Legal-Ethical Oversight*. New York: Springer (ISBN: 9789400713567).

Hicks, D., and Beaudry, M.C. (2010) *The Oxford Handbook of Material Culture Studies*. Oxford, UK: Oxford University Press (ISBN: 9780199218714).

Hoffman, D.L., Standish, C.D., García-Diez, M., Pettitt, P.B., Milton, J.A., Zilhão, J., Alcolea-González, J.J., Cantalejo-Duarte, P., Collado, H., de Balbín, R., Lorblanchet, M., Ramos-Muñoz, J., Weniger, G-Ch., and Pike, A.W.G. (2018) U-Th dating of carbonate crusts reveals Neandertal origin of Iberian cave art. *Science*, Vol. 359, Issue 6378, pp. 912–915 (DOI: 10.1126/science.aap7778).

Hogben, L. (1952) Astraglossa or first steps in celestial syntax. *Journal of British Interplanetary Society*, Vol. 11, Issue 6, pp. 258–274.

Hogben, L. (1961) Cosmical language. *Nature*, Vol. 192, Issue 4805, pp. 826–827 (DOI: 10.1038/192826a0).

Högberg, A., Holtorf, C., May, S., and Wollentz, G. (2017) No future in archaeological heritage management? *World Archaeology*, Vol. 49, Issue 5, pp. 639–647 (DOI: 10.1080/00438243.2017.1406398).

Holtorf, C., and Högberg, A. (2021) *Cultural Heritage and the Future*. London: Routledge (ISBN: 978113882 9015).

Hubel, D.H., and Wiesel, T.N. (1959) Receptive fields of single neurons in the cat's striate cortex. *Journal of Physiology*, Vol. 148, Issue 3, pp. 574–591 (DOI: 10.1113/jphysiol.1959.sp006308).

Hubel, D.H., and Wiesel, T.N. (1962) Receptive fields, binocular interaction and functional architecture in cats' visual cortex. *Journal of Physiology*, Vol. 160, Issue 1, pp. 106–154 (DOI: 10.1113/jphysiol.1962.sp006837).

Hubel, D.H. and Wiesel, T.N. (1968) Receptive fields and functional architecture of monkey striate cortex. *Journal of Physiology*, Vol. 195, Issue 1, pp. 215–243 (DOI: 10.1113/jphysiol.1968.sp008455).

IAA (2010) Declaration of Principles Concerning the Conduct of the Search for Extraterrestrial Intelligence. International Academy of Astronautics (IAA): SETI Permanent Committee. Retrieved on 30 September 2015, available at http://resources.iaaseti.org/protocols_rev2010.pdf

International Space University (2007) *Phoenix*. Illkirch, France: International Space University. Retrieved on 28 September 2017, available at https://isulibrary.isunet.edu/doc_num.php?explnum_id=103.

The Interstellar Beacon (2020) The Interstellar Beacon Project: Backup Humanity. Interstellar Beacon website. Retrieved on 05 May 2020, available at https://www.interstellarbeacon.org/.

Jack, C.E., and Thurlow, W.R. (1973) Effects of degree of visual association and angle of displacement on the "ventriloquism" effect. *Perceptual and Motor Skill*, Vol. 37, Issue 3, pp. 967–979 (DOI: 10.2466/pms.19 73.37.3.967).

Jacobs, G.H. (2018) Photopigments and the dimensionality of animal color vision. *Neuroscience and Biobehavioral Reviews*, Vol. 86, pp. 108–130 (DOI: 10.1016/j.neubiorev.2017.12.006).

Jarvis, W.E. (1992) Modern time capsules: Symbolic repositories of civilization. *Libraries & Culture*, Vol. 27, Issue 3, pp. 279–295.

Jarvis, W.E. (2003) *Time Capsules: A Cultural History*. Jefferson, NC: McFarland (ISBN: 9780786412617).

Jonas, D., and Jonas, D.M. (1976) *Other Senses, Other Worlds*. New York: Littlehampton Book Services (ISBN: 9780304297627).

Judge, A. (2000) Test challenges for alien encounter: Communicating with aliens series. *Laetus in Praesens* (personal) website. Retrieved on 14 January 2014, available at https://www.laetusinpraesens.org/docs/alien1.php.

Kahn, D. (1997) John Cage: Silence and silencing. *The Musical Quarterly*, Vol. 81, Issue 4, pp. 556–598 (DOI: 10.1093/mq/81.4.556).

Kay, P., and Kempton, W. (1984) What is the Sapir-Whorf Hypothesis? *American Anthropologist*, Vol. 86, Issue 1, pp. 65–79 (DOI: 10.1525/aa.1984.86.1.02a00050).

Kazansky, P., Cerkauskaite, A., Beresna, M., Drevinskas, R., Patel, A., Zhang, J., and Gecevicius, M. (2016) Eternal 5D data storage by ultrafast laser writing in glass. Conference paper. Researchgate (DOI: 10.1117/2.1201 603.006365).

Kershenbaum, A. (2021) *The Zoologist's Guide to the Galaxy: What Animals on Earth Reveal About Aliens—and Ourselves*. London: Penguin/ Randomhouse (ISBN: 9780241986844).

Kirschvink, J.L. (1982) Birds, bees and magnetism—A new look at the old problem of Magnetoreception. *Trends in Neurosciences*, Vol. 5, Issue 5, pp. 160–167 (DOI: 10.1016/0166-2236(82)90090-X).

Kitaoka, A., and Ashida, H. (2003) Phenomenal characteristics of the peripheral drift illusion. *Vision*, Vol. 15, Issue 4, pp. 261–262 (DOI: 10.11247/jssdj.58.69).

Kminek, G., Conley, C., Hipkin, V., and Yano, H. (2017) COSPAR's planetary protection policy. Committee on Space Research (COSPAR) publications. Retrieved on 16 August 2019, available at https://cosparhq.cnes.fr/sites/default/files/pppolicydecember_2017.pdf.

Knappett, C. (2005) *Thinking Through Material Culture: An Interdisciplinary Perspective*. Philadelphia: University of Pennsylvania Press (ISBN: 9780812237887).

Koffka, K. (1935) *Principles of Gestalt Psychology*. London: Lund Humphries (ISBN: 9788857523934).

Kolb, H., Helga, H., Ripps, H., and Wu, S. (2001) *Concepts and Challenges in Retinal Biology*. Amsterdam: Elsevier (ISBN: 9780444514844).

Korbitz, A. (2014) Towards understanding the Active SETI debate: Insights from risk communication and perception. *Acta Astronautica*, Vol. 105, Issue 2, pp. 517–520 (DOI: 10.1016/j.actaastro.2014.07.005).

Koren, M. (2018) How to build a museum in outer space. *The Atlantic*. Retrieved on 19 February 2019, available at https://www.theatlantic.com/science/archive/2018/12/museum-orbit-space-heritage/576922/.

Kornmeier, J., and Bach, M. (2005) The Necker cube—An ambiguous figure disambiguated in early visual processing. *Vision Research*, Vol. 45, Issue 8, pp. 955–960 (DOI: 10.1016/j.visres.2004.10.006).

Kornmeier, J., and Bach, M. (2012) Ambiguous figures—What happens in the brain when perception changes but not the stimulus. *Frontiers in Human Neuroscience*, Vol. 6, Issue 51 (DOI: 10.3389/fnhum.2012.00051).

Kovács, I., and Julesz, B. (1993) A closed curve is much more than an incomplete one—Effect of closure in figure ground segmentation. *Proceedings of the National Academy of Sciences of the United States of America*, Vol. 90, Issue 16, pp. 7495–7497 (DOI: 10.1073/pnas.90.16.7495).

Kramer, W. (2014) Extraterrestrial environmental impact assessments—A foreseeable prerequisite for wise decisions regarding outer space exploration, research and development. *Space Policy*, Vol. 30, Issue 4, pp. 215–222 (DOI: 10.1016/j.spacepol.2014.07.001).

Krznaric, R. (2021) *The Good Ancestor: How to Think Long Term in a Short-Term World.* London: W.H. Allen (ISBN: 978–0753554517).

Kuffler, S.W. (1953). Discharge patterns and functional organization of mammalian retina. *Journal of Neurophysiology*, Vol. 16, Issue 1, pp. 37–68 (DOI: 10.1152/jn.1953.16.1.37).

Kuhn, T.S. (1962) *The Structure of Scientific Revolutions.* Chicago: University of Chicago Press (ISBN: 9780226458083).

Langston, M.C. (2014) The accidental altruist: Inferring altruism from an extraterrestrial signal. In: Vakoch, D.A. (ed.), *Extraterrestrial Altruism: Evolution and Ethics in the Cosmos.* New York: Springer (ISBN: 9783642377495).

Lemarchand, G.A., and Lomberg, J. (2011) Communication among interstellar intelligent species: A search for universal cognitive maps. In: Vakoch, D.A. (ed.), *Communication with Extraterrestrial Intelligence.* Albany, NY: State University of New York Press (ISBN: 9781438437941).

Lestel, D. (2014) Ethology, ethnology, and communication with extraterrestrial intelligence. In: Vakoch, D.A. (ed.), *Archaeology, Anthropology and Interstellar Communication.* Washington, D.C.: The NASA History series (ISBN: 9781501081729).

Levi-Setti, R. (1993) *Trilobites.* Chicago: University of Chicago Press (ISBN: 9780226474526).

Lewis-Williams, D.J. (2002) *The Mind in the Cave: Consciousness and the Origins of Art.* London: Thames & Hudson (ISBN: 9780500284650).

Loeb, A., and Turner, E.L. (2012) Detection technique for artificially illuminated objects in the outer solar system and beyond. *Astrobiology*, Vol. 12, Issue 4, pp. 290–294 (DOI: 10.1089/ast.2011.0758).

Loffler, G. (2008) Perception of contours and shapes: Low and intermediate stage mechanisms. *Vision Research*, Vol. 48, Issue 20, pp. 2106–2127 (DOI: 10.1016/j.visres.2008.03.006).

Lomberg, J. (2004) A portrait of humanity. *Contact in Context* (online). Retrieved on 13 August 2015, available at www.jonlomberg.com/articles.html.

Lomberg, J. and Hora, S.C. (1997) Very long-term communication intelligence: The case of markers for nuclear waste sites. *Technological Forecasting and Social Change*, Vol. 56, Issue 2, pp. 171–188 (DOI: 10.1016/S0040–1625(97)00057–7).

Lovelock, J.E. (1972) Gaia as seen through the atmosphere. *Atmospheric Environment*, Vol. 6, Issue 8, pp. 579–580 (DOI: 10.1016/0004–6981(72)90076–5).

Lower, T.A., Vakoch, D.A., Clearwater, Y., Niles, B.A., and Scanlin, J.E. (2011) What the world needs to know: Identifying the relative degree of specific Maslovian needs and degree of species-level self-identification in interstellar messages submitted by a multinational sample. In: Vakoch, D.A. (ed.), *Communication with Extraterrestrial Intelligence.* Albany: State University of New York Press (ISBN: 9781438437934).

Macauley, W.R. (2006) Inscribing scientific knowledge: NASA's Pioneer plaque and the history of interstellar communication. In: Geppert, A.C.T. (ed.), *Imagining Outer Space: European Astroculture in the Twentieth Century.* London: Palgrave Macmillan (ISBN: 9781349312153).

Macauley, W.R. (2010) Picturing knowledge: NASA's Pioneer plaque, Voyager record and the history of interstellar communication, 1957–1977. PhD Thesis for the Centre for the History of Science at the University of Manchester. Retrieved on 09 September 2018, available at https://ethos.bl.uk/OrderDetails.do?uin=uk.bl.ethos.529240.

Machilsen, B., Pauwels, M., and Wagemans, J. (2009) The role of vertical mirror symmetry in visual shape detection. *Journal of Vision*, Vol. 9, Issue 12, pp. 1–11 (DOI: 10.1167/9.12.11).

Malafouris, L. (2004) The cognitive basis of material engagement: Where brain, body and culture conflate. In: Demarrais, E., Gosden, C., and Renfrew, C. (eds.), *Rethinking Materiality: The Engagement of Mind with the Material World.* Cambridge, UK: McDonald Institute for Archaeological Research (ISBN: 9781902937304).

Malafouris, L. (2007) Before and beyond representation: Towards an enactive conception of the palaeolithic image. In: Renfrew, C. and Morley, I. (eds.), *Image and Imagination: A Global History of Figurative Representation.* Cambridge, UK: McDonald Institute for Archaeological Research (ISBN: 9781902937489).

Malafouris, L. (2018) Mind and material engagement. *Phenomenology and the Cognitive Sciences*, Vol. 18, pp. 1–17 (DOI: 10.1007/s11097–018–9606–7).

Malafouris, L. (2019) Thinking as "thinging": Psychology with things. *Current Directions in Psychological Science*, Vol. 29, Issue 1, pp. 3–8 (DOI: 10.1177/0963721419873349).

Mallove, E. (1987) *The Quickening Universe: Cosmic Evolution and Human Destiny.* New York: St. Martin's Press (ISBN: 9780312000622).

Manz, A. (2010) The Human Document Project and Challenges. *Challenges*, Vol. 1, Issue 1, pp. 3–4 (DOI: 10.3390/challe1010003).

Marchant, G.E. (2011) The growing gap between emerging technologies and the law. In: Marchant, G.E., Allenby, B.R., and Herkert, J.R. (eds.), *The Growing Gap Between Emerging Technologies and Legal-Ethical Oversight*. New York: Springer (ISBN: 9789400713567).

Margenau, H. (1952) Physics and ontology. *Philosophy of Science*, Vol. 19, Issue 4, pp. 342–345 (DOI: 10.1086/287217).

Margenau, H. (1977) *The Nature of Physical Reality*. Woodbridge, CT: Ox Bow Press (ISBN: 9780918024022).

Marr, D. (1982). *Vision: A Computational investigation into the Human Representation and Processing of Visual Information*. San Francisco: W.H. Freeman. (ISBN: 9780716712848).

Martin, E. (1991) The egg and the sperm: How science has constructed a romance based on stereotypical male-female roles. *Signs Journal*, Vol. 16, Issue. 3, pp. 485–501.

Martinez, C.L.F. (2014) SETI in the light of cosmic convergent evolution. *Acta Astronautica*, Vol. 104, Issue 1, pp. 341–349 (DOI: 10.1016/j.actaastro.2014.08.013).

Maximov, V.V. (2000) Environmental factors which may have led to the appearance of color vision. *Philosophical Transactions of the Royal Society of London—B-Biological Sciences Series*, Vol. 355, Issue 1401, pp. 1239–1242 (DOI: 10.1098/rstb.2000.0675).

May, S. (2020) Micro-messaging/space messaging: A comparative exploration of #GoodbyePhilae and #MessageToVoyager. In: Harrison, R., DeSilvey, C., Holtorf,. C., Macdonald, S., Bartolini, N., Breithoff, E., Fredheim, H., Lyons, A., May, S., Morgan, J., and Penrose, S. (eds.), *Heritage Futures: Comparative Approaches to Natural and Cultural Heritage Practices*. London: UCL Press (ISBN: 9781787356016).

McCall, G.J.H. (2006) The Vendian (Ediacaran) in the geological record: Enigmas in geology's prelude to the Cambrian explosion. *Earth-Science Reviews*, Vol. 77, Issue 1–3, pp. 1–229 (DOI: 10.1016/j.earscirev.2005.08.004).

McCarthy, J. (2019) Sónar calling GJ273B: The argumentative issue of METI. *Theology and Science*, Vol. 17, Issue 1, pp. 59–68 (DOI: 10.1080/14746700.2019.1557803).

McCray, P.W. (2012) *The Visioneers: How a Group of Elite Scientists Pursued Space Colonies, Nanotechnologies, and a Limitless Future*. New Jersey: Princeton University Press (ISBN: 9780691139838).

McNeill, J.R. (2000) *Something New Under the Sun: An Environmental History of the Twentieth-Century World*. New York: WW Norton (ISBN: 0393049175).

McNiven, I.J., and Lynette, R. (2005) *Appropriated Pasts: Indigenous Peoples and the Colonial Culture of Archaeology*. Oxford: AltaMira Press (ISBN: 9780759109070).

McTier, M.A.S., and Kipping, D.M. (2018) Finding mountains with molehills: The detectability of exotopography. *Monthly Notices of the Royal Astronomical Society*, Vol. 475, Issue 4, pp. 4978–4985 (DOI: 10.1093/mnras/sty143).

Mehoke, T.S. (2009) Technology and material culture in science fiction. In: Darrin, A., and O'Leary, B.L. (eds.), *The Handbook of Space Engineering, Archaeology, and Heritage*. Boca Raton: CRC Press (ISBN: 9781420084313).

Meisinger, H. (2003) Christian love and biological altruism. *Journal of Religion & Science*, Vol. 35, Issue 4, pp. 745–782 (DOI: 10.1111/1467-9744.00312).

Michaud, M.A.G. (1982) Towards a Grand Strategy for the Species. *Earth-Oriented Applications of Space Technology*, Vol. 2, pp. 213–219.

Michaud, M.A.G. (1992) An international agreement concerning the detection of extraterrestrial intelligence. *Acta Astronautica*, Vol. 26, Issue 3-4, pp. 291–294 (DOI: 10.1016/0094-5765(92)90114-X).

Michaud, M.A.G. (1993) SETI and Diplomacy. In: Shostak, S.G. (ed.), *Progress in the Search for Extraterrestrial Life: 1993 Bioastronomy Symposium*, Vol. 74, pp. 551–556 (ISBN: 0937707937).

Michaud, M.A.G. (2003) Ten decisions that could shake the world. *Space Policy*, Vol. 19, Issue 2, pp. 131–136 (DOI: 10.1016/S0265-9646(03)00019-5).

Michaud, M.A.G. (2007) *Contact with Alien Civilizations: Our Hopes and Fears About Encountering Extraterrestrials*. New York: Springer-Verlag (ISBN: 9780387285986).

Miller, D., Abed Rabho, L., Awondo, P., de Vries, M., Duque, M., Garvey, P., Haapio-Kirk, L., Hawkins, C., Otaegui, A., Walton, S., and Wang, X. (2021) *The Global Smartphone: Beyond a Youth Technology*. London: UCL Press (ISBN: 9781787359611).

Minsky, M. (1985) Communication with alien intelligence. *Byte Magazine*, Vol. 10, Issue 4, pp. 127–138.

Mitchell, F.J., and Ellis, W.L. (1972) Microbe survival analyses, part A, Surveyor 3: Bacterium isolated from lunar retrieved television camera. In: *Analysis of Surveyor 3 Material and Photographs Returned by Apollo 12*. Washington, D.C.: NASA Scientific and Technical Information Office. Retrieved on 14 June 2017, available at https://www.lpi.usra.edu/lunar/documents/NTRS/collection2/NASA_SP_284.pdf.

Mitchison, N. (1962) *Memoirs of a Spacewoman*. England: Kennedy & Boyd (ISBN: 9781849210355).

Morrison, P. (1963) Interstellar communication. In: Cameron, A.G.W., *Interstellar Communication a Collection of Reprints and Original Contributions*. New York: W.A. Benjamin (ANIS: B000GSNRBU).

Musso, P. (2012) The problem of active SETI: An overview. *Acta Astronautica*, Vol. 78, pp. 43–54 (DOI: 10.1016/j.actaastro.2011.12.019).

NASA (1969) *Soil Mechanics Investigation* in *Apollo 11 Preliminary Science Report (SP-214)*. Washington,

D.C.: NASA Scientific and Technical Information Office. Retrieved on 22 July 2018, available at https://www.hq.nasa.gov/alsj/a11/as11psr.pdf.

NASA (2011) NASA's recommendations to space-faring entities on how to protect and preserve the historic and scientific value of US government artifacts, July 20, 2011. Retrieved on 23 August 2017, available at http://www.nasa.gov/directorates/heo/library/reports/lunar-artifacts.html.

NASA (2019) *NASA Planetary Protection Independent Review Board (PPIRB)*. Washington, D.C.: NASA Independent Review Board (Report to NASA/ SMD). Retrieved on 23 October 2019, available at https://www.nasa.gov/sites/default/files/atoms/files/planetary_protection_board_report_20191018.pdf.

NASA History Program Office (2012) *Catalogue of Manmade Material on the Moon*. Washington, D.C.: NASA History Program Office. Retrieved on 14 September 2016, available at https://history.nasa.gov/FINAL%20Catalogue%20of%20Manmade%20Material%20on%20the%20Moon.pdf.

Nelson, S., and Polansky, L. (1993) The music of the voyager interstellar record. *Journal of Applied Communication Research*, Vol. 21, Issue 4, pp. 358–376 (DOI: 10.1080/00909889309365379).

Neri, P., Morrone, M.C., and Burr, D.C. (1998) Seeing biological motion. *Nature*, Vol. 395, Issue 6705, pp. 894–896 (DOI: 10.1038/27661).

Newton, I. (1704) *Opticks: Or, a Treatise of the Reflexions, Refractions, Inflexions and Colours of Light. Also Two Treatises of the Species and Magnitude of Curvilinear Figures*. London: Smith and Walford.

Noble, D.F. (1997) *The Religion of Technology: The Divinity of Man and the Spirit of Invention*. New York: Random House (ISBN: 9780679425649).

Oberhaus, D. (2019) *Extraterrestrial Languages*. Cambridge, MA: MIT Press (ISBN: 9780262043069).

O'Connor, A. (2003) Geology, archaeology, and "the raging vortex of the 'eolith' controversy." *Proceedings of the Geologists' Association*, Vol. 114, Issue 3, pp. 255–262 (DOI: 10.1016/S0016-7878(03)80018-4).

O'Leary, B.L. (2009a) Evolution of space archaeology and heritage. In: Darrin, A., and O'Leary, B.L. (eds.), *The Handbook of Space Engineering, Archaeology, and Heritage*. Boca Raton, FL: CRC Press (ISBN: 9781420084313).

O'Leary, B.L. (2009b) One giant leap: Preserving cultural resources on the Moon. In: Darrin, A., and O'Leary, B.L. (eds.), *The Handbook of Space Engineering, Archaeology, and Heritage*. Boca Raton, FL: CRC Press (ISBN: 9781420084313).

O'Leary, B.L. (2009c) Plan for the future: Preservation of space. In: Darrin, A., and O'Leary, B.L. (eds.), *The Handbook of Space Engineering, Archaeology, and Heritage*. Boca Raton, FL: CRC Press (ISBN: 9781420084313).

Oliver, B.M., and Billingham, J. (1972) *Project Cyclops: A Design Study for a System for Detecting Extraterrestrial Life*. Washington, D.C.: NASA (NASA-CR-114445).

Ollongren, A. (1999) Large-size message construction for ETI: Typing static relations. Proceedings of the 50th International Astronautical Congress. Amsterdam, Netherlands, 4–8 October 1999.

Ollongren, A. (2001) Large-size message construction for ETI: An experiment in CETI. Proceedings of the 52nd International Astronautical Congress. Toulouse, France, 1–5 October 2001.

Ollongren, A. (2004) Large-size message construction for ETI: Non-deterministic typing and symbolic computation in LINCOS. Proceedings of the 55th International Astronautical Congress. Vancouver, Canada, 4–8 October 2004 (DOI: 10.2514/6.IAC-04-IAA.1.1.2.07).

Ollongren, A. (2010) On the signature of LINCOS. *Acta Astronautica*, Vol. 67, Issue 11–12, pp. 1440–1442 (DOI: 10.1016/j.actaastro.2010.04.006).

Ollongren, A. (2011) Large-size message construction for ETI: Aristotelian syllogisms. *Acta Astronautica*, Vol. 83, pp. 6–9 (DOI: 10.1016/j.actaastro.2012.09.003).

Ollongren, A. (2013) *Astrolinguistics: Design of a Linguistic System for Interstellar Communication Based on Logic*. New York: Springer (ISBN: 9781461454670).

Olson, V.A. (2010) American extreme: An ethnography of astronautical visions and ecologies. PhD Thesis for William Marsh Rice University, Texas.

Osborne, H. (2016) Alpha Centauri mission: Stephen Hawking and Yuri Milner plan to send tiny spacecrafts to the stars. *International Business Times UK* (online). Retrieved on 11 July 2017, available at https://www.ibtimes.co.uk/alpha-centauri-mission-stephen-hawking-yuri-milner-plan-send-tiny-spacecrafts-stars-1554511.

Osiander, R. (2009) From Vengeance 2 to Sputnik 1: The beginnings. In: Darrin, A., and O'Leary, B.L. (eds.), *The Handbook of Space Engineering, Archaeology, and Heritage*. Boca Raton, FL: CRC Press (DOI: 10.1201/9781420084320).

Paglen, T. (2012) *The Last Pictures*. Berkeley: University of California Press (ISBN: 9780520275003).

Palmeri, J. (2009) Bringing cosmos to culture: Harlow Shapley and the uses of cosmic evolution. In: Dick, S.J., and Lupisella, M.L. (eds.), *Cosmos & Culture: Cultural Evolution in a Cosmic Context*. Washington, D.C.: The NASA History series (ISBN: 9780160831195).

Panecasio, S. (2020) Witches on TikTok have apparently tried to "hex" the Moon. CNET. Retrieved on 19 July 2020, available at https://www.cnet.com/news/witches-on-tiktok-have-apparently-tried-to-hex-the-moon/.

Paterson, C. (2014) Making scents of life on Earth: Embedding olfactory information into multi-channel interstellar messages. In: *Communicating Across the Cosmos: How Can We Make Ourselves Understood by Other Civilizations in the Galaxy?* SETI Institute, Mountain View, California, 10–11 November 2014. Retrieved on 16 November 2014, available at https://www.youtube.com/watch?v=FWQlop5Yedk.

Patton, P. (2018) Language in the Cosmos I: Is universal grammar really universal? Universe Today website. Retrieved on 03 July 2018, available at https://www.universetoday.com/139326/universal-grammar-really-universal/.

Peirce, C.S. (1931–1935) *The Collected Papers of Charles Sanders Peirce, Volumes I and II: Principles of Philosophy and Elements of Logic.* Cambridge, MA: Harvard University Press (ISBN: 978–0674138001).

Peirce, C.S., and Buchler, J. (2011) *Philosophical Writings of Peirce* (Originally published as *The Philosophy of Peirce, Selected Writings* by Routledge and Kegan Paul Ltd. in 1940). Mineola, NY: Dover Publications (ISBN: 9780486202174).

Pentland, A. (1989) Shape information from shading: A theory about human perception. *Spatial Vision*, Vol. 4, Issue 2, pp. 165–182 (DOI: 10.1163/156856889X00103).

Perkins, S. (2012) Dear future Earthlings: A message in a bottle won't be enough to communicate with distant generations. *Science News*, Vol. 182, Issue. 12, pp. 26–28 (DOI: 10.1002/scin.5591821223).

Persson, E. (2013) Philosophical aspects of astrobiology. In: Dunér, D., Parthemore, J., Persson, E., and Holmberg, G. (eds.), *The History and Philosophy of Astrobiology.* Newcastle upon Tyne: Cambridge Scholars Publishing (ISBN: 9781443850353).

Peters, T. (2011) The implications of the discovery of extra-terrestrial life for religion. *Philosophical Transactions of the Royal Society A: Mathematical, Physical and Engineering Sciences*, Vol. 369, Issue 1936, pp. 644–655 (DOI: 10.1098/rsta.2010.0234).

Peters, T. (2017) Projecting Earth on to heaven: The hopes of Frank Drake and the fears of Stephen Hawking. *Theology and Science*, Vol. 15, Issue 2, pp. 160–161 (DOI: 10.1080/14746700.2017.1299387).

Peters, T. (2019) Does extraterrestrial life have intrinsic value? An exploration in responsibility ethics. *International Journal of Astrobiology*, Vol. 18, Issue 4, pp. 304–310 (https://doi.org/10.1017/S1473550041700057X).

Petersen, S. (2009) *Space-Age Aesthetics: Lucio Fontana, Yves Klein, and the Postwar European Avant-Garde.* Philadelphia: Penn State University Press (ISBN: 9780271033426).

Pichalakkattu, B. (2018) SETI & METI: An Indian perspective. *Theology and Science*, Vol. 17, Issue 1, pp. 49–58 (DOI: 10.1080/14746700.2019.1557801).

Pike, A.W.G., Hoffmann, D.L., García-Diez, M., Pettitt, P.B., Alcolea, J., De Balbín, R., González-Sainz, C., de las Heras, C., Lasheras, J.A., Montes, R., and Zilhão, J. (2012) U-series dating of Paleolithic art in 11 caves in Spain. *American Association for the Advancement of Science; Science Magazine*, Vol. 336, Issue 6087, pp. 1409–1413 (DOI: 10.1126/science.1219957).

Pike, A.W.G., Hoffmann, D.L., Pettitt, P.B., García-Diez, M., and Zilhão, J. (2016) Dating Paleolithic cave art: Why U—Th is the way to go. *Quaternary International*, Vol. 432, pp. 41–49 (DOI: 10.1016/j.quaint.2015.12.013).

Pitt, J.C. (1982) Will a rubber ball still bounce? In: Smith, N.D. (ed.), *Philosophers Look at Science Fiction.* Chicago: Nelson-Hall (ISBN: 0882298070).

Pizon-Rodriguez, A., Bensch, S., and Mulheim, R. (2018) Expression patterns of cryptochrome genes in avian retina suggest involvement of Cry4 in light-dependent magnetoreception. *Royal Society*, Vol. 15, Issue 140 (DOI: 10.1098/rsif.2018.0058).

Planck Collaboration (2015) Planck 2015 results. XIII. Cosmological parameters. *Astronomy & Astrophysics*, Vol. 594 (DOI: 10.1051/0004–6361/201525830).

Platoff, A.M. (1993) Where no flag has gone before: Political and technical aspects of placing a flag on the Moon. NASA Contractor Report 188251. Retrieved on 18 September 2019, available at https://ntrs.nasa.gov/archive/nasa/casi.ntrs.nasa.gov/19940008327.pdf.

Plog, F. (1974) *The Study of Prehistoric Change.* New York: Academic Press (9780127856452).

Plumwood, V. (2001) Nature as agency and the prospects for a progressive naturalism. *Capitalism Nature Socialism*, Vol. 12, Issue 4, pp. 3–32 (DOI: 10.1080/104557501101245225).

Ponce, C.R. and Born, R.T. (2008) Stereopsis. *Current Biology*, Vol. 18, Issue 8, pp. 845–850 (DOI: 10.1016/j.cub.2008.07.006).

Pope, M. (1999) *The Story of Decipherment: From Egyptian Hieroglyphs to Maya Script.* London: Thames and Hudson (ISBN: 9780500281055).

Putnam, H. (1982) *Reason, Truth, and History.* Cambridge, UK: Cambridge University Press (ISBN: 978–0521297769).

Quast, P.E. (2017) A human perspective of Earth: An overview of dominant themes to emerge from global "A Simple Response…" messages (DOI: 10.13140/RG.2.2.36341.78563).

Quast, P.E. (2018a) A profile of humanity: The cultural signature of Earth's inhabitants beyond the atmosphere. *International Journal of Astrobiology*, Vol. 20, Issue 3, pp. 194–214 (DOI: 10.1017/S147355041800290).

Quast, P.E. (2018b) Beyond the Earth; Schematics for "Companion Guide for Earth" archival elements residing within geosynchronous orbit (DOI: 10.13140/RG.2.2.14177.97127).

Quast, P.E. (2019) A profile of humanity: How "green" are our postcards from Earth? In: *Proceedings of the UN/ GREEN: Naturally Artificial Intelligences Conference*, RIXC Art Science Festival, Riga, Latvia, 4–6 July 2019.

Quast, P.E. (2021) Re-encountering signs of agency: Surveying the appearance of "layering" patterns within our interstellar messaging record as representational signs for Earth. In: Crawford, I. (ed.), *Expanding Worldviews: Astrobiology, Big History and Cosmic Perspectives, Astrophysics and Space Science Proceedings 58.* Switzerland: Springer.

Quast, P.E. (2022) Remembering the conversation? Discussions about the challenges and next steps in establishing a long-term METI archive. In: *Proceedings from the UK SETI Research Network (UKSRN) Conference 2022*, University of Durham, United Kingdom. 7–8 July 2022.

Ramachandran, V.S., and Gregory, R.L. (1991) Perceptual filling in of artificially induced scotomas in human vision. *Nature,* Vol. 350, Issue 6320, pp. 699–702 (DOI: 10.1038/350699a0).

Rand, A. (1971) *The Romantic Manifesto.* New York: Signet Books (ISBN: 9780451149169).

Rathje, W.L. (1999) An archaeology of space garbage. *Discovering Archaeology,* Vol. 1, Issue 5, pp. 108–111.

Rathje, W.L., and Murphy, C. (1992) *Rubbish! The Archaeology of Garbage.* Tucson: University of Arizona Press (ISBN: 9780816521432).

Raulin-Cerceau, F. (2010) The pioneers of interplanetary communication: From Gauss to Tesla. *Acta Astronautica,* Vol. 67, Issue 11–12, pp. 1391–1398 (DOI: 10.1016/j.actaastro.2010.05.017).

Raulin-Cerceau, F., and Cyrille-Olou, D. (2019) *A la recherche d'intelligence extraterrestre.* Paris: Nouveau Monde Editions (ISBN: 9782369428152).

Reeves, R. (1994) *The Superpower Space Race: An Explosive Rivalry through the Solar System.* New York: Plenum Publishing Corporation (ISBN: 0306447681).

Reid, J.J., Schiffer, M.B., and Rathje, W.L. (1975) Behavioral archaeology: Four strategies. *American Anthropologist,* Vol. 77, Issue 4, pp. 864–869. Retrieved on 16 April 2019, available at https://www.jstor.org/ stable/674794.

Reijnen, G.C.M. (1990) Basic elements of an international terrestrial reply following the detection of a signal from extraterrestrial intelligence. *Acta Astronautica,* Vol. 21, Issue 2, pp. 143–148 (DOI: 10.1016/ 0094–5765(90)90142–8).

Rein, H., Tamayo, D., and Vokrouhlický, D. (2018) The random walk of cars and their collision probabilities with planets. *Aerospace,* Vol. 5, Issue 2, p. 57 (DOI: 10.3390/aerospace5020057).

Reynolds, E., Murray, J., Stronge, C., McCullough, K., Cranstoun, T., Hickey, J.B., Gurnett, D.A., Stone, E., Kohlhase, C., Bagenal, F., and Harmen, R. (2019) *The Farthest* [video file]. Arlington, VA: Public Broadcasting Service (ISBN: 9781531702427). Retrieved on 06 November 2017, available at https://www.youtube.com/watch?v=VQZPpFuQdTk.

Riegl, A. (1903) The modern cult of monuments: Its character and origin. Republished in: Stanley-Price, N., Kirby-Talley, M. and Vaccaro, A.M. (eds.), *Historical and Philosophical Issues in the Conservation of Cultural Heritage.* Los Angeles: Getty Conservation Institute (ISBN: 9780892363988).

RK&M (2019) *Preservation of Records, Knowledge, and Memory (RK&M) Across Generations: Developing a Key Information File for a Radioactive Waste Depository.* Boulogne-Billancourt, France: Nuclear Energy Agency (Report 7377). Retrieved on 8 October 2019, available at https://www.oecd-nea.org/rwm/ pubs/2019/7377-rkm-kif.pdf.

Rogers, T.F. (2004) Safeguarding Tranquility Base: Why the Earth's Moon base should become a World Heritage Site. *Space Policy,* Vol. 20, Issue 1, pp. 5–6 (DOI: 10.1016/j.spacepol.2003.11.001).

Rose, A. (2008) Long-term materials testing on the ISS. The Long Now Foundation (blog). Retrieved on 11 November 2017, available at https://blog.longnow.org/02008/11/19/long-term-materials-testing-on-the-iss/.

Rosenboom, D. (2003) The imperative of co-creation in interstellar communication: Lessons from experimental music. *Leonardo* (Electronic Almanac), Vol. 7, Issue 11 (ISBN: 9780983357100).

Ruxton, G.D. (2008) Non-visual crypsis: A review of the empirical evidence for camouflage to senses other than vision. *Philosophical Transactions of the Royal Society of London—B-Biological Sciences Series,* Vol. 364, Issue 1516, pp. 549–557 (DOI: 10.1098/rstb.2008.0228).

Ruxton, G.D., Sherratt, T.N., and Speed, M.P. (2004) Countershading and counterillumination. In: Ruxton, G.D., Sherratt, T.N. and Speed, M.P. (eds.), *Avoiding Attack: The Evolutionary Ecology of Crypsis, Warning Signals and Mimicry.* Oxford, UK: Oxford University Press (ISBN: 9780198528609).

Sagan, C. (1973) *Communication with Extraterrestrial Intelligence (CETI).* Cambridge, MA: MIT Press (ISBN: 0262690373).

Sagan, C. (1980) *Cosmos: A Personal Voyage, Episode 13: Who Speaks for Earth?* [video file]. Arlington, VA: Public Broadcasting Service (ASIN: B071H28346). Retrieved on 16 December 2018, available at https:// www.youtube.com/watch?v=luTxCri2bf4.

Sagan, C. (1985) *Contact.* New York: Simon & Schuster. (ISBN: 9781857235807).

Sagan, C. (1994) *Pale Blue Dot: A Vision of the Human Future in Space.* London: Headline Book Publishing (ISBN: 0747215537).

Sagan, C., Drake, F.D., Druyan, A., Ferris, T., Lomberg, J., and Salzman-Sagan, L. (1978) *Murmurs of Earth: The Voyager Interstellar Record.* New York: Ballantine Books (ISBN: 9780345315366).

Saint-Gelais, R. (2014) Beyond linear B: The metasemiotic challenge of communication with extraterrestrial intelligence. In: Vakoch, D.A. (ed.), *Archaeology, Anthropology and Interstellar Communication*. Washington, D.C.: The NASA History series, pp. 79–94 (ISBN: 9781501081729).

Sample, I. (2019) Tardigrades may have survived spacecraft crashing on Moon. *The Guardian* (online). Retrieved on 07 August 2019, available at https://www.theguardian.com/science/2019/aug/06/tardigrades-may-have-survived-spacecraft-crashing-on-moon.

Samuels, D. (2006) Alien tongues. In: Battaglia, D. (ed.), *ET Culture: Anthropology in Outerspaces*. Durham, NC: Duke University Press (ISBN: 9780822336211).

Sandberg, A. (1999) The physics of information processing superobjects: Daily life among the Jupiter brains. *Journal of Evolution and Technology*, Vol. 5, Issue 1, pp. 1–34.

Sassi, M., Demeyer, M., and Wagemans, J. (2014) Peripheral contour grouping and saccade targeting: The role of mirror symmetry. *Symmetry-Basel*, Vol. 6, Issue 1, pp. 1–22 (DOI: 10.3390/sym6010001).

Sassi, M., Machilsen, B., and Wagemans, J. (2012) Shape detection of gaborized outline versions of everyday objects. *i-perception journal*, Vol. 3, Issue 10, 745–764 (DOI: 10.1068/i0499).

Sassi, M., Vancleef, K., Machilsen, B., Panis, S., and Wagemans, J. (2010) Identification of everyday objects on the basis of gaborized outline versions. *i-perception journal*, Vol. 1, Issue 3, pp. 121–142 (DOI: 10.1068/i0384).

Saunders, B. (1995) Disinterring basic color terms: A study in the mystique of cognitivism. *History of the Human Sciences*, Vol. 8, Issue 7, pp. 19–38 (DOI: 10.1177/095269519500800402).

Sautuola, M.S. (1880) *Breves apuntes sobre algunos objetos prehistóricos de la provincia de Santander*. Real Academia de la Historia. Retrieved on 01 December 2019, available at http://centrodeestudiosmontaneses.com/wp-content/uploads/DOC_CEM/BIBLIOTECA/EDICION_OTROS/breves-apuntes-objetos-prehistoricos_sautuola_1880.pdf.

Sawicka, E., Stramski, D., Darecki, M., and Dubranna, J. (2012) Power spectral analysis of wave induced fluctuations in downward irradiance within the near-surface ocean under sunny conditions. In: *Proceedings of Ocean Optics XXI*. Glasgow, Scotland, 8–12 October 2012.

Scheffer, L. (2004) Aliens can watch "I Love Lucy." *Contact in Context* (online), Vol. 2, Issue 1.

Schiffer, M.B. (1975) Archaeology as behavioral science. *American Anthropologist*, Vol. 77, Issue 4, pp. 836–848 (DOI: 10.1525/aa.1975.77.4.02a00060).

Schiffer, M.B. (1976) *Behavioral Archeology (Studies in Archeology)*. New York: Academic Press (ISBN: 9780126241501).

Schiffer, M.B. (1987) *Formation Processes of the Archaeological Record*. Salt Lake City: University of Utah Press (ISBN: 9780874805130).

Schiffer, M.B. (1999) *The Material Life of Human Beings. Artifacts, Behavior, and Communication*. London: Routledge (ISBN: 9781134637256).

Schmitt, R.M. (2017) Archiving the "best of ourselves" on the Voyager Golden Records: Rhetorics of the frontier, memory, and technology. Master of Arts Thesis for the Department of Communication at the University of North Texas. Retrieved on 16 November 2019, available at https://scholar.colorado.edu/comm_gradetds/73.

Schmitz, H., and Bleckmann, H. (1998) The photomechanic infrared receptor for the detection of forest fires in the beetle *Melanophila acuminata*. *Journal of Comparative Physiology A*, Vol. 182, Issue 5, pp. 647–657 (DOI: 10.1007/s003590050).

Schneider, J., Léger, A., Fridlund, M., White, G.J., Eiroa, C., Henning, T., Herbst, T., Lammer, H., Liseau, R., Paresce, F., Penny, A., Quirrenbach, A., Röttgering, H., Selsis, F., Beichman, C., Danchi, W., Kaltenegger, L., Lunine, J., Stam, D., and Tinetti, G. (2010) The far future of exoplanet direct characterization. *Astrobiology*, Vol. 10, Issue 1, pp. 121–126 (DOI: 10.1089/ast.2009.0371).

Schuyler, W.M., Jr. (1982) Could anyone here speak Babel-17? In: Smith, N.D. (ed.), *Philosophers Look at Science Fiction*. Chicago: Nelson-Hall (ISBN: 0882298070).

Schwab, I.R. (2017) The evolution of eyes: Major steps. *The Keeler lecture 2017: Centenary of Keeler Ltd. Eye*, Vol. 32, Issue 2, pp. 302–313 (DOI: 10.1038/eye.2017.226).

Schwartz, J.S.J. (2016) Where no planetary protection policy has gone before. *International Journal of Astrobiology*, Vol. 18, Issue 4, pp. 1–9 (DOI: 10.1017/S1473550418000228).

Schwartz, R.N., and Townes, C.H. (1961) Interstellar and interplanetary communication by optical masers. *Nature*, Vol. 190, Issue 4772, pp. 205–208 (DOI: 10.1038/190205a0).

Scott, J. (2019) *The Vinyl Frontier: The Story of the Voyager Golden Record*. London: Bloomsbury Publishing (ISBN: 9781472956132).

Sefler, G.F. (1982) Alternative linguistic frameworks: Communication with extraterrestrial beings. In: Smith, N.D. (ed.), *Philosophers Look at Science Fiction*. Chicago: Nelson-Hall (ISBN: 0882298070).

Seligman, M.E.P., Railton, P., Baumeister, R.F., and Sripada, C. (2016) *Homo Prospectus*. Oxford: Oxford University Press (ISBN: 9780199374472).

SETI@Home (2016) Regarding messaging to extraterrestrial intelligence (METI)/Active SETI searches for extraterrestrial intelligence (Active SETI). UC Berkeley: SETI@Home website. Retrieved on 16 April 2016, available at http://setiathome.berkeley.edu/meti_statement_0.html.

Shamay-Tsoory, S.G. (2011) The neural bases for empathy. *The Neuroscientist*, Vol. 17, Issue 1, pp. 18–24 (DOI: 10.1177/1073858410379268).

Shams, L., Kamitani, Y., and Shimojo, S. (2002) Visual illusion induced by sound. *Brain Research Cognitive Brain Research*, Vol. 14, Issue 1, pp. 147–152 (DOI: 10.1016/s0926–6410(02)00069–1).

Sharp, S.R. (2016) A theatrical release party for a reimagined golden record. Hyperallergic website. Retrieved on 07 January 2020, available at https://hyperallergic.com/337116/a-theatrical-release-party-for-a-reimagined-golden-record/

Sherrington, C.S. (1906) Observations on the scratch-reflex in the spinal dog. *Journal of Physiology*, Vol. 34, Issue 1–2, pp. 1–50 (DOI: 10.1113/jphysiol.1906.sp001139).

Shostak, S.G. (2005) SETI and intelligent design. Space.com (online). Retrieved on 12 January 2018, available at https://www.space.com/1826-seti-intelligent-design.html.

Shostak, S.G. (2011) Limits on interstellar messages. *Acta Astronautica*, Vol. 68, Issue 3, pp. 366–371 (DOI: 10.1016/j.actaastro.2009.10.021).

Smil, V. (2006) *Energy*. London: Oneworld Publications (ISBN: 9781780741512).

Smith, K.C. (2017) Hawking and the METI Hawks: Right for the wrong reasons. *Theology and Science*, Vol. 15, Issue 2, pp. 147–149 (DOI: 10.1080/14746700.2017.1299383).

Smith, K.C. (2019) METI or REGRETTI: Informed consent, scientific paternalism and alien intelligence. In: Smith, K.C., and Mariscal, C. (eds.), *Social and Conceptual Issues in Astrobiology*. Oxford: Oxford University Press.

Smith, K.C., Abney, K., Anderson, G., Billings, L., DeVito, C.L., Green, B.P., Johnson, A.R., Marino, L., Munevar, G., Oman-Reagan, M.P., Potthast, A., Schwartz, J.S.J., Tachibana, K., and Jensen-Wells, S. (2019) The great colonization debate. *Futures Journal*, Vol. 110, pp. 4–14 (DOI: 10.1016/j.futures.2019.02.004).

Smith, R.F.W. (1963) Communication with extraterrestrial beings called improbably unless man can signal in two systems of thought. *Science Fortnightly*, 30 October 1963 Edition.

SNACI—The Staff at the National, A. and C. Ionosphere (1975) The Arecibo Message of November 1974. *Icarus*, Vol. 26, Issue 4, pp. 462–466 (DOI: 10.1016/0019-1035(75)90116-5).

Socas-Navarro, H. (2018) Possible photometric signatures of moderately advanced civilizations: The Clarke Exobelt. *The Astrophysics Journal*, Vol. 855, Issue 2 (DOI: 10.3847/1538–4357/aaae66).

Sonesson, G. (2013) Preparations for discussing constructivism with a Martian (the Second Coming). In: Dunér, D., Parthemore, J., Persson, E. and Holmberg, G. (eds.), *The History and Philosophy of Astrobiology*. Newcastle upon Tyne: Cambridge Scholars Publishing (ISBN: 9781443850353).

Spennemann, D.H.R. (2004) The ethics of treading on Neil Armstrong's footprints. *Space Policy*, Vol. 20, Issue 4, pp. 279–290 (DOI: 10.1016/j.spacepol.2004.08.005).

Spennemann, D.H.R. (2007) Extreme cultural tourism from Antarctica to the Moon. *Annals of Tourism Research*, Vol. 34, Issue 4, pp. 898–918 (DOI:10.1016/j.annals.2007.04.003).

Spennemann, D.H.R. (2009) On the nature of the cultural heritage values of spacecraft crash sites. In: Darrin, A. and O'Leary, B.L. (eds.), *The Handbook of Space Engineering, Archaeology, and Heritage*. Boca Raton, FL: CRC Press (ISBN: 9781420084313).

Spivack, N. (2003) *The Genesis Project*. Nova Spivack typepad (blog). Retrieved on 24 August 2018, available at https://novaspivack.typepad.com/nova_spivacks_weblog/2003/08/the_genesis_pro.html.

Spivack, N. (2019) The Lunar Library: Genesis Mission. Arch Mission website. Retrieved on 9 March 2019, available at https://www.archmission.org/lunar-library-overview.

Staski, E. (2009) Archaeology: The basics. In: Darrin, A. and O'Leary, B.L. (eds.), *The Handbook of Space Engineering, Archaeology, and Heritage*. Boca Raton, FL: CRC Press (ISBN: 9781420084313).

Staski, E., and Gerke, R. (2009) Natural formation processes and their effects on exoatmospheric objects, structures, and sites. In: Darrin, A. and O'Leary, B.L. (eds.), *The Handbook of Space Engineering, Archaeology, and Heritage*. Boca Raton, FL: CRC Press (ISBN: 9781420084313).

Steffen, W., Grinevald, J., Crutzen, P., and McNeill, J. (2011) The Anthropocene: Conceptual and historical perspectives. *Philosophical Transactions of The Royal Society*, Vol. 369, Issue 1938, pp. 842–867 (DOI: 10.1098/rsta.2010.0327).

Steffen, W., and Hughes, L. (2013) *The Critical Decade 2013: Climate Change Science, Risks and Responses*. Canberra: A.C.T. Climate Commission Secretariat (ISBN: 9781925006216).

Sterling, P., and Demb, J.B. (2004) Retina. In: Shepherd, G.M. (ed.), *The Synaptic Organization of the Brain*. Oxford, UK: Oxford University Press (DOI: 9780195159561).

Stevens, M., and Merilaita, S. (2011) *Animal Camouflage: Mechanisms and Function*. Cambridge: Cambridge University Press (ISBN: 9780521152570).

Stiegler, B. (2009) *Technics and Time 2: Disorientation* (translated by Stephen Barker). Stanford: Stanford University Press (ISBN: 9780804730143).

Stoneley, J., and Lawton, A.T. (1976) *CETI: Communication with Extra-Terrestrial Intelligence*. London: Star Books (ISBN: 9780352397812).

Stooke, P.J. (2007) *The International Atlas of Lunar Exploration*. Cambridge: Cambridge University Press (ISBN: 9780521819305).

Stothard, M. (2016) Nuclear waste: Keep out for 100,000 years. *Financial Times* (online). Retrieved on 16 September 2016, available at https://www.ft.com/content/db87c16c-4947-11e6-b387-64ab0a67014c.

Szondy, D. (2018) The Tesla Roadster could be the dirtiest manmade object in space. New Atlas (online). Retrieved on 29 September 2018, available at https://newatlas.com/tesla-roadster-dirtiest-space/535 90/.

Teichert, M., and Bolz, J. (2018) How senses work together: Cross-modal interactions between primary sensory cortices. *Neural Plast,* Vol. 2018 (DOI: 10.1155/2018/5380921).

Tolkien, J.R.R. (1940) On translating *Beowulf.* In: Hall, J.R.C. (1967) *Beowulf and the Finnesburg Fragment.* Australia: George Allen & Unwin (ASIN: B004J02OQ0).

Tough, A. (2000) How to achieve contact: Five promising strategies. In: Tough, A. (ed.), *When SETI Succeeds: The Impact of High-Information Contact.* Bellevue, WA: Foundation for the Future (ISBN: 97809 67725222).

Traphagan, J.W. (2010) Ritual, meaningfulness, and interstellar message construction. *Acta Astronautica,* Vol. 67, Issue 7–8, pp. 954–960 (DOI: 10.1016/j.actaastro.2010.06.011).

Traphagan, J.W. (2014) Anthropology at a distance: SETI and the production of knowledge in the encounter with an extraterrestrial other. In: Vakoch, D.A. (ed.), *Archaeology, Anthropology and Interstellar Communication.* Washington, D.C.: The NASA History series (ISBN: 9781501081729).

Traphagan, J.W. (2015) Culture, intelligence, and ETI. In: Traphagan, J.W. (ed.), *Extraterrestrial Intelligence and Human Imagination: SETI at the Intersection of Science, Religion, and Culture.* New York: Springer (ISBN: 9783319105505).

Traphagan, J.W. (2016) *Science, Culture and the Search for Life on Other Worlds.* New York: Springer Science + Business Media (ISBN: 9783319417455).

Traphagan, J.W. (2017) Do no harm? Cultural imperialism and the ethics of active SETI. *Journal of the British Interplanetary Society,* Vol. 70, Issue 5–6, pp. 219–224 (JBIS Refcode: 2017.70.219).

Traphagan, J.W. (2018) Cargo cults and the ethics of active SETI. *Space Policy,* Vol. 46, pp. 18–22 (DOI: 10.1016/j.spacepol.2018.04.001).

Traphagan, J.W. (2019) Active SETI and the problem of research ethics. *Theology and Science,* Vol. 17, Issue 1, pp. 69–78 (DOI: 10.1080/14746700.2019.1557806).

Traphagan, J.W. (2000) *Taming Oblivion: Aging Bodies and the Fear of Senility in Japan.* Albany, New York: State University of New York Press (ISBN: 978-0791445006).

Traphagan, J.W., and Traphagan, W. (1987) Music, meaning, history. *Phenomenological Inquiry,* Vol. 11, pp. 60–71.

Trauth, K.M., Hora, S.C., and Guzowski, R.V. (1993) *Sandia Report: Expert Judgement on Markers to Deter Inadvertent Human Intrusion into the Waste Isolation Pilot Plant.* Albuquerque, New Mexico: Sandia National Laboratories (Report SAND92–1382/ UC-721). Retrieved on 15 November 2015, available at https://prod-ng.sandia.gov/techlib-noauth/access-control.cgi/1992/921382.pdf.

Triscott, N. (2016) Transmissions from the noosphere: Contemporary art and outer space. In: Dickens, P. and Ormrod, J. (eds.), *The Palgrave Handbook of Society, Culture & Outer Space.* London: Palgrave Macmillan (ISBN: 9781349577057).

Troscianko, T., Benton, C.P., Lovell, P.G., Tolhurst, D.J., and Pizlo, Z. (2009) Camouflage and visual perception. *Philosophical Transactions of the Royal Society of London—B-Biological Sciences Series,* Vol. 364, Issue 1516, pp. 449–461 (DOI: 10.1098/rstb.2008.0218).

Turner, D. (2006) Church's thesis and functional programming. In: Olszewski, A. (ed.), *Church's Thesis After 70 Years.* Berlin: Logos Verlag (ISBN: 9783110324945).

United Nations (1967) *Treaty on Principles Governing the Activities of States in the Exploration and Use of Outer Space, including the Moon and Other Celestial Bodies (Outer Space Treaty).* United Nations Office for Outer Space Affairs (UNOOSA): Resolution 2222 XXI. Retrieved on 02 April 2016, available at https://www.unoosa.org/oosa/en/ourwork/spacelaw/treaties/introouterspacetreaty.html.

Vakoch, D.A. (1996) An iconic approach to communicating chemical concepts to extraterrestrials. *Proceedings of SPIE - The International Society for Optical Engineering,* Vol. 2704, pp. 140–149 (DOI: 10.1117/12.243431).

Vakoch, D.A. (1998a) Signs of life beyond Earth: A semiotic analysis of interstellar messages. *Leonardo,* Vol. 31, Issue 4, pp. 313–319 (DOI: 10.2307/1576671).

Vakoch, D.A. (1998b) The Dialogic Model: Representing human diversity in messages to extraterrestrials. *Acta Astronautica,* Vol. 42, Issue 10–12, pp. 705–710 (DOI: 10.1016/S0094-5765(98)00030-7).

Vakoch, D.A. (2008) Representing culture in interstellar messages. *Acta Astronautica,* Vol. 63, Issues 5–6, pp. 657–664 (DOI: 10.1016/j.actaastro.2008.05.011).

Vakoch, D.A. (2009) Honest exchanges with ET. *New Scientist,* Vol. 202, Issue 2705, pp. 22–23 (DOI: 10.1016/ S0262-4079(09)61112-X).

Vakoch, D.A. (2010a) An iconic approach to communicating musical concepts in interstellar messages. *Acta Astronautica,* Vol. 67, Issue 11, pp. 1406–1409 (DOI: 10.1016/j.actaastro.2010.01.006).

Vakoch, D.A. (2010b) Integrating active and passive SETI programs: Prerequisites for multigenerational research. In: *Proceedings of the Astrobiology Science Conference 2010, Evolution and Life: Surviving*

Catastrophes and Extremes on Earth and Beyond. League City, Texas, 20–26 April 2010. Issue 1538, pp. 5213 (Bibcode: 2010LPICo1538.5213V).

Vakoch, D.A. (2011a) The art and science of interstellar message composition: A report on international workshops to encourage multidisciplinary discussion. *Acta Astronautica*, Vol. 68, Issue 3–4, pp. 451–458 (10.1016/j.actaastro.2010.01.011).

Vakoch, D.A. (2011b) *Communication with Extraterrestrial Intelligence.* Albany, NY: State University of New York Press (ISBN: 9781438437941).

Vakoch, D.A. (2012) What would you say to an extraterrestrial? [video file]. Nashville, TX: TEDX Talk, March 2012. Retrieved on 19 February 2016, available at https://www.youtube.com/watch?v=hx 9i-KRMCCc.

Vakoch, D.A. (2014) *Archaeology, Anthropology and Interstellar Communication.* Washington, D.C.: The NASA History series (ISBN: 9781501081729).

Vakoch, D.A. (2016) In defense of METI. *Nature Physics*, Vol. 12, Issue 10, pp. 890–890 (DOI: 10.1038/nphys3897).

Vakoch, D.A. (2011c) Asymmetry in active SETI: A case for transmissions from Earth. *Acta Astronautica*, Vol. 68, pp. 476–488 (DOI: 10.1016/j.actaastro.2009.03.008).

Vakoch, D.A., and Ollongren, A. (2003) Large-size message construction for ETI: Self-interpretation in LINCOS. In: Norris, R. and Stootman, F. (eds.), *Bioastronomy: Life Amongst the Stars.* San Francisco: Astronomical Society of the Pacific (ISBN: 9781583811719).

Vakoch, D.A., and Ollongren, A. (2011) Processes in Lingua Cosmica. In: Vakoch, D.A. (ed.), *Communication with Extraterrestrial Intelligence.* Albany: State University of New York Press (ISBN: 9781438437934).

Vakoch, D.A. and Ollongren, A. (2013) Large-size message construction for ETI, Logical existence expressed in Lingua Cosmica. *Acta Astronautica*, Vol. 83, Issue 3–4, pp. 6–9 (DOI: 10.1016/j.actaastro.2012.09.003).

Vakoch, D.A., Lower, T.A., Niles, B.A., Rast, K.A., and DeCou, C. (2013) What should we say to extraterrestrial intelligence? An analysis of responses to "Earth Speaks." *Acta Astronautica*, Vol. 86, pp. 136–148 (DOI: 10.1016/j.actaastro.2011.05.022).

Van Gennep, A. (1960 [1909]) *The Rites of Passage.* 7th Edition. Chicago: University of Chicago Press (ISBN: 9780415330237).

Vishwanath, D. (2016) Induction of monocular stereopsis by altering focus distance: A test of Ames's Hypothesis. *i-Perception*, Vol. 7, Issue 2 (DOI: 10.1177/2041669516643236).

Vishwanath, D., and Hibbard, P.B. (2013) Seeing in 3-D with just one eye: Stereopsis in the absence of binocular disparities. *Psychological Science*, Vol. 24, Issue 9, pp. 1673–1685 (DOI: 10.1177/0956797613477867).

Wagemans, J. (1995) Detection of visual symmetries. *Spatial Vision*, Vol. 9, Issue 1, pp. 9–32 (DOI: 10.1163/156856895x00098).

Wald, G., and Brown, P.K. (1965) Human color vision and color blindness. In: *Cold Spring Harbour Symposia on Quantitative Biology*, Vol. 30, pp. 345–361 (DOI: 10.1101/sqb.1965.030.01.035).

Ward, B. (1966) *Spaceship Earth.* New York: Columbia University Press (ASIN: B003NIC9WS).

Ward, B. (1973) *Who Speaks for Earth?* New York: W.W. Norton (ISBN: 9780393093414).

Warner, M., and Collective, S.T. (1993) *Fear of a Queer Planet: Queer Politics and Social Theory.* Minneapolis: University of Minnesota Press (ISBN: 9780816623341).

Webb, S. (2002) *If the Universe Is Teeming with Aliens...Where Is Everybody? Fifty Solutions to the Fermi Paradox and the Problem of Extraterrestrial Life.* New York: Springer (ISBN: 9780387955018).

Welcher, L. (2018) Building a long-term archive of messages to ET. In: *SoCIA: Social and Contemporary Issues in Astrobiology* annual conference. University of Nevada, Reno, 13–15 April 2018.

Wells-Jensen, S. (2001) ENG 480/580; Extraterrestrial Language. Bowling Green State University, Course Outline. Retrieved on 14 March 2016, available at http://personal.bgsu.edu/~swellsj/xenolinguistics/.

Wells-Jensen, S. (2018) Could we learn E.T.'s language? (2nd part). METI International (blog). Retrieved on 16 February 2018, available at http://meti.org/en/blog/could-we-learn-ets-language-2nd-part.

Westwood, L., O'Leary, B.L., and Donaldson, M.W. (2017) *The Final Mission: Preserving NASA's Apollo Sites.* Gainesville: University Press of Florida (ISBN: 9780813064741).

Wheatstone, C. (1838) Contributions to the physiology of vision—Part the first. On some remarkable, and hitherto unobserved, phenomena of binocular vision. *Philosophical Transactions of the Royal Society of London—B-Biological Sciences Series*, Vol. 128, Issue 1838, pp. 371–394 (DOI: 10.1098/rspl.1837.0035).

White, F. (1987) *The Overview Effect: Space Exploration and Human Evolution.* Reston, VA: American Institute of Aeronautics and astronautics (ISBN: 9781624102622).

White, F. (1990) *The SETI Factor: How the Search for Extraterrestrial Intelligence Is Revolutionizing Our View of the Universe and Ourselves.* New York: Walter Publishing Company (ISBN: 0802711057).

Whitley, D.S. (2006) *Introduction to Rock Art Research.* Walnut Creek, CA: Left Coast Press (ISBN: 9781598740004).

Wikimedia Deutschland (2016) *Wikipedia to the Moon.* Wikimedia: Meta-Wiki Project Log. Retrieved on 10 September 2017, available at https://meta.wikimedia.org/wiki/Wikipedia_to_the_Moon/About.

Wilson, P. (2015) *Translation After Wittgenstein.* New York: Routledge (ISBN: 978-1138799875).

Wolfram, S. (2018) Showing off to the universe: Beacons for the afterlife of our civilization. Stephen

Wolfram Writings (blog). Retrieved on 21 February 2018, available at https://writings.stephenwolfram.com/2018/01/showing-off-to-the-universe-beacons-for-the-afterlife-of-our-civilization/.

Wong, S. (2019) First moon plants sprout in China's Chang'e 4 biosphere experiment. *New Scientist* (online). Retrieved on 20 January 2019, available at https://www.newscientist.com/article/2190704-first-moon-plants-sprout-in-chinas-change-4-biosphere-experiment/.

Wright, J.T., and Oman-Reagan, M.P. (2018) Visions of human futures in space and SETI. *International Journal of Astrobiology*, Vol. 17, Issue 2, pp. 177–188 (DOI: 10.1017/S1473550417000222).

Wright, R. (2004) *A Short History of Progress*. Edinburgh: Canongate Books (ISBN: 9781841958309).

Zagaeski, M., and Moss, C.F. (1994) Target surface texture discrimination by the echolocating bat, Eptesicus fuscus. *The Journal of the Acoustical Society of America*, Vol. 95, Issue 2881 (DOI: 10.1121/1.409387).

Zaitsev, A.L. (2006) Transforming SETI to METI. SETI League website. Retrieved on 17 January 2017, available at http://www.setileague.org/editor/metitran.htm.

Zaitsev, A.L. (2008) Sending and searching for interstellar messages. *Science Direct*, Vol. 63, Issue 5-6, pp. 614–617 (DOI: 10.1016/j.actaastro.2008.05.014).

Zaitsev, A.L. (2011) METI: Messaging to Extraterrestrial Intelligence. In: Schuch, P.H. (ed.), *Searching for Extraterrestrial Intelligence: SETI Past, Present, and Future*. New York: Springer (ISBN: 9783642131967).

Zaitsev, A.L. (2012) Classification of interstellar radio messages. *Acta Astronautica*, Vol. 78, pp. 16–19 (DOI: 10.1016/j.actaastro.2011.05.026).

Zerubavel, E. (2003) *Time Maps: Collective Memory and the Social Shape of the Past*. Chicago: University of Chicago Press (ISBN: 9780226981536).

Zhang, J. (2013) 5D data storage by ultrafast laser nanostructuring in glass. In: *Proceedings of the Conference on Lasers and Electro-Optics*, Optical Society of America. San Jose, California, 9–14 June 2013 (DOI: 10.1364/CLEO_SI.2013.CTh5D.9).

Zhang, S. (2017) Has a mysterious medieval code really been solved? *Atlantic* (online). Retrieved on 22 March 2018, available at https://www.theatlantic.com/science/archive/2017/09/has-the-voynich-manuscript-really-been-solved/539310/.

Zubrin, R. (2017) Interstellar communication using microbial data storage: Implications for SETI. *Journal of the British Interplanetary Society*, Vol. 70, Issue 5-6, pp. 163–174.

About the Contributors

Klara Anna **Capova** is a socio-cultural anthropologist specializing in science and technology studies. She is primarily interested in the roles of space sciences and space technologies and the social changes and societal impacts they make. This includes transformations in human relations to outer space; developments in contemporary worldviews; and the cultural, economic, geo-political and public dimension of space exploration programs, and their relevance for the UN Sustainable Development Goals 2030.

David **Dunér** is a professor of history of science and ideas at Lund University, Sweden. He is a board member and director of METI International, and member of the working group on the historical, philosophical, societal and ethical issues in astrobiology within the European Astrobiology Institute. He has edited *The History and Philosophy of Astrobiology: Perspectives on Extraterrestrial Life and the Human Mind* (2013) and special issues on the history and philosophy of astrobiology in *Astrobiology* (2012) and on the history and philosophy of the origin of life in *International Journal of Astrobiology* (2016).

Christopher **Gillespie** has a PhD in visual perception from St. Andrews University and a MSc in acoustics and music technology and a BSc (Hons) in astrophysics from the University of Edinburgh. His research focuses on contour integration and shape perception and he is also interested in more broad theoretical issues and sensory ecology. He works at Edinburgh Napier University as senior technician where he is involved in supporting and teaching scientific and experimental programming, statistical analysis and methodology to psychologists.

Alice **Gorman** is an internationally recognized leader in the field of space archaeology and author of the award-winning book *Dr. Space Junk vs the Universe: Archaeology and the Future* (2019). She is an associate professor in archaeology at Flinders University in Adelaide and a member of the advisory council of the Space Industry Association of Australia. Asteroid 551014 Gorman is named after her in recognition of her work in establishing space archaeology.

Cornelius **Holtorf** is professor of archaeology and holds an UNESCO Chair on Heritage Futures at Linnaeus University, Sweden. He was an advisor to the New Horizons: One Earth Message initiative and a co-investigator of the Heritage Futures research program, and he has led the Memory Across Generations project involving the Swedish nuclear waste sector. He has published extensively, including *Cultural Heritage and the Future* (2021, coedited with A. Högberg).

Paul E. **Quast** is an interdisciplinary scholar based in Scotland. He holds an MFA from the College of Arts, Humanities and Social Sciences (University of Edinburgh), and is a director for the Beyond the Earth Foundation. The APOH catalogue is the first catalogue arising from this long-term archival work. He is a member of SoCIA and UKSRN, and an advisor for the Memory of Mankind and for All Moonkind consortia.

Kelly **Smith** is a professor of philosophy and biological sciences as well as the chair of the Department of Philosophy and Religion at Clemson University. His research includes work on the concept of genetic disease; the relationship between religious faith and scientific reasoning; the ethical implications of technology; and the social and conceptual issues in astrobiology and space exploration. He has published in a wide array of professional journals, done consulting work for NASA and other organizations, and given numerous public presentations on his work.

John W. **Traphagan** is an anthropologist and visiting research scholar in the Interplanetary Initiative at Arizona State University, visiting professor in the Center for International Education at Waseda University and professor emeritus in the Program in Human Dimensions of Organizations at the University of Texas at Austin. His research has appeared in numerous journals including *The International Journal of Astrobiology, Acta Astronautica, Futures, and Space Policy.* He is the author of *Extraterrestrial Intelligence and Human Imagination: SETI at the Intersection of Science, Religion, and Culture* (2015) and *Culture, Science, and the Search for Life on Other Worlds* (2016).

Index

Numbers in **bold italics** indicate pages with illustrations